CAMBRIDGE MONOGRAPHS ON PHYSICS

Liquid crystals

Liquid crystals

S. CHANDRASEKHAR

Raman Research Institute, Bangalore

CAMBRIDGE UNIVERSITY PRESS
CAMBRIDGE
LONDON · NEW YORK · MELBOURNE

Published by the Syndics of the Cambridge University Press
The Pitt Building, Trumpington Street, Cambridge CB2 1RP
Bentley House, 200 Euston Road, London NW1 2DB
32 East 57th Street, New York, NY 10022, USA
296 Beaconsfield Parade, Middle Park, Melbourne 3206, Australia

First published 1977

Set on Linotron filmsetter by J. W. Arrowsmith Ltd., Bristol, England
and printed in Great Britain at the University Press, Cambridge

Library of Congress Cataloguing in Publication Data

Chandrasekhar, Sivaramakrishna, 1930–
Liquid crystals
(Cambridge monographs on physics)
Includes bibliographical references and index
1. Liquid crystals I. Title
QD923.C46 548′.9 75-32913
ISBN 0 521 21149 2

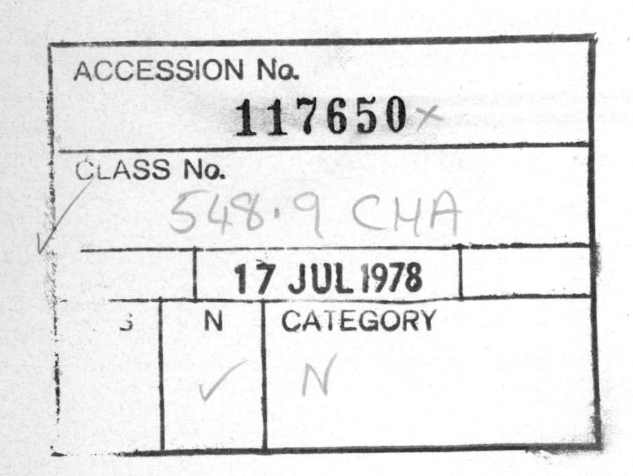

Contents

Preface

'I am at a loss to give a distinct idea of the nature of this liquid, and cannot do so without many words.'

The Narrative of Arthur Gordon Pym of Nantucket,
Edgar Allan Poe

The foundations of the physics of liquid crystals were laid in the 1920s but, surprisingly, interest in these substances died down almost completely during the next three decades. The situation was summarized by F. C. Frank in his opening remarks at a Discussion of the Faraday Society in 1958: 'After the Society's successful Discussion on liquid crystals in 1933, too many people, perhaps, drew the conclusion that the major puzzles were eliminated, and too few the equally valid conclusion that quantitative experimental work on liquid crystals offers powerfully direct information about molecular interactions in condensed phases.' In the last few years there has been a resurgence of activity in this field, owing partly to the realization that liquid crystals have important uses in display technology.

An exposition of the physics of liquid crystals involves many disciplines: continuum mechanics, optics of anisotropic media, statistical physics, crystallography etc. In covering such a wide field it is difficult to define what precisely the reader is expected to know already. An attempt is made to present as far as possible a self-contained treatment of each of these different aspects of the subject. Naturally, discussion of some topics has had to be curtailed for reasons of space. For example, we have not dealt with lyotropic systems, whose complex structures are only just beginning to be elucidated; or the special applications of magnetic resonance techniques, as these have been adequately reviewed elsewhere; or the very recent results of neutron scattering experiments. The primary aim of this monograph is to provide an insight into the variety of new phenomena exhibited by these intermediate states of matter.

This book would probably never have been completed without the unstinting cooperation of my young colleagues in Bangalore. I am particularly indebted to N. V. Madhusudana, G. S. Ranganath, R.

Shashidhar, U. D. Kini, R. Nityananda and T. G. Ramesh, whose notes and critical comments were of inestimable value at every stage of the writing, and to K. A. Suresh and B. R. Ratna for their assistance in preparing the diagrams and the list of references. Finally, it is a pleasure to express my thanks to Michael Woolfson for his encouraging interest and advice.

Raman Research Institute
August 1975 S. CHANDRASEKHAR

1
Introduction

The term *liquid crystal* signifies a state of aggregation that is intermediate between the crystalline solid and the amorphous liquid. As a rule, a substance in this state is strongly anisotropic in some of its properties and yet exhibits a certain degree of fluidity, which in some cases may be comparable to that of an ordinary liquid. The first observations of liquid crystalline or *mesomorphic* behaviour were made towards the end of the last century by Reinitzer[1] and Lehmann.[2] Several thousands of organic compounds are now known to form liquid crystals.[3,4] An essential requirement for mesomorphism to occur is that the molecule must be highly geometrically anisotropic, usually long and relatively narrow. Depending on the detailed molecular geometry, the system may pass through one or more mesophases before it is transformed into the isotropic liquid. Transitions to these intermediate states may be brought about by purely thermal processes (*thermotropic* mesomorphism) or by the influence of solvents (*lyotropic* mesomorphism).

1.1. Structure and classification of the mesophases

1.1.1. Thermotropic liquid crystals

Following the nomenclature proposed originally by Friedel,[5] thermotropic liquid crystals are classified broadly into three types: *nematic*, *cholesteric* and *smectic*.

The nematic liquid crystal has a high degree of long range orientational order of the molecules, but no long range translational order (fig. 1.1.1). Thus it differs from the isotropic liquid in that the molecules are spontaneously oriented with their long axes approximately parallel. The preferred direction usually varies from point to point in the medium, but a homogeneously aligned specimen is optically uniaxial, positive and strongly birefringent. The mesophase owes its fluidity to the ease with which the molecules slide past one another while still retaining their parallelism. Recent X-ray studies[6,7] indicate that some nematics are composed of groups of about 10^2 molecules – called *cybotactic* groups[8] – the molecular centres in each group arranged in layers. In ordinary

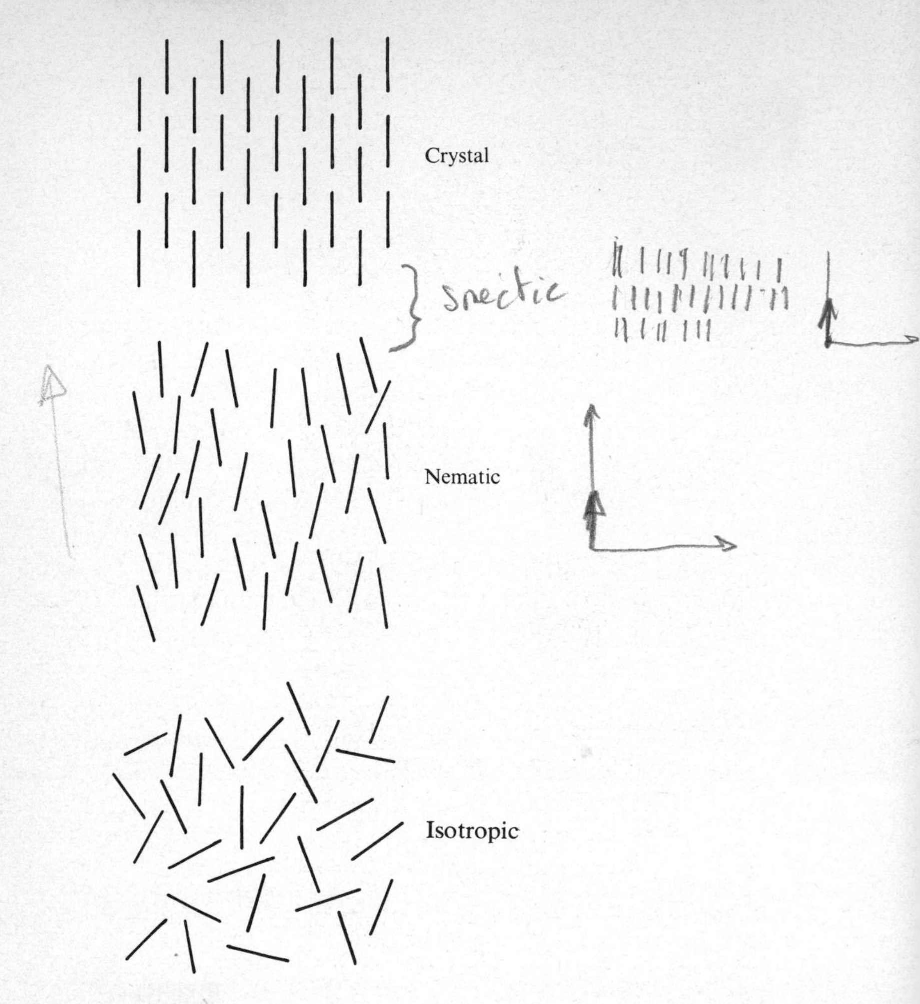

Fig. 1.1.1. Schematic representation of molecular order in the crystalline, nematic and isotropic phases.

nematics the cybotactic groups, if they do exist, are smaller than can be detected by X-ray methods (see plates 1 and 2).

The cholesteric mesophase is also a nematic type of liquid crystal except that it is composed of optically active molecules. As a consequence the structure has a screw axis superimposed normal to the preferred molecular direction (fig. 1.1.2). Optically inactive molecules or racemic mixtures result in a helix of infinite pitch which corresponds to the true nematic. Thermodynamically, the cholesteric is very similar to the nematic as the energy of twist forms only a minute part ($\sim 10^{-5}$) of the total

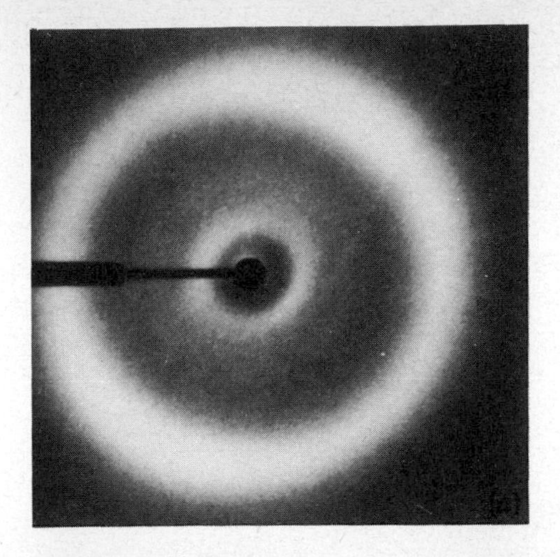

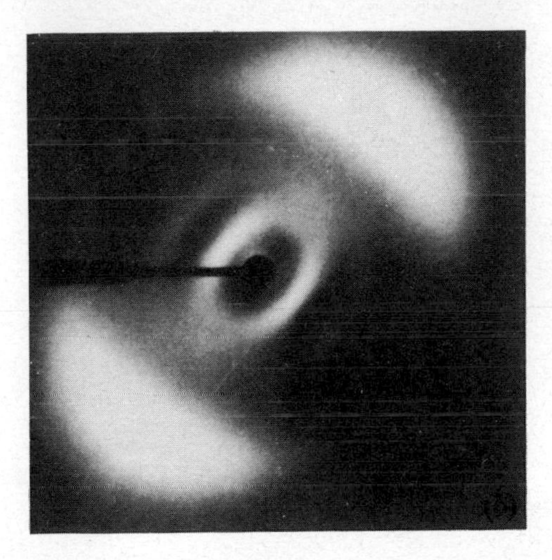

Plate 1. X-ray diffraction patterns from an 'ordinary' nematic liquid crystal: 4-n-propyloxybenzylidene-4′-n-propylaniline. (*a*) A fairly randomly oriented sample, (*b*) a well aligned sample. (De Vries.[7])

energy associated with the parallel alignment of the molecules.[9] This is well illustrated by the fact that when a small quantity of a cholesteric substance,[5] or even a non-mesomorphic optically active substance,[10] is added to a nematic the mixture adopts a helical configuration. The spiral arrangement of the molecules in this mesophase is responsible for its unique optical properties, viz, selective reflexion of circularly polarized light and a rotatory power about a thousand times greater than that of an ordinary optically active substance.

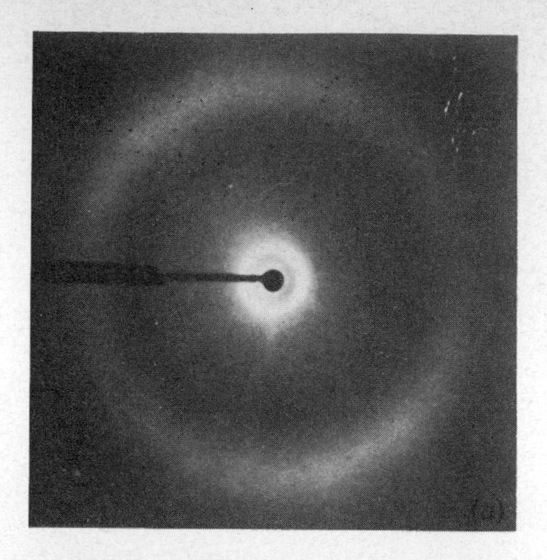

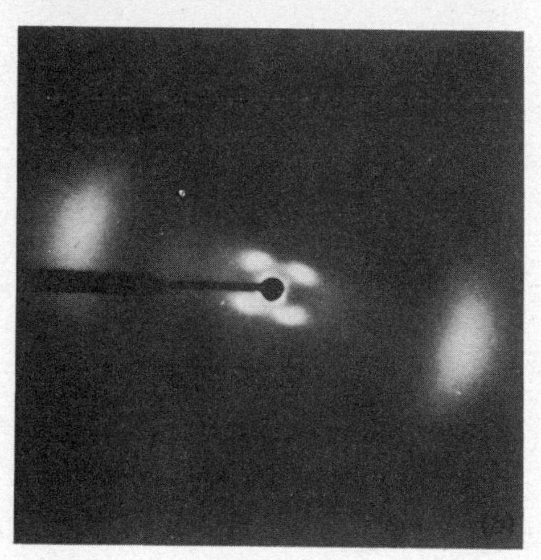

Plate 2. X-ray diffraction patterns from a 'cybotactic' nematic: bis-(4'-n-octyloxybenzal)-2-chloro-1,4-phenylenediamine. (*a*) A randomly oriented sample, (*b*) a well aligned sample. (De Vries.[7])

Smectic liquid crystals have stratified structures but a variety of molecular arrangements are possible within each stratification. In smectic *A* the molecules are upright in each layer with their centres irregularly spaced in a 'liquid-like' fashion (fig. 1.1.3(*a*)). The thickness of the layer is of the order of the length of the free molecule. The interlayer attractions are weak as compared with the lateral forces between molecules and in consequence the layers are able to slide over one another relatively easily. Thus this mesophase has fluid properties, though as a rule it is very much more viscous than the nematic mesophase. Smectic *B* differs from *A* in

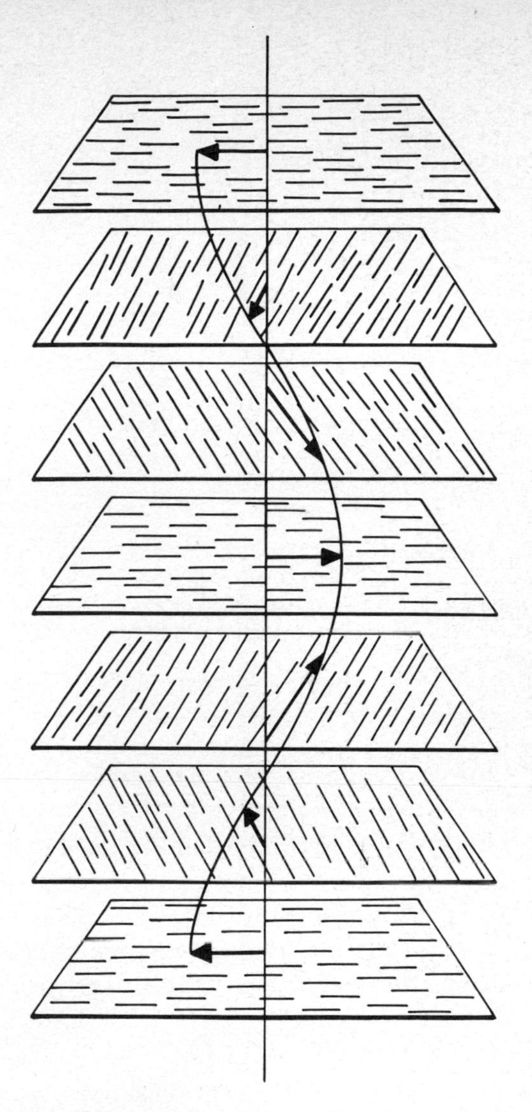

Fig. 1.1.2. The cholesteric liquid crystal: schematic representation of the helical structure.

that the molecular centres in each layer are hexagonal close packed. Smectic C is a tilted form of smectic A, i.e., the molecules are inclined with respect to the layers (fig. 1.1.3(b)). If in addition to the tilt there is an ordered arrangement within each layer, it is labelled B_c. At least four other distinct smectic modifications have been identified(11,12) but their structures are not yet known with any certainty. Smectic D has been

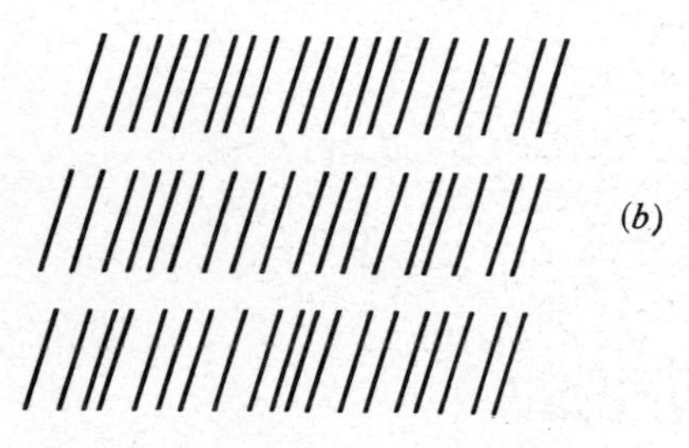

Fig. 1.1.3. Normal and tilted smectic structures.

reported to be cubic[13] and would appear to be an exception to the rule that smectics have well defined stratifications.

The energy required to deform a liquid crystal is so small that even the slightest perturbation, caused say by a dust particle or a surface inhomogeneity, can distort the structure quite profoundly. Thus when a liquid crystal is taken between glass plates and examined under a polarizing microscope one rarely sees the familiar interference figures expected from the equilibrium structures shown in figs. 1.1.1–1.1.3. Instead, one usually obtains a rather complex optical pattern. For example, a nematic film shows a characteristic threaded texture from which this mesophase derives its name (plate 3) and a smectic *A* film a focal conic texture (plate 4). These textures are useful in the optical identification of the mesophases[14] and their nature is generally well understood, as we shall see in later chapters. 'Single crystal' films with the molecules aligned perpendicular to the plates (*homeotropic* structure) or parallel to them (*homogeneous* or *planar* structure) can be prepared by suitable prior treatment of the glass surfaces.

From purely geometrical arguments, Herrmann[15] concluded that there should be 18 distinct mesomorphic groups between the perfectly ordered crystalline arrangement and the truly amorphous one. Examples

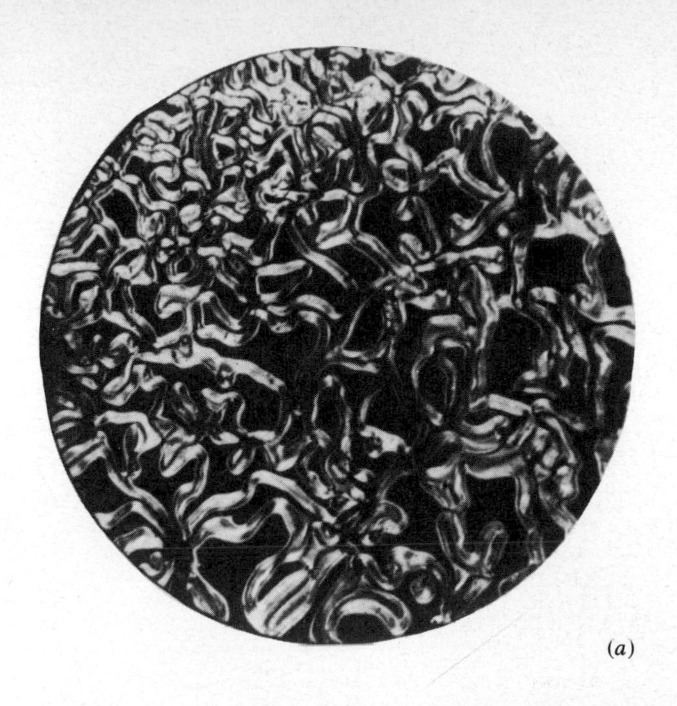

(a)

(b)

Plate 3. (a) Threads in a nematic liquid crystal. Crossed polarizers. Film thickness ~100 μm. (b) Schlieren texture in a nematic film of thickness ~10 μm. Crossed polarizers. (Sackmann and Demus.[14])

Plate 4. Focal conic textures in smectic *A*. (*a*) The polygonal texture. Crossed polarizers. (Friedel.[5]) (*b*) Simple fan-shaped texture. Crossed polarizers. (Sackmann and Demus.[14])

of some of these groups have been found in plant virus preparations[16] and in surfactant–water compositions,[17] but it is not yet clear whether in fact all of them can give rise to energetically feasible configurations. Thus Friedel's nomenclature offers a convenient basis for the classification of thermotropic liquid crystals and is universally adopted at present.

1.1.2. Lyotropic liquid crystals

Lyotropic liquid crystals are made up of two or more components.[18] Generally, one of the components is an amphiphile (containing a polar head group attached to one or more long hydrocarbon chains) and another is water. A familiar example of such a system is soap (sodium dodecyl sulphate) in water. As the water concentration is increased several mesophases are obtained. The types of molecular packing in these mesophases are represented schematically in figs. 1.1.4 and 1.1.5.

In the *lamellar* or *neat* phase, water is sandwiched between the polar heads of adjacent layers, while the hydrocarbon tails, which are disordered or in a liquid-like configuration, are in a non-polar environment (fig. 1.1.4). In the *cubic* or *viscous isotropic* phase the layers are bent to

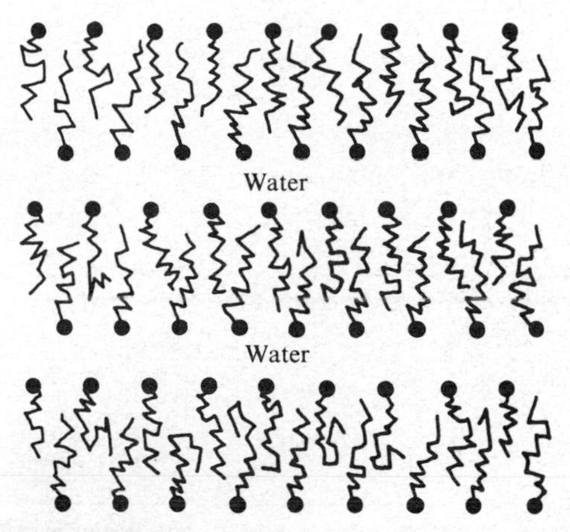

Fig. 1.1.4. The lamellar or neat phase of soaps.

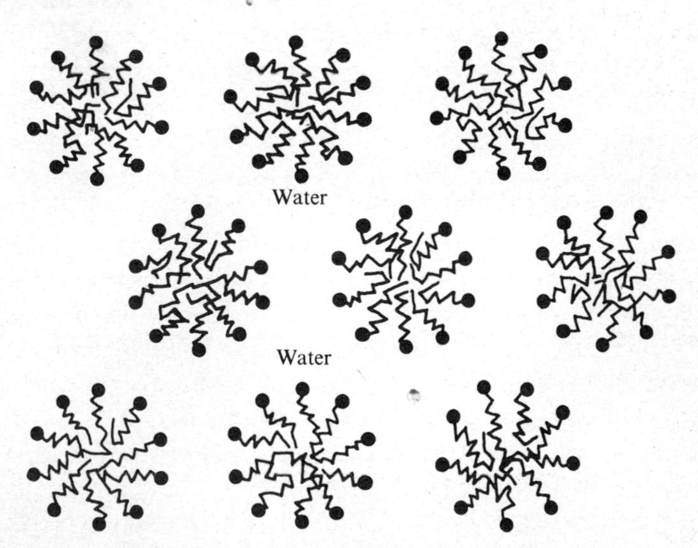

Fig. 1.1.5. The hexagonal or middle phase of soaps.

form spherical units, the polar heads lying on the surface of the sphere and the hydrocarbon chains filling up the inside. The spherical units form a body-centred cubic arrangement,[19] water taking up the space between the units. In the *hexagonal* or *middle* phase, the layers are rolled up into cylinders. The cylindrical units of indefinite length are arranged parallel to one another in a hexagonal array (fig. 1.1.5). A nematic type of ordering has also been observed in some soap systems.[20] It is believed that there exists a cylindrical superstructure similar to the hexagonal phase, but this has not been proved conclusively.[21]

In hydrophobe-dominated compositions, such as aerosol OT-water system,[22] *inverted middle* and *inverted viscous isotropic* phases can occur in which the tails point outward towards the hydrophobic medium while water is trapped inside.

Two mesophases may coexist over small ranges of composition and temperature. Phase diagrams have been constructed for a large number of binary systems. Transitions may be brought about from any one of the mesophases directly to the isotropic solution at appropriate temperatures. Ternary and higher-component systems exhibit essentially the same types of structures, the phase diagrams being of course much more complicated.

Cholesteric liquid crystals are formed by solutions of synthetic polypeptides, e.g., poly-γ-benzyl-L-glutamate, in organic liquids when the concentration exceeds a certain critical value.[23]

Lyotropic liquid crystals occur abundantly in nature, particularly in living systems.[24] Their structures are quite complex and are only just beginning to be elucidated. However, in this monograph we shall be confining our attention mainly to the physics of thermotropic liquid crystals and do not propose to deal with the structures of lyotropic systems in any further detail.

1.2. Polymorphism in thermotropic liquid crystals

Some typical examples of mesophase transitions in thermotropic materials are given below. The latent heats (in kilojoules per mole) are also presented for those cases for which data are available. All the transitions listed here, except one, are *enantiotropic*, i.e., they take place reversibly on heating and cooling, though the reversal to the solid phase is usually accompanied by supercooling. The one exception is cholesteryl nonanoate (see (1.2.6) below) which shows a *monotropic* transition – the smectic A phase occurs only on cooling.

A considerable body of experimental data has accumulated over the years in regard to the relationship between mesomorphism and chemical constitution, but we shall not be discussing this aspect of the problem. For a concise review of current trends in the synthetic chemistry of mesogenic compounds with emphasis on the special requirements for technological applications reference may be made to a recent article by Gray.[32]

4,4′-Di-methoxyazoxybenzene (*p*-azoxyanisole)[25]

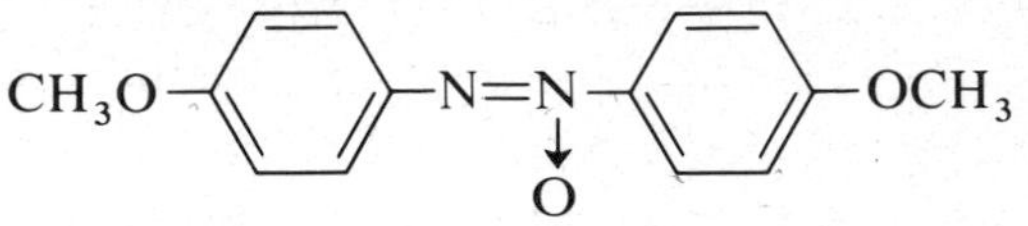

$CH_3O-C_6H_4-N{=}N(\rightarrow O)-C_6H_4-OCH_3$

$$\text{Solid} \underset{29.57\ \text{kJ}}{\overset{118.2\ ^\circ\text{C}}{\longleftrightarrow}} \text{nematic} \underset{0.57\ \text{kJ}}{\overset{135.3\ ^\circ\text{C}}{\longleftrightarrow}} \text{isotropic} \tag{1.2.1}$$

4′-n-Pentyl-4-cyanobiphenyl[26]

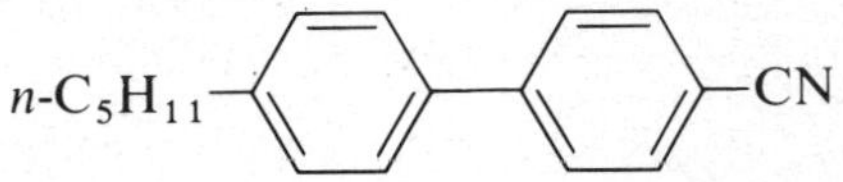

$n\text{-}C_5H_{11}-C_6H_4-C_6H_4-CN$

$$\text{Solid} \underset{17.2\ \text{kJ}}{\overset{22.5\ ^\circ\text{C}}{\longleftrightarrow}} \text{nematic} \underset{0.8\ \text{kJ}}{\overset{35\ ^\circ\text{C}}{\longleftrightarrow}} \text{isotropic} \tag{1.2.2}$$

p-Quinquephenyl[27]

$$\text{Solid} \overset{401\ ^\circ\text{C}}{\longleftrightarrow} \text{nematic} \overset{445\ ^\circ\text{C}}{\longleftrightarrow} \text{isotropic} \tag{1.2.3}$$

Cholesteryl benzoate[28]

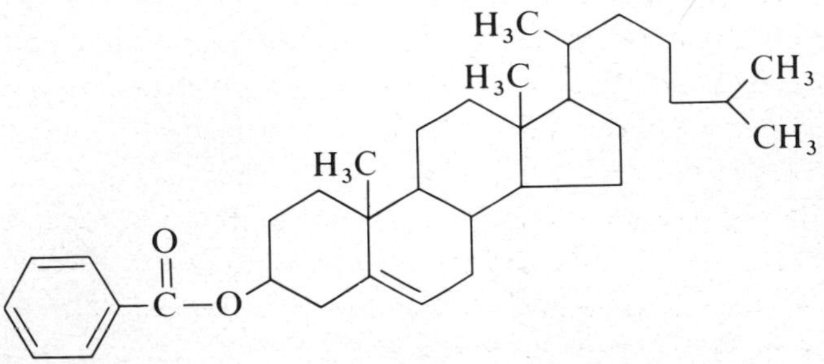

H_3C, H_3C, H_3C, CH_3, CH_3, O, C–O

$$\text{Solid} \overset{147\ ^\circ\text{C}}{\longleftrightarrow} \text{cholesteric} \overset{186\ ^\circ\text{C}}{\longleftrightarrow} \text{isotropic} \tag{1.2.4}$$

Ethyl *p*-azoxycinnamate[29]

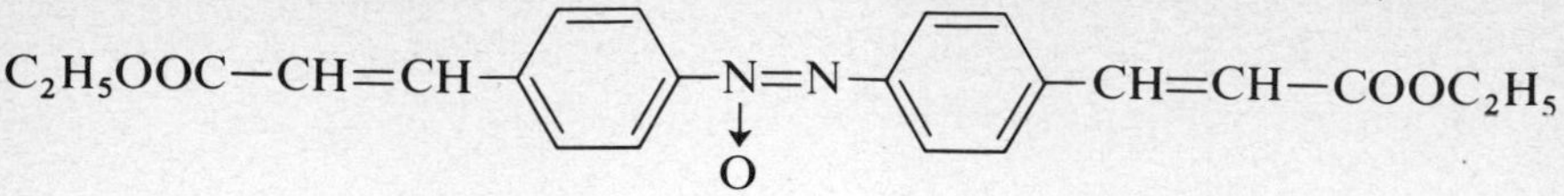

$$C_2H_5OOC-CH=CH-C_6H_4-N(\rightarrow O)=N-C_6H_4-CH=CH-COOC_2H_5$$

$$\text{Solid} \xleftrightarrow{140\,^\circ\text{C}} \text{smectic } A \xleftrightarrow{250\,^\circ\text{C}} \text{isotropic} \qquad (1.2.5)$$

Cholesteryl nonanoate (monotropic)[30]

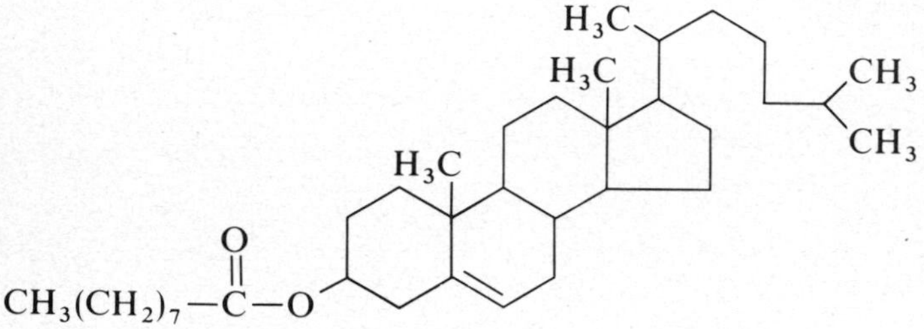

H_3C H_3C CH_3 CH_3 H_3C O

$CH_3(CH_2)_7-C(=O)-O$

$$\text{Solid} \xrightarrow{78.6\,^\circ\text{C}} \text{cholesteric} \xleftrightarrow[0.66\text{ kJ}]{91.2\,^\circ\text{C}} \text{isotropic}$$

cholesteric $\xrightarrow[0.25\text{ kJ}]{75.5\,^\circ\text{C}}$ smectic A $\rightarrow$ Solid

(1.2.6)

Cholesteryl myristate[30]

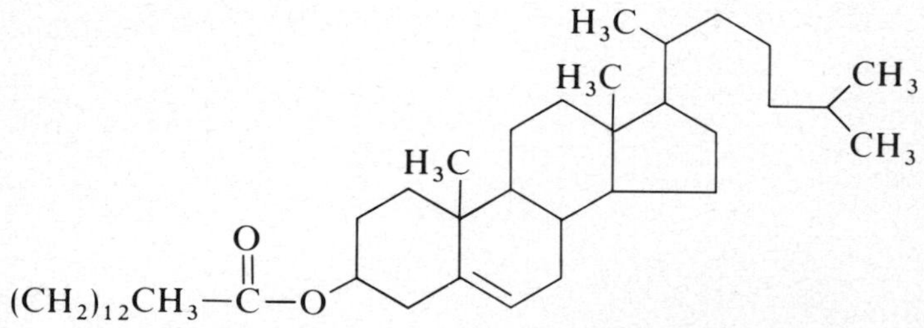

H_3C H_3C CH_3 CH_3 H_3C O

$(CH_2)_{12}CH_3-C(=O)-O$

$$\text{Solid} \xleftrightarrow[17.5\text{ kJ}]{71.0\,^\circ\text{C}} \text{smectic } A \xleftrightarrow[1.78\text{ kJ}]{79.1\,^\circ\text{C}} \text{cholesteric} \xleftrightarrow[1.51\text{ kJ}]{84.6\,^\circ\text{C}} \text{isotropic} \qquad (1.2.7)$$

4,4′-Di-heptyloxyazoxybenzene[25]

$$C_7H_{15}O-C_6H_4-N(\rightarrow O)=N-C_6H_4-OC_7H_{15}$$

$$\text{Solid} \xleftrightarrow[40.92\ \text{kJ}]{74.4\ ^\circ\text{C}} \text{smectic } C \xleftrightarrow[1.59\ \text{kJ}]{95.4\ ^\circ\text{C}} \text{nematic} \xleftrightarrow[1.02\ \text{kJ}]{124.2\ ^\circ\text{C}} \text{isotropic} \quad (1.2.8)$$

Terephthal-bis-4-n-butylaniline[31]

$$H_9C_4-C_6H_4-N=HC-C_6H_4-CH=N-C_6H_4-C_4H_9$$

$$\text{Solid} \xleftrightarrow[18.17\ \text{kJ}]{113.0\ ^\circ\text{C}} \text{smectic } B \xleftrightarrow[3.76\ \text{kJ}]{144.1\ ^\circ\text{C}} \text{smectic } C \xleftrightarrow[\sim 0\ \text{kJ}]{172.5\ ^\circ\text{C}} \text{smectic } A \xleftrightarrow[0.29\ \text{kJ}]{199.6\ ^\circ\text{C}} \text{nematic} \xleftrightarrow[0.75\ \text{kJ}]{236.5\ ^\circ\text{C}} \text{isotropic} \quad (1.2.9)$$

2-(4-n-Pentylphenyl)-5-(4-n-pentyloxyphenyl)-pyrimidine[12]

$$C_5H_{11}O-C_6H_4-C_4H_2N_2-C_6H_4-C_5H_{11}$$

$$\text{Solid} \xleftrightarrow[34.98\ \text{kJ}]{79\ ^\circ\text{C}} \text{smectic } G \xleftrightarrow[0.58\ \text{kJ}]{102.7\ ^\circ\text{C}} \text{smectic } F \xleftrightarrow[0.50\ \text{kJ}]{113.8\ ^\circ\text{C}} \text{smectic } C \xleftrightarrow[0.20\ \text{kJ}]{144\ ^\circ\text{C}} \text{smectic } A \xleftrightarrow[11.45\ \text{kJ}]{210\ ^\circ\text{C}} \text{isotropic} \quad (1.2.10)$$

2

Statistical theories of nematic order

2.1. Melting of molecular crystals: the Pople–Karasz model

X-ray analyses of the crystal structures of typical nematogenic compounds [1–3] have established that the long narrow molecules are more or less parallel and interleave one another to form what was described by Bernal and Crowfoot[1] as an 'imbricated' arrangement (fig. 2.1.1). The transformation from the solid to the nematic phase is characterized by the breakdown of the positional order of the molecules but not of the orientational order. The mesophase is fluid and at the same time anisotropic because of the facility with which the molecules slide over one another while still preserving their parallelism. The degree of orientational order in the liquid crystal drops gradually on heating, whereas certain other thermodynamic properties, such as the specific heat, thermal expansion and isothermal compressibility, increase rapidly as the temperature approaches the nematic–isotropic point T_{NI}. At T_{NI} a 'weak' first order (discontinuous) phase transition takes place, accompanied by the complete breakdown of the long range orientational order. The changes of entropy and volume attending this transition are typically only a few per cent of the corresponding values for the solid–nematic transition.

In this section, we shall consider an elementary model[4–6] which illustrates the principal mechanisms responsible for this unusual type of melting phenomenon.

2.1.1. Order–disorder in positions and orientations

The simplest treatment of the melting of inert gas crystals is that due to Lennard-Jones and Devonshire[7] (LJD) who regarded the mechanism of fusion as a positional order–disorder phenomenon. They postulated that the atoms may occupy sites on one of two equivalent interpenetrating lattices, which we shall refer to as A-sites and B-sites. The lowest energy configuration (corresponding to the state of perfect order) is that in which all atoms occupy the same kind of sites, say the A-sites. With increasing temperature, the number of atoms in the interstitial B-sites increases till a critical stage is reached when there is complete collapse of the long range

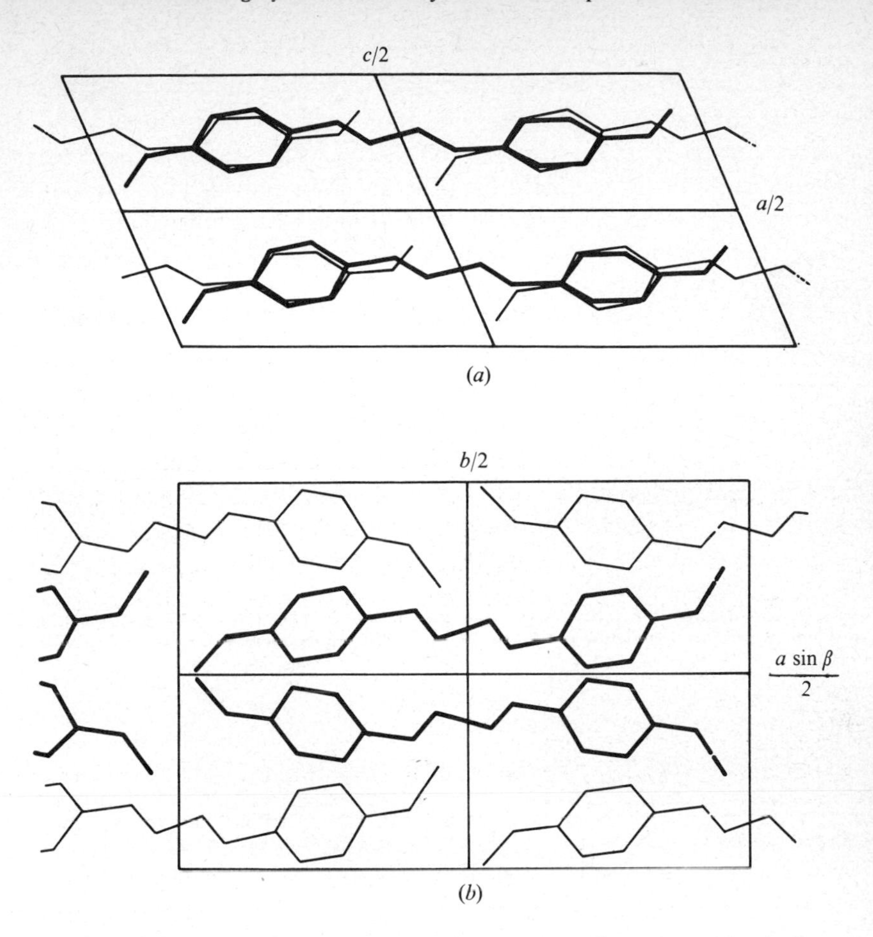

Fig. 2.1.1. Molecular arrangement in the crystalline phase of anisaldazine: (*a*) projection of the structure along [010], (*b*) projection along [001]. Crystals are monoclinic with $a = 17.46$ Å, $b = 10.76$ Å, $c = 8.45$ Å, $\beta = 113°48'$. Space group *Cc* with 4 molecules per unit cell. (After Galigne and Falgueirettes.[3])

order and both kinds of sites become equally populated. The system in which there is an equal number of occupied and unoccupied A-sites corresponds to a liquid, for under such circumstances migration from one site to another becomes easily possible. By assuming an appropriate volume dependence for the energy of interaction between atoms in A and B sites, Lennard-Jones and Devonshire showed that the melting parameters predicted by the theory agree well with the data for a number of spherical or nearly spherical molecules.

However, the theory fails for anisotropic molecules where the effects of orientational disorder become important. The thermodynamic data on melting suggest that there are two classes of molecular crystals: those which undergo phase transitions associated with rotational motions at temperatures below the melting point and those in which the rotational and melting transitions coalesce. The former have entropies of fusion lower than the inert gas crystals, while the latter have much higher entropies of fusion (table 2.1.1).

TABLE 2.1.1. *Melting entropy of molecular crystals*[8] (*in entropy units,* $\Delta S/R$)

Inert gas crystals				
	Ne	3.26	Kr	3.36
	A	3.35	Xe	3.40
Crystals in which the rotational transition precedes the melting transition				
	O_2	2.0	PH_3	1.92
	SiH_4	1.8	CD_4	2.42
	CF_4	2.0	CH_4	2.47
Crystals in which the two transitions coalesce				
	CO_2	9.25	$SiCl_4$	9.06
	N_2O	8.58	HCN	7.73
	C_2H_4	7.70	CS_2	6.51
	SO_2	8.95	$(CN)_2$	7.90

Pople and Karasz[4] proposed a simple extension of the LJD model which at once provides an interpretation of these variations in the melting properties. They assumed that the molecule can take up one of two orientations on any site, so that it now has four possibilities, A_1, A_2, B_1 and B_2. The state of perfect order (or the solid at zero temperature) may then be regarded as one in which all molecules occupy sites and orientations of the same type, say A_1, and the state of complete disorder (or the liquid phase) as one in which all four configurations are equally populated. Clearly, there can also be states with positional order and no orientational order and vice versa.

When a molecule is turned into an unfavourable orientation, local strains are set up and consequently there is an increased tendency for the molecules in the vicinity to move to interstitial sites. To allow for this coupling between orientational and positional ordering, Pople and Karasz made the simple assumption that the orientational component of the AB interactions is negligible compared with that of the AA or BB interactions, so that the B-sites near a misorientation in an A-lattice are favoured. Accordingly the energy required for a molecule to diffuse to an interstitial B-site in an A-lattice is determined only by the AB interactions regardless of orientation, and that required for a molecule to assume an A_2-orientation in an A_1-lattice (or a B_2-orientation in a B_1-lattice) is determined by the A_1A_2 (or B_1B_2) interactions only.

Let W be the energy of an AB interaction and W' the orientational energy of an A_1A_2 or B_1B_2 interaction. We neglect interactions between more distant neighbours. If for any configuration of the system of N molecules, N_{AB} is the number of neighbouring AB pairs, $N_{A_1A_2}$ the number of relative misorientations on neighbouring A-sites and $N_{B_1B_2}$ is similarly defined, the partition function for the whole assembly is

$$
\begin{aligned}
Z &= f^N \Omega, \\
&= f^N \sum \exp[-(N_{AB}W + N_{A_1A_2}W' + N_{B_1B_2}W')/k_B T]
\end{aligned}
$$

where the sum is over all configurations taking into account all possible arrangements in positions and orientations; f is the partition function per molecule in a state of perfect order which is treated as a function of volume per molecule and temperature only, and k_B is the Boltzmann constant.

Approximate solution of the cooperative order–disorder problem We define the degree of positional order as $q = 2\mathcal{Q} - 1$, where $\mathcal{Q}$ is the fraction of molecules in the A-sites, and the degree of orientational order as $s = 2\mathcal{S} - 1$, where $\mathcal{S}$ is the fraction of molecules in 1-orientations. If N is the total number of molecules in the system, there are evidently

$\frac{1}{4}N(1+q)(1+s)$ molecules in A_1 positions,

$\frac{1}{4}N(1+q)(1-s)$ molecules in A_2 positions,

$\frac{1}{4}N(1-q)(1+s)$ molecules in B_1 positions,

$\frac{1}{4}N(1-q)(1-s)$ molecules in B_2 positions.

Let there be z B-sites adjacent to each A-site and z' A-sites closest to each A-site. This implies, of course, that there are also z A-sites adjacent to each B-site and z' B-sites closest to each B-site.

Applying the Bragg–Williams (or zeroth order) approximation,[9] the configurational partition function $\Omega = \gamma(q, s) \exp[-\{\frac{1}{4}NzW(1-q^2) + \frac{1}{8}Nz'W'(1-s^2)(1+q^2)\}/k_BT]$, where

$$\gamma(q, s) = \left(\frac{N!}{[\frac{1}{2}N(1+q)]![\frac{1}{2}N(1-q)]!}\right)^2$$

$$\times\left(\frac{[\frac{1}{2}N(1+q)]!}{[\frac{1}{4}N(1+q)(1+s)]![\frac{1}{4}N(1+q)(1-s)]!}\right)$$

$$\times\left(\frac{[\frac{1}{2}N(1-q)]!}{[\frac{1}{4}N(1-q)(1+s)]![\frac{1}{4}N(1-q)(1-s)]!}\right).$$

Using Stirling's theorem

$$N^{-1} \log \Omega = (1+q) \log\left(\frac{1+q}{2}\right) - (1-q) \log\left(\frac{1-q}{2}\right)$$

$$-\frac{1+s}{2} \log\left(\frac{1+s}{2}\right)$$

$$-\left(\frac{1-s}{2}\right) \log\left(\frac{1-s}{2}\right) - \frac{zW}{k_BT}\left(\frac{1-q^2}{4}\right)$$

$$-\frac{z'W'}{k_BT}\left(\frac{1-s^2}{4}\right)\left(\frac{1+q^2}{2}\right). \tag{2.1.1}$$

To evaluate the contribution of the disorder to the various thermodynamic quantities it is necessary to specify the dependence of W and W' on volume. It is found[5,6] that the empirical trends in the properties, especially for large orientational barriers which is the main region of interest in the present discussion, are reproduced well by the model when $W = W_0(V_0/V)^4$ and $W' = W'_0(V_0/V)^3$ where the suffix zero denotes the value corresponding to the equilibrium intermolecular separation as defined by the Lennard-Jones potential. When $W' = 0$, the theory reduces identically to the LJD model for spherical molecules. Applying the equilibrium conditions,

$$\frac{\partial \log \Omega}{\partial q} = \frac{\partial \log \Omega}{\partial s} = 0,$$

we obtain

$$\log\left(\frac{1+q}{1-q}\right) = \frac{zW}{k_BT}\left[\frac{1}{2} - \frac{V\nu}{V_0}\left(\frac{1-s^2}{4}\right)\right] q, \tag{2.1.2}$$

$$\log\left(\frac{1+s}{1-s}\right) = \frac{zW}{k_BT}\nu\frac{V}{V_0}\left(\frac{1+q^2}{2}\right) s, \tag{2.1.3}$$

where $\nu = z'W'_0/zW_0$ is a measure of the relative barriers for the rotation of a molecule and for its diffusion to an interstitial site.

The Helmholtz free energy may be conveniently expressed as $F = F' + F''$, where $F' = -Nk_BT \log f$ is the contribution due to the perfectly ordered system and $F'' = -k_BT \log \Omega$ is that due to disorder. The pressure due to disorder $p'' = -(\partial F''/\partial V)_T$, so that from (2.1.1),

$$\frac{p''V_0}{Nk_BT} = \frac{zW}{k_BT}\frac{V_0}{V}\left[(1-q^2)+\tfrac{3}{8}\nu\frac{V}{V_0}(1-s^2)(1+q^2)\right], \qquad (2.1.4)$$

where q and s are now the equilibrium values determined by (2.1.2) and (2.1.3) If p' is the ordered part of the pressure, the total pressure $p \doteqdot p' + p''$. Extensive calculations of $p'V/Nk_BT$ and other thermodynamic parameters have been carried out as functions of V/V_0 and k_BT/ε for a face-centred cubic lattice in terms of the Lennard-Jones 6–12 potential, ε being the minimum energy of interaction of a pair of particles.[10] Using these values and (2.1.4) the complete isotherm can be obtained. In the actual calculations the ratio W_0/ε was adjusted empirically to give the correct melting temperature of argon. The parameters used were $z = 6$ for an interpenetrating face-centred cubic lattice, and $W_0/\varepsilon = 0.977$. The melting properties of a range of materials can then be investigated as the *non-dimensional* parameter ν increases from zero to a high value.

2.1.2. Plastic crystals and liquid crystals

Hereafter q and s will be understood to refer to those values that satisfy (2.1.2) and (2.1.3). For low zW/k_BT, $q = s = 0$ is the only solution that minimizes the free energy for any given k_BT/ε. The behaviour for higher zW/k_BT depends critically on the strength of the orientational barrier. The variation of the equilibrium values of q and s with zW/k_BT for three typical values of ν and the corresponding isotherms are shown in figs. 2.1.2–2.1.4. The sigmoid portions of the isotherms represent phase transitions, i.e., the two phases will be in equilibrium at a given pressure when the areas enclosed above and below the pressure line are equal. For small ν (less than about 0.3), the theory predicts two transitions, a solid state rotational transition followed by a melting transition. The intermediate phase corresponds to the orientationally disordered crystal, sometimes referred to as the *plastic crystal*. For ν between 0.3 and 0.975, the two transitions coalesce and there occurs a single transition with a much greater entropy of fusion. For $\nu > 0.975$ there are again two

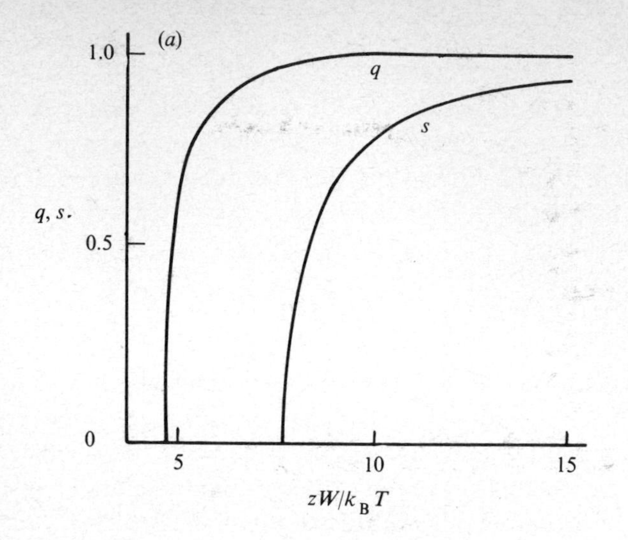

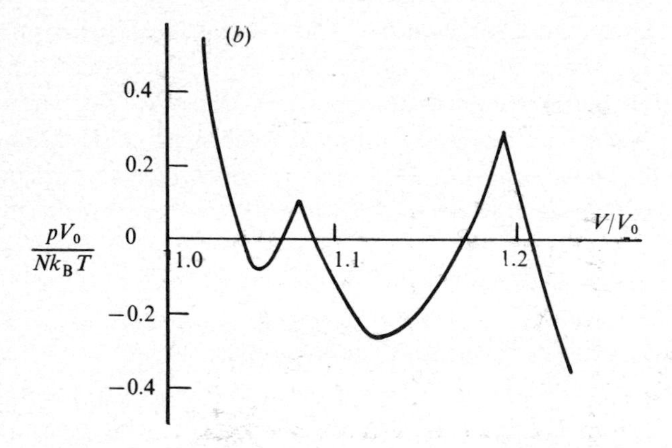

Fig. 2.1.2. (a) Variation of the equilibrium value of the positional order parameter q and the orientational order parameter s with zW/k_BT for $\nu = 0.3$. (b) Theoretical isotherm for $\nu = 0.3$ showing a rotational transition in the solid state followed by the melting transition. (The isotherm is drawn for the solid rotational transition temperature $k_BT/\varepsilon = 0.593$.)

transitions; the positional melting now precedes the rotational melting, and there occurs an intermediate phase similar to the nematic liquid crystal with orientational order but no translational order. (Over a small range of ν, $0.975 < \nu < 1.047$, the theory predicts a second order (continuous) nematic–isotropic transition, whereas experimentally it is found

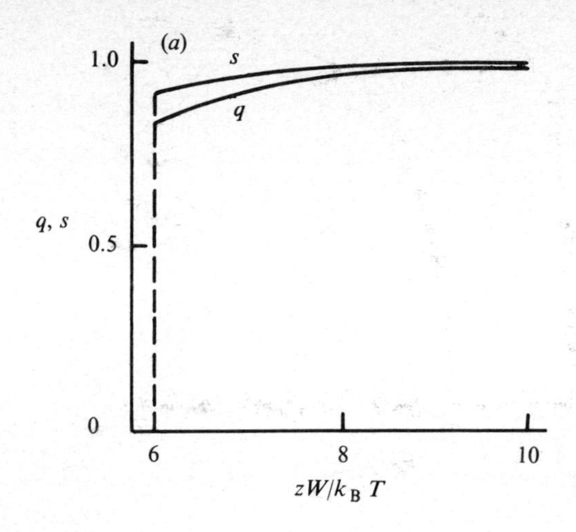

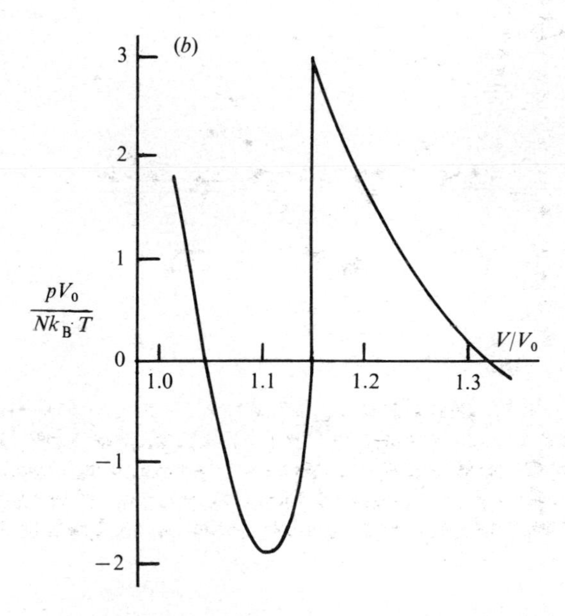

Fig. 2.1.3. (*a*) Variation of the equilibrium values of q and s with $zW/k_B T$ for $\nu = 0.8$. (*b*) Theoretical isotherm for $\nu = 0.8$ showing a single transition in which both positional and orientational orders collapse simultaneously ($k_B T/\varepsilon = 0.625$).

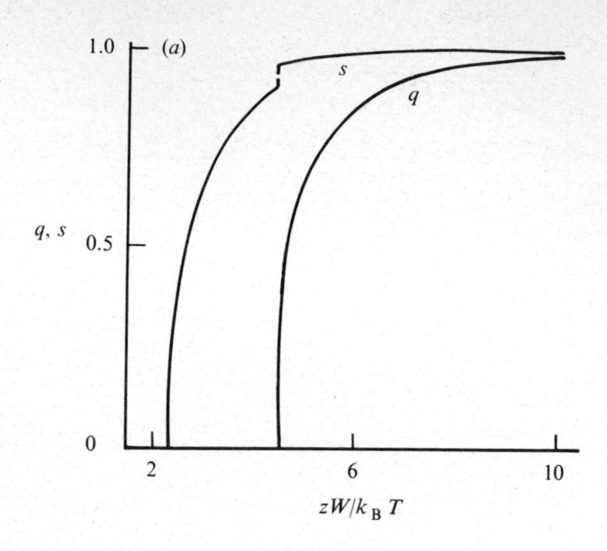

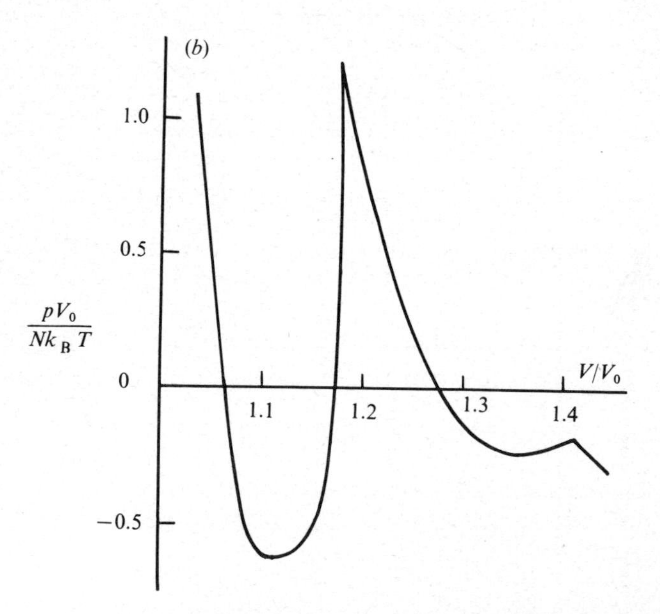

Fig. 2.1.4. (*a*) Variation of the equilibrium values of q and s with zW/k_BT for $\nu = 1.3$. (*b*) Theoretical isotherm for $\nu = 1.3$ showing solid–nematic and nematic–isotropic transitions. (The isotherm is drawn for the solid–nematic transition temperature $k_BT/\varepsilon = 0.678$.) (After reference 5.)

that this transition is always discontinuous, though weakly so. This limitation of the model probably arises from the restriction that the molecule can take up only two orientations. An extension of the treatment to more than two discrete orientations has been given by Amzel and Becka[11] for low ν.)

The entropy S, specific heat c_V and isothermal compressibility β can be evaluated from the thermodynamic relations:

$$S'' = -\left(\frac{\partial F''}{\partial T}\right)_V,$$

$$c''_V = k_B\left[T^2\left(\frac{\partial^2 \log \Omega}{\partial T^2}\right)_V + 2T\left(\frac{\partial \log \Omega}{\partial T}\right)_V\right],$$

$$\frac{1}{\beta''} = -k_B TV\left(\frac{\partial^2 \log \Omega}{\partial V^2}\right)_T.$$

Figs. 2.1.5–2.1.7 give the reduced temperatures of transition and the corresponding entropy and volume changes as functions of ν. For a

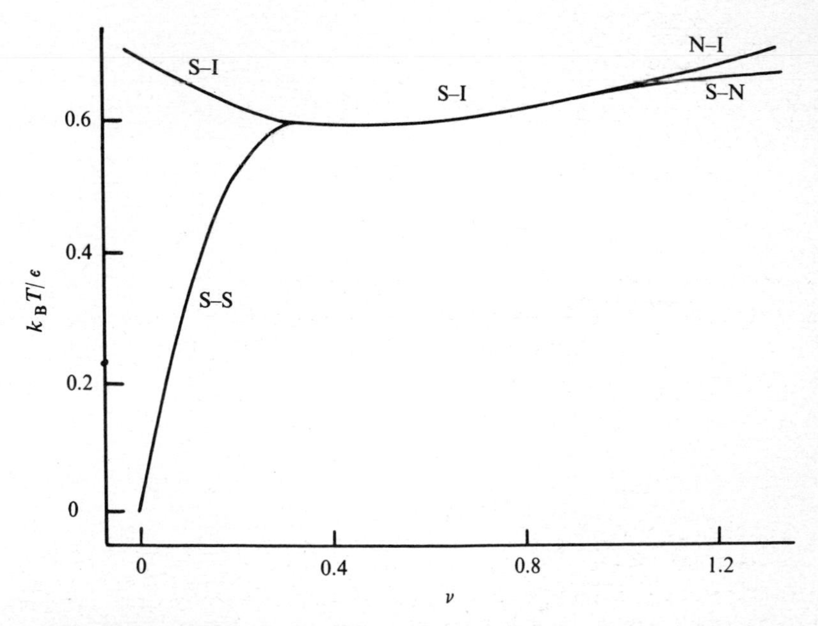

Fig. 2.1.5. Reduced transition temperature versus ν. S–S, solid–solid; S–I, solid–isotropic liquid; S–N, solid–nematic; N–I, nematic–isotropic. (After reference 6.)

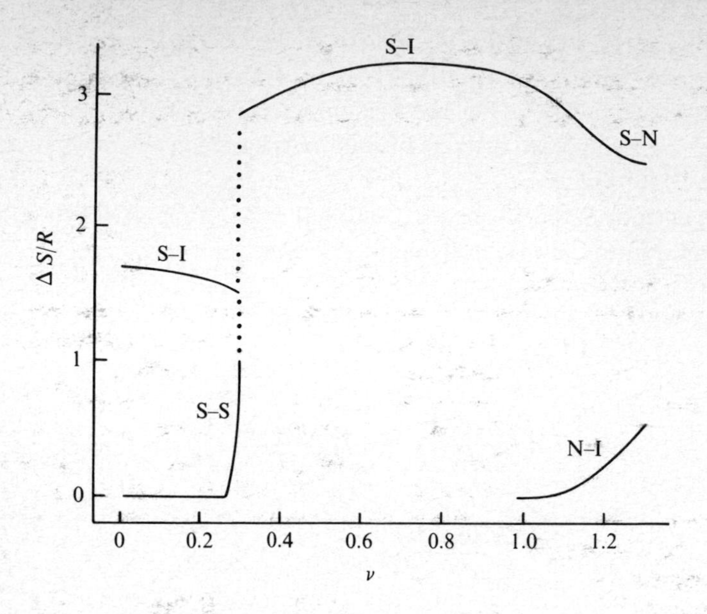

Fig. 2.1.6. Entropy of transition versus ν (see legend of fig. 2.1.5).

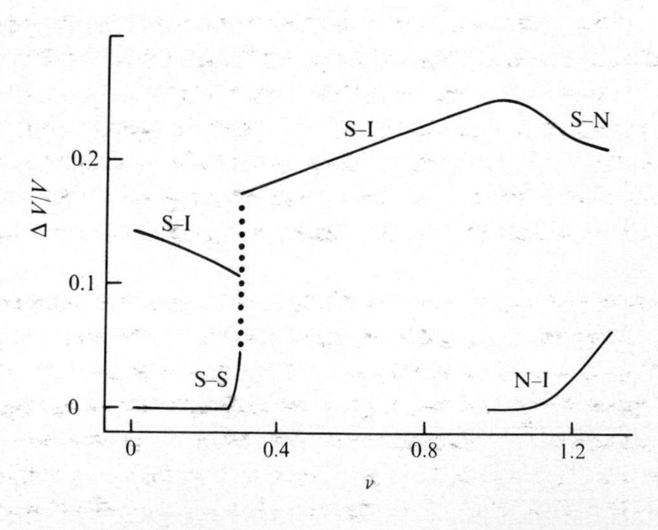

Fig. 2.1.7. Relative change of volume on transition versus ν (see legend of fig. 2.1.5).

certain range of ν, $\Delta S/R$ and $\Delta V/V$ for the nematic–isotropic transition are only very small fractions of the values for the solid–nematic transition. This is indeed a distinctive feature of such transitions in general (table 2.1.2).

TABLE 2.1.2

Compound	Solid–nematic transition		Nematic–isotropic transition	
	Latent heat (kJ mole^{-1})	Volume change $\Delta V/V$ (%)	Latent heat (kJ mole^{-1})	Volume change $\Delta V/V$ (%)
4-(4-n-Propylmercapto benzalamino)-azobenzene[12]	31.30	5.07	0.368	0.09
p-Azoxyanisole	29.57[14]	11.03[13]	0.574[14]	0.36[15]
p-Azoxyphenetole	26.87[14]	8.4[16]	1.366[14]	0.60[16]
4-Methoxybenzylidene-4′-butylaniline	18.06[17]	–	0.582[17]	0.40[18]

Curves for the order parameter, thermal expansion, specific heat and isothermal compressibility derived from the theory are presented in figs. 2.1.8–2.1.11, and it can be seen that they reproduce quite well the observed trends in the properties of the nematic phase. Thus while a simple model of this type cannot be expected to be applicable in detail to any particular substance, it does serve to bring out the effects of orientational disordering on the thermodynamics of the melting process.

However, as near-neighbour correlations are neglected completely in the Bragg–Williams approximation, the theory is unable to account for the specific heat and other anomalies in the isotropic phase. An attempt has been made to extend the theory by using the quasi-chemical or first approximation.[21] This results in a very weak anomaly in the specific heat above T_{NI}, but not in the thermal expansion or isothermal compressibility. It is clear that the model is far too elementary for describing these and other short range order effects in the isotropic phase (see § 2.4).

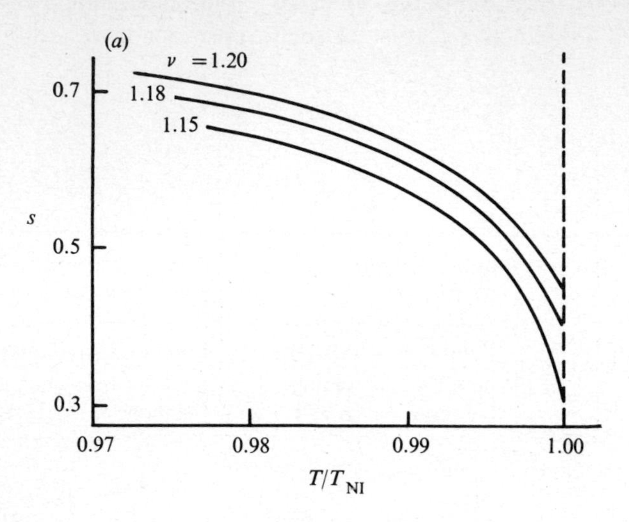

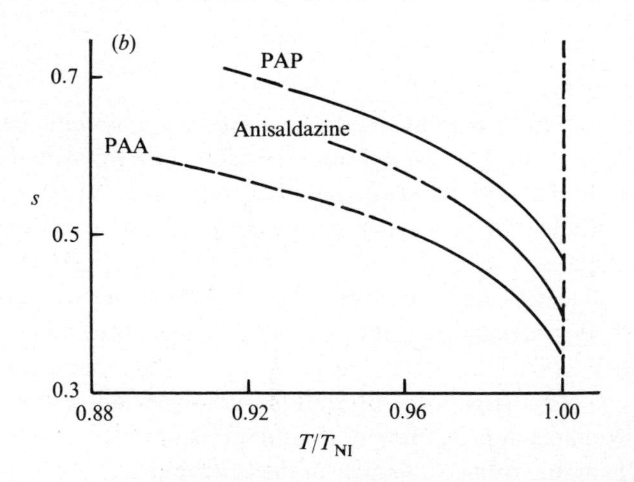

Fig. 2.1.8. Orientational order parameters versus temperature: (*a*) theoretical curves for $\nu = 1.15$, 1.18 and 1.20, (*b*) experimental curves of s defined by (2.3.1) for *p*-azoxyanisole (PAA), anisaldazine and *p*-azoxyphenetole (PAP).[19] The dashed portions of the curves represent supercooled regions. (After reference 5.)

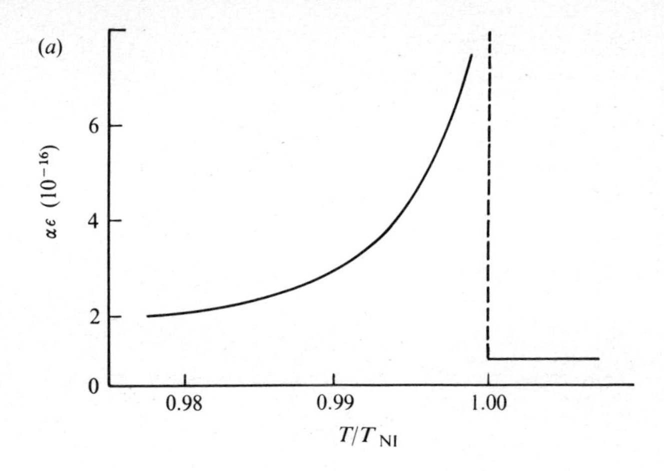

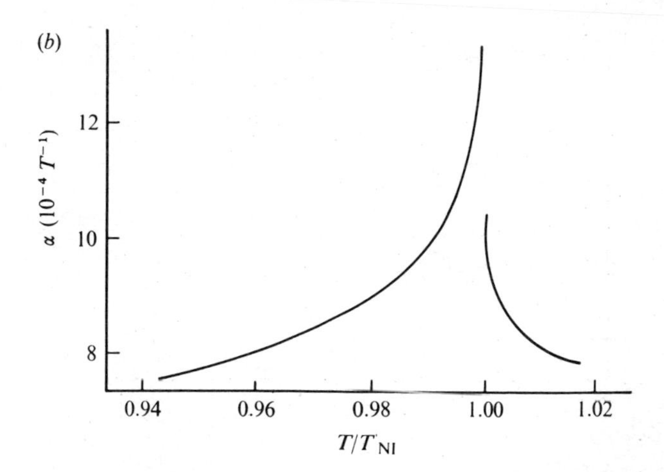

Fig. 2.1.9. Coefficient of thermal expansion in the nematic and isotropic phases: (*a*) theoretical curve for $\nu = 1.15$, (*b*) experimental curve for PAA.[15] (After reference 5.)

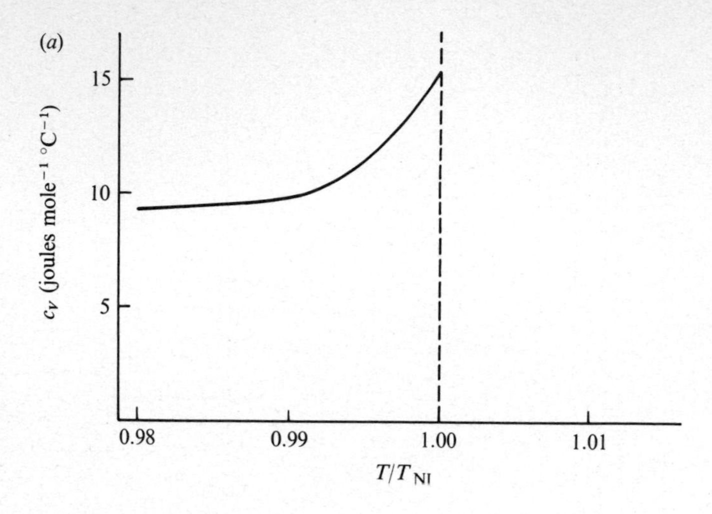

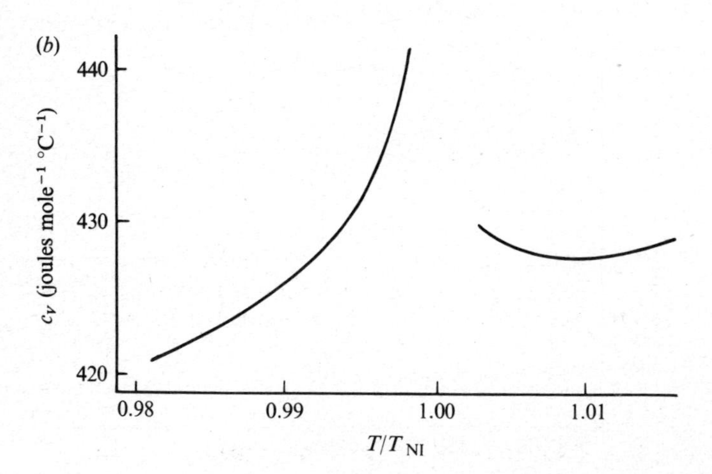

Fig. 2.1.10. Specific heat at constant volume in the nematic and isotropic phases: (*a*) theoretical contribution due to disorder for $\nu = 1.15$, (*b*) experimental curve for PAA derived from the observed values of c_p[14] and β.[20] (After reference 5.)

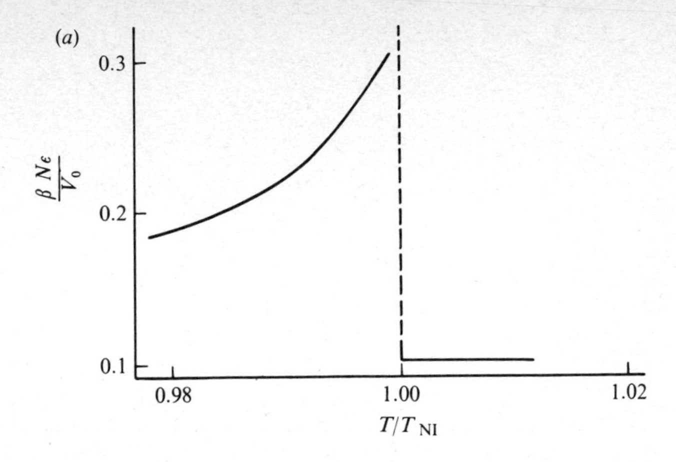

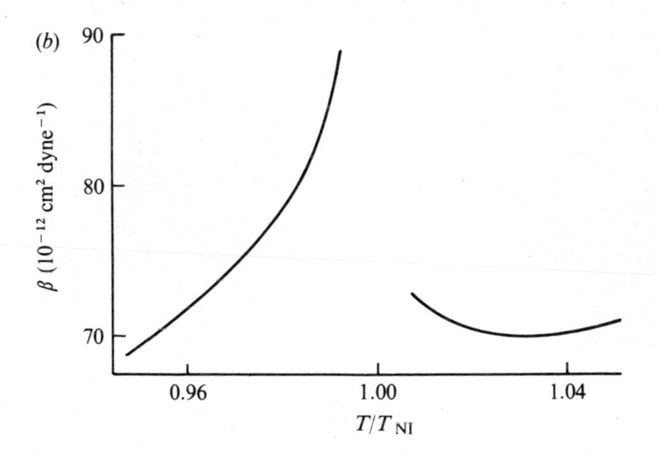

Fig. 2.1.11. Isothermal compressibility in the nematic and isotropic phases: (*a*) theoretical contribution due to disorder for $\nu = 1.15$, (*b*) experimental data for PAA.[20]

2.1.3. Pressure-induced mesomorphism

Theoretical phase diagrams for two representative values of ν are illustrated in fig. 2.1.12. In fig. 2.1.12(*a*) the orientational barrier is large enough for the nematic phase to occur at zero pressure. As the pressure is raised, both the solid–nematic and the nematic–isotropic transition temperatures increase, the slope $\mathrm{d}T/\mathrm{d}p$ for the latter being greater in accordance with the experimental data (fig. 2.1.13). Of course, such a behaviour is also to be expected from the Clausius–Clapeyron equation.

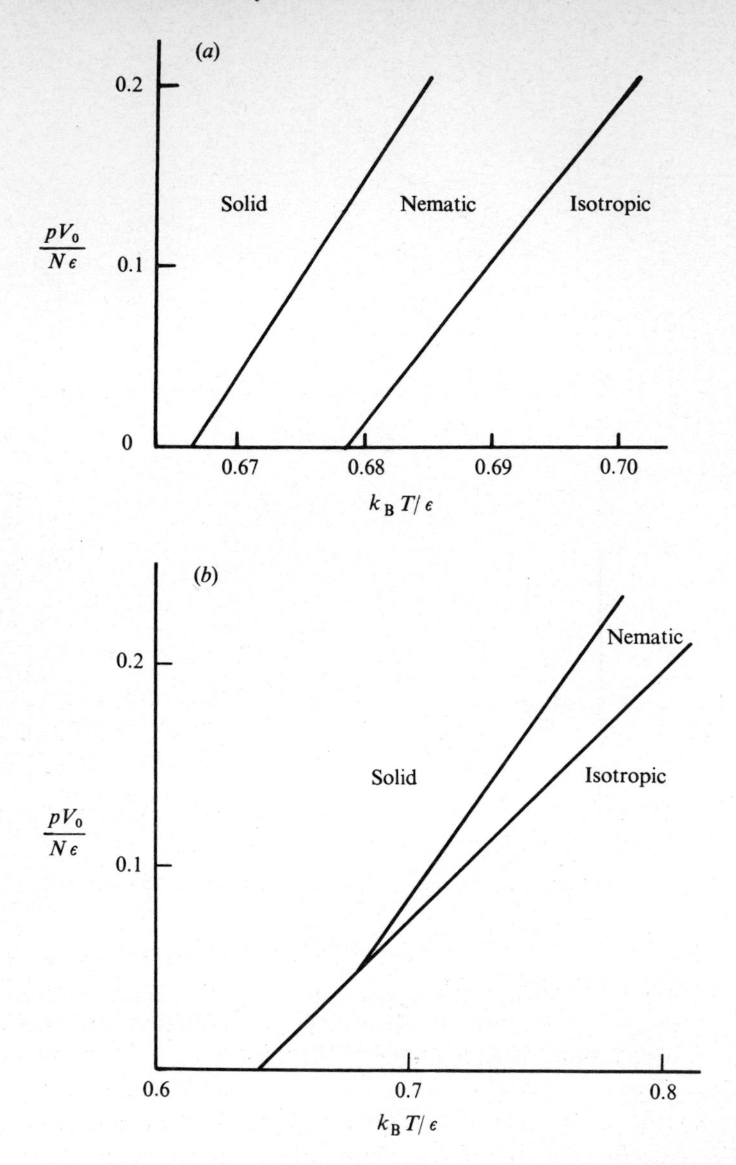

Fig. 2.1.12. Theoretical variation of the transition temperatures with pressure: (*a*) $\nu = 1.15$, (*b*) $\nu = 0.95$.

Fig. 2.1.12(*b*) represents a more interesting case. Here ν is just below the critical value for the nematic phase to occur at zero pressure. As the pressure is raised, there is initially a single transition, viz, the solid–liquid melting transition, but at higher pressures branching takes place and

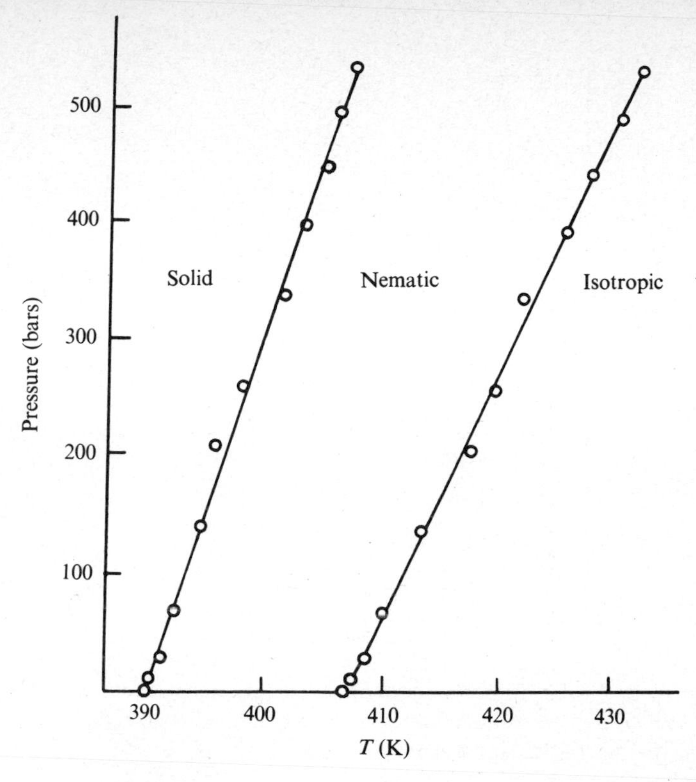

Fig. 2.1.13. Experimental phase diagram for PAA. (After reference 22.)

there are two transitions. The branching point represents the solid–nematic–isotropic *triple* point.

Experiments were done on the first two members of the *p*-n-alkoxybenzoic acid series to verify this prediction.[22] These two compounds, methoxy- and ethoxybenzoic acids, do not form liquid crystals at atmospheric pressure, whereas propoxybenzoic acid and the higher homologues show at least one liquid crystalline phase. As the pressure is raised, both compounds exhibit mesophases, initially a nematic phase and then, at higher pressures, a smectic phase as well (fig. 2.1.14). (The smectic phase is not included in the framework of the theory in its present form. It would require the introduction of another order parameter, viz, a lamellar order parameter (see chapter 5).) The transitions were detected by differential thermal analysis and the mesophases identified by microscopic observations using a high pressure optical cell (plate 5). These studies also established for the first time the existence of the solid–

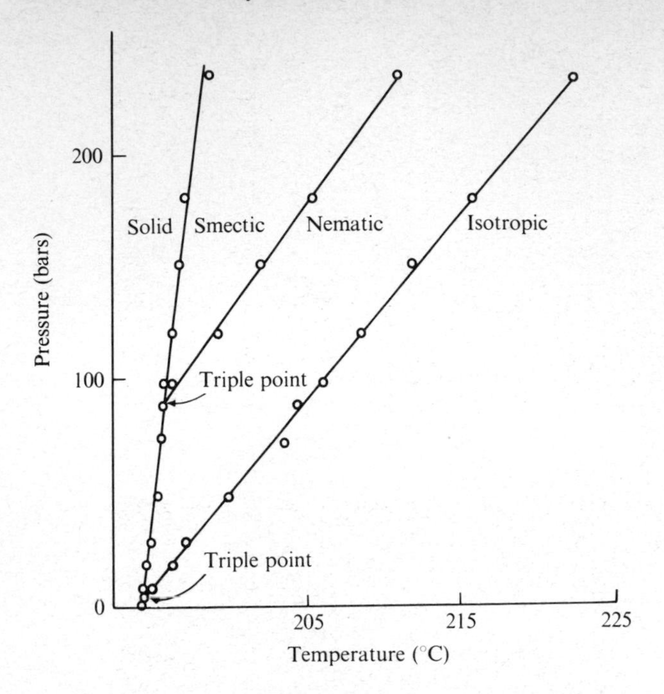

Fig. 2.1.14. Experimental phase diagram for *p*-ethoxybenzoic acid showing the solid–nematic–isotropic and solid–smectic–nematic triple points. (After reference 22.)

nematic–isotropic and the solid–smectic–nematic triple points in single component systems.

2.2. Phase transition in a fluid of hard rods

We have emphasized that the asymmetry of the molecular shape is an important factor in determining whether or not a substance exhibits the liquid crystalline phase. Consider a fluid of long thin rods without any forces between them other than the one preventing their interpenetration. At sufficiently low densities the rods can assume all possible orientations and the fluid will be isotropic. As the density increases, it becomes increasingly difficult for the rods to point in random directions and intuitively one may expect the fluid to undergo a transition to a more ordered anisotropic phase having uniaxial symmetry. That this does indeed happen was first proved by Onsager.[23] Other hard rod theories have since been proposed, which we shall refer to briefly at the end of this section, but Onsager's approach, based on an exact density expansion for the free energy, is still probably the most satisfying. However, the

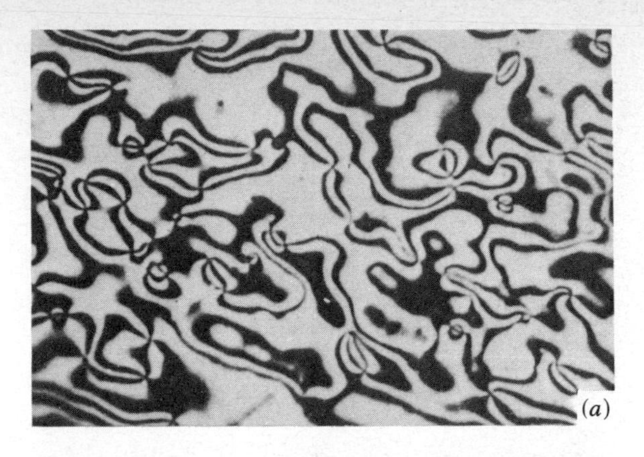

Plate 5. Pressure-induced mesophases in *p*-ethoxybenzoic acid: (*a*) nematic schlieren textures, (*b*) focal conics and bâtonnets in the smectic phase. (Reference 22.)

mathematical analysis involved in the theory is rather complex – even in the lowest order it leads to a non-linear integral equation. Truncation of the series after the linear term, as was done by Onsager, would appear to be satisfactory only for very long rods and not for shorter ones, but attempts to evaluate the coefficients in the expansion beyond that of the linear term have so far been unsuccessful because of computational difficulties.

Zwanzig[24] proposed a simplified version of Onsager's theory which enables the virial coefficients to be evaluated to a much higher order. The simplification consists in the following assumptions: (*a*) the rods can take up only discrete orientations along three mutually perpendicular axes, whereas in the original theory they can point in any direction, and (*b*) the length to breadth ratio of the rods is very large (tending to infinity). Though these assumptions may be somewhat too drastic as far as real liquid crystal systems are concerned[25] the model does serve to illustrate the essential principles of Onsager's method and may prove to be useful as a starting point for possible extensions and applications of the general theory.

In Zwanzig's model the rods are rectangular parallelopipeds of length l and square cross-sectional area d^2. Assuming that the potential energy of interaction of N such hard rods is a sum of pair potentials, we may write

$$U_N = \tfrac{1}{2} \sum_{i \neq j}^{N} U_{ij} \tag{2.2.1}$$

where U_{ij} is ∞ if the molecules i and j intersect and zero if they do not. The intersection may occur either in a parallel or perpendicular configuration as shown in fig. 2.2.1. Let the position vector of the jth rod be

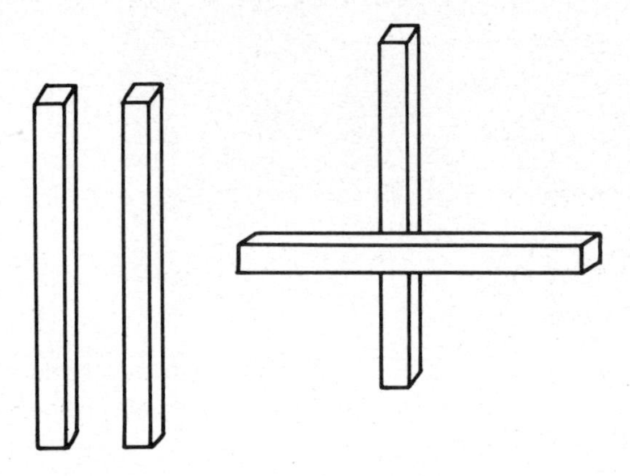

Fig. 2.2.1. Intersection of rods in the parallel and perpendicular configurations.

represented by $\mathbf{R}_j$ and its orientation by $\mathbf{u}_j$, and let the set of all positions be abbreviated by $\mathbf{R}$ and the set of all orientations by $\mathbf{u}$. The configurational integral for the system consisting of N rods in the volume V can then be expressed as

$$Q_N(T, V) = \frac{1}{N!} \frac{1}{3^N} \sum_{\mathbf{u}} \int \mathrm{d}\mathbf{R} \exp(-U_N/k_B T)$$

where $N!$ allows for the indistinguishability of the particles and 3^N for the total number of orientational states. The potential energy of the system depends on the sets **R** and **u**. Since the configurational space available for each molecule is V,

$$\int d\mathbf{R}\exp(-U_N/k_BT) = V^N \exp(-\varphi_N(\mathbf{u})/k_BT),$$

where the new function φ_N depends only on the orientations. Thus

$$Q_N = \frac{V^N}{N!3^N}\sum_{j_1=1}^{3}\ldots\sum_{j_N=1}^{3}\exp[-\varphi_N\{u_1(j_1)\ldots u_N(j_N)\}/k_BT].$$

Now for any given configuration of the system, there will be $N(1)$ molecules pointing in the direction $\mathbf{u}(1)$, $N(2)$ in the direction $\mathbf{u}(2)$ and $N(3)$ in the direction $\mathbf{u}(3)$. φ_N therefore becomes a function of the occupation numbers in the three allowed directions, i.e.,

$$\varphi_N = \varphi_N(N(1), N(2), N(3)).$$

The indistinguishability of the molecules requires that for any given set of occupation numbers $N(1)$, $N(2)$ and $N(3)$ there are $N!/N(1)!N(2)!N(3)!$ equivalent configurations, so that

$$Q_N = \frac{V^N}{N!3^N}\sum_{N(1)=0}^{N}\sum_{N(2)=0}^{N}\sum_{N(3)=0}^{N}\frac{N!\exp(-\varphi_N/k_BT)}{N(1)!N(2)!N(3)!},$$

where $N(1)+N(2)+N(3)=N$. Putting

$$t(N(1), N(2), N(3)) = \frac{V^N}{3^N}\frac{\exp(-\varphi_N/k_BT)}{N(1)!N(2)!N(3)!},$$

we have

$$Q_N = \sum_{N(1)=0}^{N}\sum_{N(2)=0}^{N}\sum_{N(3)=0}^{N} t(N(1), N(2), N(3)). \qquad (2.2.2)$$

To evaluate this configurational integral, we make use of the 'maximum term method' of statistical mechanics,[26] i.e., in the limit $N\to\infty$, $V\to\infty$, but $N/V=\rho=$ constant, Q_N can be approximated by the largest term t_{max} in the sum. The problem therefore reduces to one of maximizing t with respect to the occupation numbers. Using Stirling's approximation and introducing the mole fraction $x_\alpha = N(\alpha)/N$ of the various components,

$$N^{-1}\log t = 1-\log\rho-\log 3-\sum_{\alpha=1}^{3}x_\alpha\log x_\alpha-(\varphi_N/Nk_BT), \qquad (2.2.3)$$

subject to the condition

$$\sum_{\alpha=1}^{3}x_\alpha = 1.$$

To proceed further, we need to know the dependence of the function φ_N on density, which can be obtained by resorting to the virial expansion. Now, the molecules having different orientations may be regarded as belonging to different species, so that in effect we have a multicomponent system. The virial expansion of the free energy of a gaseous mixture in ascending powers of the density is well known:[27] for our present purpose we use the form

$$-\frac{\varphi_N}{Nk_{\mathrm{B}}T}=\sum_{N(1)}\sum_{N(2)}\sum_{N(3)}B(N(1),N(2),N(3))$$
$$\times\prod_{\alpha=1}^{3}(x_\alpha)^{N(\alpha)}\rho^{\Sigma N(\alpha)-1}, \tag{2.2.4}$$

where the virial coefficient $B(N(1), N(2), N(3))$ is defined as

$$B(N(1),N(2),N(3))=\frac{1}{VN(1)!N(2)!N(3)!}\int\Sigma\Pi f.$$

f is the Mayer function given by

$$f_{ij}=\exp(-U_{ij}/k_{\mathrm{B}}T)-1,$$
$$=\begin{cases}-1 & \text{if } i \text{ and } j \text{ intersect,}\\ 0 & \text{otherwise.}\end{cases}$$

$\int\Sigma\Pi f$ is the standard abbreviation in the cluster theory of imperfect gases which signifies the integral over all positions of a sum of products of Mayer f functions taken over all irreducible clusters (or graphs) involving $N(1)$ molecules of species 1, $N(2)$ molecules of species 2 and $N(3)$ molecules of species 3.

The maximum value of t is obtained from the condition

$$\frac{\partial}{\partial x_\alpha}(N^{-1}\log t)=0,\quad \alpha=1,2,3.$$

The configurational free energy of the system is given by

$$F=-k_{\mathrm{B}}T\log Q_N$$

and the pressure by

$$\frac{p}{Nk_{\mathrm{B}}T}=\frac{\partial}{\partial V}(N^{-1}\log t_{\mathrm{max}})_T. \tag{2.2.5}$$

Thus by determining t_{max}, the pressure can be derived as a function of the density.

Calculation of the virial coefficients The simplification introduced by restricting the rods to three discrete orientations will now be apparent. The rods must intersect only in those three directions, so that the cluster integrals in three dimensions can be factorized into products of cluster integrals in one dimension. Consider for example the second virial coefficient for parallel rods. By definition

$$B(2,0,0)=\frac{1}{V(2!0!0!)}\int d^3R_1\int d^3R_2 f_{12}(2,0,0).$$

The first volume integral is evaluated directly so that

$$B(2,0,0)=\tfrac{1}{2}\int d^3R_2 f_{12}(2,0,0).$$

We observe that the Mayer function f_{12} is zero everywhere except in the range $-l$ to l in the x direction and $-d$ to d in the other two directions. In the range where f is non-zero, its value is -1. Hence

$$B(2,0,0)=-4ld^2.$$

Similarly, the second virial coefficient for two rods in an orthogonal configuration

$$B(1,1,0)=\frac{1}{V(1!1!0!)}\int d^3R_1\, d^3R_{12} f_{12}(1,1,0)$$

$$=\int_{-(l+d)/2}^{(l+d)/2}\int_{-(l+d)/2}^{(l+d)/2}\int_{-d}^{d} d^3R_2=-2d(l+d)^2.$$

When the rods are very long and very thin, $l \gg d$,

$$B(1,1,0)\simeq -2l^2d[1+2(d/l)]$$

$$B(2,0,0)\simeq -4l^2d(d/l).$$

Thus, the virial coefficient for parallel configuration is smaller than that for orthogonal configuration by an order d/l. Higher order virial coefficients exhibit the same behaviour. To simplify the calculation Zwanzig neglected all terms of order d/l, i.e., he confined himself to the limit $l\to\infty$, $d\to 0$, $l^2d=$ constant, thereby reducing considerably the number of coefficients to be considered. It is readily verified that only coefficients of the type $B(m,n,0)$, $B(0,m,n)$ and $B(m,0,n)$ survive, and that by symmetry $B(m,0,n)=B(n,0,m)=B(0,m,n)=B(0,n,m)$.

Let us now choose the z axis ($i=3$) as a preferred direction. By symmetry the number of molecules in the x and y directions are equal, so that the mole fractions along the three directions can be expressed in terms of a single order parameter x:

$$x_1 = x,$$

$$x_2 = x,$$

$$x_3 = 1 - 2x.$$

Therefore, (2.2.4) reduces to

$$-\frac{\varphi_N}{Nk_\mathrm{B}T} = \sum_m \sum_n B(m, n, 0)\rho^{m+n-1} \times [x^{m+n} + x^m(1-2x)^n + x^n(1-2x)^m]. \tag{2.2.6}$$

The mole fraction x is then determined by

$$\frac{\mathrm{d}}{\mathrm{d}x}(N^{-1} \log t) = 0,$$

i.e.,

$$2 \log\left(\frac{x}{1-2x}\right) = \frac{\mathrm{d}}{\mathrm{d}x}(-\varphi_N/Nk_\mathrm{B}T).$$

For example, in the second virial approximation this yields

$$\log\left(\frac{x}{1-2x}\right) = -(2-6x). \tag{2.2.7}$$

The solution $x = \frac{1}{3}$ corresponding to the isotropic distribution is trivially satisfied at all densities and in any approximation. Below a certain critical density, (2.2.7) has only the isotropic root. Above this density two new roots appear; the one which corresponds to low x gives $t_{\max}$ and the others are discarded. Using (2.2.5), the pressure can then be evaluated as a function of density or volume. The transition predicted by Onsager's theory in the second virial approximation is confirmed by this model even when all virial coefficients up to the seventh are included but the properties of the anisotropic phase depend rather sensitively on the order of the approximation. Fig. 2.2.2 gives the isotherm obtained in the sixth approximation, and it can be seen that the shape of the curve is characteristic of a first order transition.

A Padé analysis of Zwanzig's model by Runnels and Colvin[28] has shown that the character of the transition is stable against increase in the order of the approximation. However, the model does not converge to the Onsager limit with increasing number of discrete orientations and

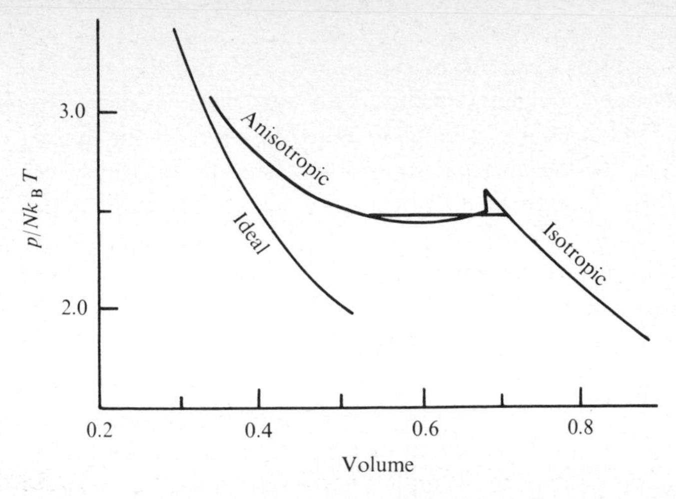

Fig. 2.2.2. Theoretical isotherm evaluated from Zwanzig's model in the sixth approximation showing a first order nematic–isotropic transition. (After Zwanzig.[24])

Straley[25] has concluded that it does not adequately represent real liquid crystal systems.

Other hard rod treatments Other statistical treatments of the hard rod fluid may be classified into two types, the lattice model and the 'scaled particle' method. Straley[29] has examined these different approaches critically and has concluded that for sufficiently long rods they are all equivalent to Onsager's theory, while for shorter rods no theory, including that of Onsager, can be considered to be accurate. However, the more recent formulation of the scaled particle approach due to Cotter[30] would appear to be quite promising in that it might be applicable even to shorter rods.

The lattice model was proposed originally by Flory[31] and has been discussed with modifications by Di Marzio[32] and by Alben.[33] In its simplest form, space is represented by a simple cubic array of cells. A rod of length L is a series of L full cells running along one of the cube axes. Let there be M cells and N rods, and let the fraction of the rods pointing in the ith direction be α_i. In the case of hard rods, no two rod segments may occupy the same cell. The aim is to find the orientational distribution α and the density N/M corresponding to the state of minimum Gibbs free energy at a given pressure and temperature in the limit N and M tending to be very large. Since there is no interaction between sites, the Gibbs free energy may be found in terms of the configurational entropy. The distribution which maximizes the entropy is isotropic or anisotropic in varying degrees depending on the density, the more highly ordered states being favoured at higher densities. The most extensive calculations based on this model are by Alben,[33] who has worked out amongst other properties, the expansivity, the compressibility and the specific heat. These three quantities do not show any pretransition effects in the ordered phase. However, when an attractive van der Waals term is added to the hard rod potential energy, pronounced pretransition effects are found which are in qualitative agreement with experiment. There are some quantitative differences which Alben has attributed to the neglect of short range order.

The scaled particle method was introduced by Reiss, Frisch and Lebowitz[34] in their treatment of hard spheres. It provides a means of deriving approximate expressions for the chemical potential and pressure of the fluid by considering the reversible work necessary to insert a scaled particle (i.e., a 'solute' particle which is a scaled replica of the 'solvent' particles) at an arbitrary point in the fluid. The method was first applied to hard rods by Cotter and Martire[35] assuming restricted orientations (as in Zwanzig's model) and extended by Lasher[36] for continuous orientations. The theory has recently been presented in a more rigorous manner by Cotter.[30] The transition parameters and the properties of the anisotropic phase predicted by this theory for a system of hard spherocylinders, i.e., cylinders capped at either end by a hemisphere, are in qualitative agreement with those given by lattice model calculations. A noteworthy feature of Cotter's theory is that for very long rods it reduces to Onsager's theory, while at the other extreme when the length to breadth ratio is 1 it is equivalent to the hard sphere theory of Reiss *et al.* This indicates that Cotter's approach is likely to give satisfactory results for shorter rods.

Hard rod models have been used to study the following problems:

(i) The effect of molecular shape (cylinders, oblate spheroids, prolate spheroids,[37] discs[38] etc.). The question of mesophases occurring in systems of disc-shaped particles is of fundamental importance. The calculations of Runnels and Colvin[38] show that an isotropic–nematic transition is possible, in principle, in such systems. So far, the only experimental evidence for the formation of mesophases of disc-shaped molecules is that reported by Brooks and Taylor.[39] They found that during the carbonization of graphitizable organic materials droplets are formed having a high degree of molecular order with the aromatic sheets stacked in a parallel array. Their optical textures are somewhat similar to those of ordinary nematic liquid crystals.

(ii) The influence of molecular flexibility on the density change at the transition.[40] A feature that is common to all rigid particle theories is that the density change at the isotropic–nematic transition is far too large to be applicable to real systems. Wulf and De Rocco[40] have shown that this discrepancy can be accounted for at least partly by considering semi-flexible rods. This has been done by assuming the rods to consist of l segments connected by $l-1$ bonds such that the bonds may lie along 3 mutually perpendicular directions. The molecule can now assume several bent configurations so that the gain in packing facility on entering the aligned phase is quite small and the density change is accordingly much less.

(iii) The contribution of the alkyl end-chain of the molecules by considering rods with semi-flexible tails[41] (see § 2.3.4).

(iv) The properties of mixtures (rods and plates,[42] rods and cubes,[43] etc.).

Further developments in the statistical mechanics of the hard rod fluid, especially computer generated experiments of idealized systems, are of considerable interest as they should yield a wealth of microscopic data of the kind that hard sphere systems have provided in elucidating the dynamical properties of the liquid state.[44]

2.3. The Maier–Saupe theory and its applications

An approach that has proved to be extremely useful in developing a theory of spontaneous long range orientational order and the related properties of the nematic phase is the molecular field method, closely analogous to that introduced by Weiss in ferromagnetism. Each molecule is assumed to be in an average orienting field due to its environment, but otherwise uncorrelated with its neighbours. The first molecular field theory of the nematic state was proposed in 1916 by Born[45] who treated the medium as an assembly of permanent electric dipoles and demonstrated the possibility of a transition from an isotropic phase to an anisotropic one as the temperature is lowered. Though this result is important qualitatively we need not discuss this particular theory because it is now well established that permanent dipole moments are not necessary for the occurrence of the liquid crystalline phase (see (1.2.3)). Moreover, the theory predicts that the aligned phase should be ferroelectric, which does not appear to be the case even when the molecules are polar. The most widely used treatment based on the molecular field approximation is that due to Maier and Saupe.[46]

2.3.1. Definition of long range orientational order

We begin by defining the long range orientational order parameter in the nematic phase. Suppose that the liquid crystal is composed of rod-like molecules in which (i) the distribution function is cylindrically symmetric about the axis of preferred orientation **n** and (ii) the directions **n** and −**n** are fully equivalent, i.e., the preferred axis is non-polar. Subject to these two symmetry properties, and assuming the rods to be cylindrically symmetric, the simplest way of defining the degree of alignment is by the parameter s, first introduced by Tsvetkov,[47]

$$s = \tfrac{1}{2}\langle 3\cos^2\theta - 1\rangle, \tag{2.3.1}$$

where θ is the angle which the long molecular axis makes with **n**, and the angular brackets denote a statistical average. For perfectly parallel alignment $s = 1$, while for random orientations $s = 0$. In the nematic phase, s has an intermediate value which is strongly temperature dependent.

The order parameter can be directly related to certain experimentally determinable quantities – say, for example, the diamagnetic anisotropy of the liquid crystal.[48] Let us choose a space-fixed cartesian coordinate system xyz with z parallel to **n**. If η_1 and η_2 are the principal diamagnetic susceptibilities of the molecule referred to its own principal axes, the average z component of the susceptibility per unit volume in the nematic phase is evidently

$$\begin{aligned}\chi_z &= n[\eta_1\langle\cos^2\theta\rangle + \eta_2\langle\sin^2\theta\rangle]\\ &= n[\bar{\eta} + \tfrac{2}{3}(\eta_1 - \eta_2)s]\end{aligned} \tag{2.3.2}$$

where $\bar{\eta} = \tfrac{1}{3}(\eta_1 + 2\eta_2)$ and n the number of molecules per unit volume. Similarly

$$\chi_x = \chi_y = n[\bar{\eta} - \tfrac{1}{3}(\eta_1 - \eta_2)s]. \tag{2.3.3}$$

Therefore

$$s = (\chi_z - \chi_x)/n(\eta_1 - \eta_2). \tag{2.3.4}$$

η_1 and η_2 can be obtained from the susceptibilities of the solid crystal if the structure is known. Thus a measurement of χ_z, χ_x or the anisotropy $\chi_z - \chi_x$, in the nematic phase gives the absolute value of s. In a similar manner, s can be related to the optical anisotropy (birefringence[49] and linear dichroism[48]) though, as is well known, there is a difficulty in this case as there is no rigorous way of correcting for the effects of the polarization field in the medium. Approximate methods have been proposed[48,50] which appear to yield satisfactory results.[19,51,52]

Another important tool for investigating molecular order is nuclear magnetic resonance spectroscopy.[53,54] The principle of this method is that if there is a pair of protons in a molecule, each spin is coupled to the external magnetic field **H** as well as to the dipole field created by its neighbour. The latter gives rise to a small perturbation of the eigenstates of the Zeeman Hamiltonian so that the signal appears as a doublet. The doublet separation is easily seen to be proportional to $(3\cos^2\theta - 1)/r_{ij}^3$ where θ is the angle which the interproton vector $\mathbf{r}_{ij}$ makes with **H**. In the isotropic phase $\mathbf{r}_{ij}$ can assume all possible orientations with respect to **H** and $\langle 3\cos^2\theta - 1\rangle$ vanishes, but in the nematic phase there is a splitting which is a measure of the order parameter s. The analysis is, of course, not so straightforward for complex molecules as it requires precise knowledge of the interproton distances and angles. The quadrupole splitting in the magnetic resonance spectrum of ^{14}N[55] or of deuterium[56] has also been employed for studying molecular order. Other methods involve the use of nuclear magnetic resonance[57] or electron spin resonance spectroscopy[58] of geometrically anisotropic molecules aligned in nematic solvents. We shall not discuss any of these techniques here as a number of excellent reviews on the subject are available.[59–64] In fig. 2.3.1 we present recent determinations of the order parameter of PAA by three different methods.

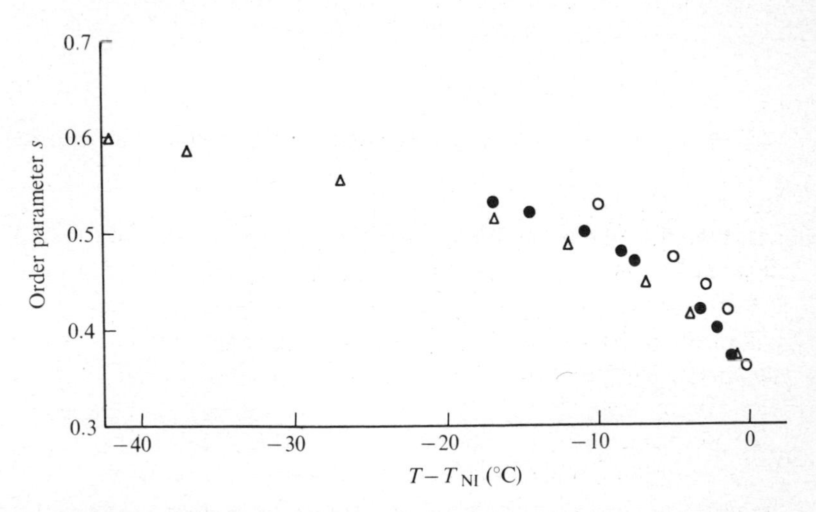

Fig. 2.3.1. The temperature variation of the long range orientational order parameter s in PAA. ○ from NMR measurements (McColl and Shih[65]), ● from diamagnetic anisotropy (Gasparoux, Regaya and Prost[66]), △ from refractive indices.[64]

NMR and dielectric measurements (see § 2.3.5) indicate that there is a fair degree of rotational freedom of the molecules about their long axes. Nevertheless the assumption that the molecules are cylindrically symmetric is not valid in general and can sometimes lead to errors in the determination of s.[67] For a molecule of arbitrary shape, (2.3.1) may be written in the generalized form[68]

$$s_{ij}^{\alpha\beta} = \tfrac{1}{2}\langle 3 i_\alpha j_\beta - \delta_{\alpha\beta}\delta_{ij}\rangle \tag{2.3.5}$$

where $\alpha, \beta = x, y, z$ refer to the space-fixed axes, $i, j = x', y', z'$ refer to the principal axes of the molecule, i_α, j_β denote the projection of the unit vectors $\mathbf{i}$, $\mathbf{j}$ along α, β, and $\delta_{\alpha\beta}$, δ_{ij} are Krönecker deltas. $s_{ij}^{\alpha\beta}$ is symmetric in i, j and in α, β. It is also a traceless tensor with respect to either pair, i.e.,

$$s_{ij}^{\alpha\alpha} = s_{ii}^{\alpha\beta} = 0,$$

where repeated tensor indices imply the usual summation convention. The generalization of (2.3.2) and (2.3.3) is

$$\chi_{\alpha\beta} - \tfrac{1}{3}\delta_{\alpha\beta}\chi_{\gamma\gamma} = n\eta_{ij}s_{ij}^{\alpha\beta},$$

where η_{ij} is the molecular susceptibility tensor which can in principle be determined from the crystal data.

$s_{ij}^{\alpha\beta}$ is a 3×3 matrix. When diagonalized the matrix has the form

$$s_{\alpha\beta} = \begin{bmatrix} s_{11} & 0 & 0 \\ 0 & s_{22} & 0 \\ 0 & 0 & -(s_{11}+s_{22}) \end{bmatrix}.$$

In the uniaxial nematic case, $s_{11} = s_{22}$.

If the molecule is flexible more parameters are required to define the degree of alignment of its different parts.[63] However, in a large part of the discussion that follows on the theory of the nematic state we shall ignore all these details and assume the molecules to be cylindrically symmetric rods.

2.3.2. The mean field approximation

We assume that each molecule is subject to an average internal field which is independent of any local variations or short range ordering. Consistent with the symmetry of the structure, viz, the cylindrical distribution about the preferred axis and the absence of polarity, we may postulate that the orientational energy of a molecule

$$u_i \propto \tfrac{1}{2}(3\cos^2\theta_i - 1)s,$$

where θ_i is the angle which the long molecular axis makes with the preferred axis and s is given by (2.3.1).

The exact nature of the intermolecular forces need not be specified for the development of the theory. However in their original presentation Maier and Saupe[46] assumed that the stability of the nematic phase arises from the dipole–dipole part of the anisotropic dispersion forces. The second order perturbed energy of the Coulomb interaction between a pair of molecules 1 and 2 is given by

$$W_{00} = \frac{1}{R^6} \sum_{\mu\nu\neq 00} \frac{1}{E_{\mu\nu} - E_{00}} \left| \sum_{X,Y,Z} \left[\left(3\frac{X^2}{R^2} - 1 \right) x_{0\mu}^{(1)} x_{0\nu}^{(2)} + \frac{3}{R^2} XY \left(y_{0\mu}^{(1)} x_{0\nu}^{(2)} + x_{0\mu}^{(1)} y_{0\nu}^{(2)} \right) \right] \right|^2, \tag{2.3.6}$$

where $x_{0\mu}^{(1)}$, $x_{0\nu}^{(2)}$ etc. are the components of the transition moments defined as

$$x_{0\mu}^{(1)} = \sum_i \int \psi_0^{(1)*} e_i^{(1)} x_i^{(1)} \psi_\mu^{(1)} \, \mathrm{d}\tau^{(1)},$$

$$x_{0\nu}^{(2)} = \sum_k \int \psi_0^{(2)*} e_k^{(2)} x_k^{(2)} \psi_\nu^{(2)} \, \mathrm{d}\tau^{(2)}, \text{ etc.,}$$

$\psi_\mu^{(1)}$, $\psi_\nu^{(2)}$ and E_μ, E_ν are respectively the eigenfunctions and eigenvalues of the unperturbed molecules corresponding to states μ and ν, $E_{\mu\nu} = E_\mu + E_\nu$, $e_i^{(1)}$, $x_i^{(1)} y_i^{(1)} z_i^{(1)}$ are the charges and their coordinates in the first molecule measured in the space-fixed coordinate system with the origin at the centre of mass of molecule 1, $\mathbf{R} = (X, Y, Z)$ is the position of the centre of mass of molecule 2, $e_k^{(2)}$, $x_k^{(2)} y_k^{(2)} z_k^{(2)}$ are the charges and their coordinates in molecule 2 measured in the coordinate system obtained by translating the one defined above by XYZ. The second summation in (2.3.6) is over terms with x, y and z changed cyclically.

We assume that the rotational motions of the molecules are independent of their translational motions. Let $\xi^{(1)}\eta^{(1)}\zeta^{(1)}$ and $\xi^{(2)}\eta^{(2)}\zeta^{(2)}$ be the two molecular coordinate systems with their origins at the centres of mass, ζ coinciding with the long axis of each molecule. We define the Eulerian angles as follows: θ = the angle between z and ζ, Θ = angle between ξ and the normal to the $z\zeta$ plane and Θ' = angle between x and the normal to the $z\zeta$ plane. The components of the transition moments are

$$\xi_{0\mu} = \sum_i \int \varphi_0^{(1)*} e_i^{(1)} \xi_i^{(1)} \varphi_\mu^{(1)} \, \mathrm{d}\tau, \text{ etc.,}$$

where φ is the eigenfunction in the $\xi\eta\zeta$ coordinate system. Suppose now that the nematic medium has cylindrical symmetry about z so that all values of Θ' are equally probable, and further that the molecules rotate freely about ζ so that all values of Θ are equally probable. Setting $\langle\cos\theta^{(1)}\cos\theta^{(2)}\rangle=0$, i.e., the molecules do not distinguish between $\theta=0$ and π, and assuming a spherical distribution of the intermolecular radius vectors, the *orientation dependent* part of the potential energy of the ith molecule can be written in the mean field approximation as

$$u_i=-\tfrac{1}{2}(3\cos^2\theta_i-1)\langle\tfrac{1}{2}(3\cos^2\theta-1)\rangle \times \sum_{\mu\nu\neq 00}\frac{\delta_{0\mu}\delta_{0\nu}}{E_{\mu\nu}-E_{00}}\sum_k\frac{M(R_{ik})}{R_{ik}^6}, \tag{2.3.7}$$

where M is a function of the intermolecular vector $\mathbf{R}_{ik}$ only and

$$\delta_{0\mu}=\tfrac{1}{3}[|\zeta_{0\mu}|^2-\tfrac{1}{2}(|\xi_{0\mu}|^2+|\eta_{0\mu}|^2)]$$

is a measure of the anisotropy of the 0–μ transition. Maier and Saupe expressed (2.3.7) in the simple form

$$u_i=-\frac{A}{V^2}s\left(\frac{3\cos^2\theta_i-1}{2}\right) \tag{2.3.8}$$

where V is the molar volume and A is taken to be a constant independent of pressure, volume and temperature.

Cotter[69] has recently examined the postulates underlying the mean field approximation in the light of Widom's analysis of this general problem and has concluded that thermodynamic consistency requires that u_i should be proportional to V^{-1} regardless of the nature of the intermolecular pair potential. However, in what follows we have assumed a V^{-2} dependence as in the original formulation of the theory by Maier and Saupe.

2.3.3. Evaluation of the order parameter

The excess thermodynamic properties of the ordered system relative to those of the disordered one can now readily be derived on the basis of (2.3.8). The internal energy per mole

$$U=\frac{N}{2}\frac{\int_0^1 u_i\exp(-u_i/k_BT)\,\mathrm{d}(\cos\theta_i)}{\int_0^1\exp(-u_i/k_BT)\,\mathrm{d}(\cos\theta_i)}=-\tfrac{1}{2}Nk_BTBs^2, \tag{2.3.9}$$

where $B=A/k_BTV^2$ and N is the Avogadro number. The partition function for a single molecule

$$f_i=\int_0^1\exp(-u_i/k_BT)\,\mathrm{d}(\cos\theta_i), \tag{2.3.10}$$

so that the entropy

$$S=-Nk_B\left[\tfrac{1}{2}Bs(2s+1)-\log\int_0^1 \exp(\tfrac{3}{2}Bs\cos^2\theta_i)\,d(\cos\theta_i)\right]. \tag{2.3.11}$$

The Helmholtz free energy

$$\begin{aligned}F&=U-TS\\&=Nk_BT\left[\tfrac{1}{2}Bs(s+1)-\log\int_0^1 \exp(\tfrac{3}{2}Bs\cos^2\theta_i)\,d(\cos\theta_i)\right].\end{aligned} \tag{2.3.12}$$

The condition for equilibrium is

$$\left(\frac{\partial F}{\partial s}\right)_{V,T}=0,$$

or

$$\frac{3s\,\partial\langle\cos^2\theta_i\rangle}{\partial s}-3\langle\cos^2\theta_i\rangle+1=0.$$

This equation is satisfied when

$$\langle\cos^2\theta_i\rangle=\langle\cos^2\theta\rangle, \tag{2.3.13}$$

which may be called the *consistency* relation.

The plot of F versus s at different temperatures evaluated from (2.3.12) is shown in fig. 2.3.2. The minimum of the free energy occurs at that value of s which satisfies the consistency relation (2.3.13). When $T<T_{NI}$, there is only one minimum which corresponds to the stable ordered phase. When T is slightly greater than T_{NI} there are two minima, but the one at $s=0$, i.e., the isotropic phase, represents the absolute minimum. At $T=T_{NI}$, there are again two minima, one at $s=0$ and the other at $s=s_c$, but the two states now have equal free energies; at this temperature therefore a discontinuous transition takes place with no change of volume but an abrupt change in the order parameter. Putting $F=0$ at the transition yields

$$A/k_BT_{NI}V_c^2=4.541, \tag{2.3.14}$$

$$s_c=0.4292, \tag{2.3.15}$$

where V_c is the molar volume of the *nematic* phase at T_{NI}. As an example, for PAA $T_{NI}=408$ K, $V_c=225$ cm^3, and $A=13.0$ erg cm^6.

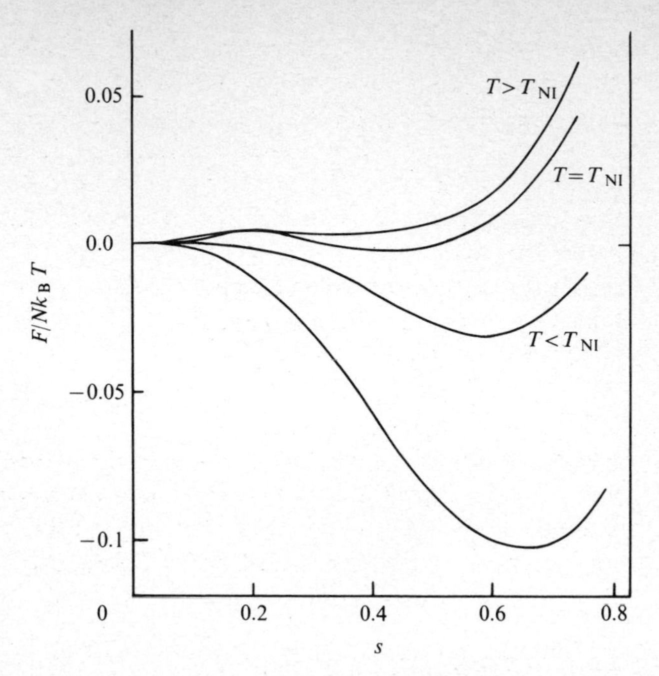

Fig. 2.3.2. Variation of the free energy with the order parameter calculated from the Maier–Saupe theory for different values of $A/k_B TV^2$. The minima in the curves occur at values of s which fulfil the consistency relation (2.3.13).[70]

Strictly speaking it is the Gibbs free energy $G = U - TS + PV$ which must be equated to zero at the transition. This leads to the following expression for the volume change at T_{NI} [46,70]:

$$\Delta V = -2F(\partial F/\partial V)_{T=T_{NI}}$$

or

$$\frac{\Delta V}{V} = \frac{1}{Bs_c^2}\left[2 \log \int_0^1 \exp(\tfrac{3}{2}Bs_c \cos^2\theta)\mathrm{d}(\cos\theta) - Bs_c(s_c+1)\right], \tag{2.3.16}$$

from which s_c can be determined if ΔV is known from experiment. However, ΔV is usually so small (see table 2.1.2) that s_c derived by this method differs hardly by 1–2 per cent from that given by (2.3.15). In effect, therefore, the theory predicts a universal value of $s_c \simeq 0.44$ for all nematic materials. While this value is in fair agreement with the observed data for a number of compounds, there are systematic deviations.[51,59] The introduction of higher order terms in the Maier–Saupe potential function to allow for contributions other than the purely dipole–dipole part of the dispersion energy has been suggested,[70–72] which enables calculations to be made for specific cases. Further, the long wavelength orientational fluctuations in the medium (see § 3.9) will diminish the effective order parameter and this has to be taken into account in the theory.[61] The assumption that the molecule is a rigid rod is also an oversimplification. A realistic treatment of the role of the flexible end-chain in the ordering process has been made by Marcelja,[73] whose theory we shall discuss presently.

Empirical data suggest that the volume dependence of u_i may be different from that assumed in (2.3.8). In general, if $u_i \propto V^{-m}$, where m is a number, then (2.3.14) gets modified to

$$A/k_B T_{NI} V_c^m = 4.541 \qquad (2.3.17)$$

but s_c still has the same value. The rate of variation of s with temperature does not depend too critically on the exponent m, as the isobaric change of volume with temperature over the entire nematic range is usually only of the order of 1–2 per cent (fig. 2.3.3). However its precise value

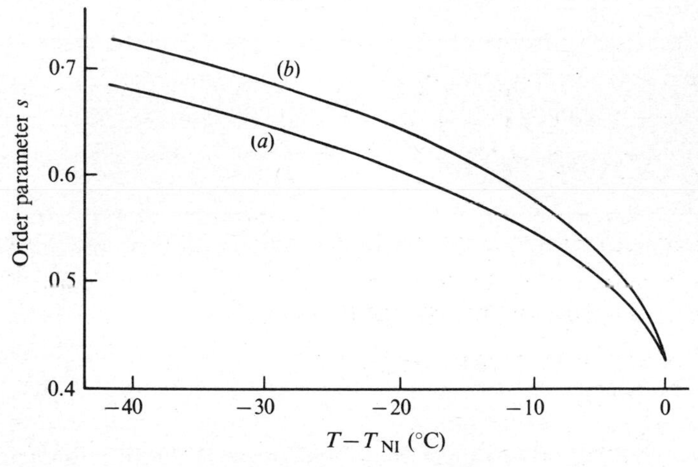

Fig. 2.3.3. Order parameter. s versus temperature predicted by the Maier–Saupe theory. (a) $u_i \propto V^{-2}$; (b) $u_i \propto V^{-4}$.

becomes rather important in accounting for the effect of pressure on the order parameter. Pressure studies[65,74] have established the following results for PAA:

(i) $(\partial \log T/\partial \log V)_s = -4$ (fig. 2.3.4). (2.3.18)

(ii) The thermal range of the nematic phase at constant volume is about 2.5 times that at constant pressure.

(iii) The order parameter s_c at the transition point is very nearly independent of the pressure.

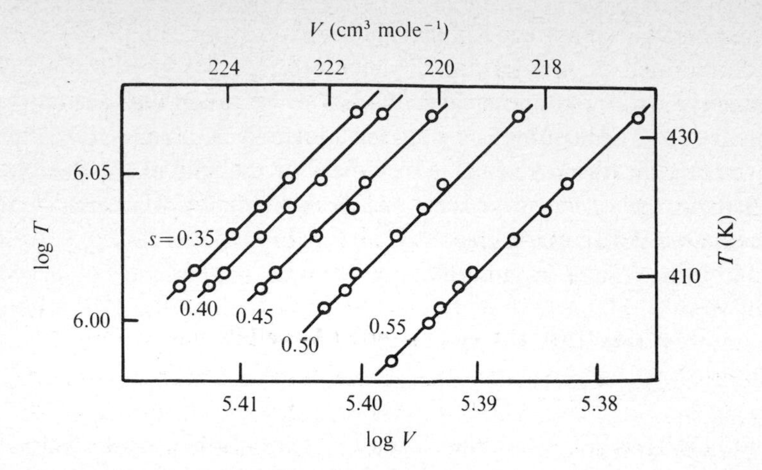

Fig. 2.3.4. Lines of constant order parameter s versus log V and log T for PAA. The slope $(\partial \log T/\partial \log V) = -4.0$. (After McColl and Shih.[65])

Now, it is clear from (2.3.12) and (2.3.13) that B should have a constant value if s is to be invariant. If B is to be kept constant while T and V are varied, as is implied in (2.3.18), TV^m should be constant, or

$$(\partial \log T/\partial \log V)_s = -m. \tag{2.3.19}$$

Empirically, therefore, $m = 4$ for PAA. It turns out that with this value of m, result (ii) follows at once from the theory; on the other hand with $m = 2$, the thermal range at constant volume is only 0.74 times the experimental value.[75] The pressure invariance of s_c is, of course, a direct consequence of (2.3.14) or (2.3.17). Thus, the experimental data can be fitted satisfactorily with $m = 4$. However, it is doubtful if this has any theoretical significance, in view of Cotter's result,[69] referred to earlier, that m should be 1 irrespective of the nature of the intermolecular pair potential.

A drawback of the theory becomes apparent when we calculate the latent heat of transition from the nematic to the isotropic phase. The heat of transition is easily shown to be given by[46,71]

$$H = T_{\mathrm{NI}}[(\alpha/\beta)\Delta V - S(V_c, T_{\mathrm{NI}})] \tag{2.3.20}$$

where α and β are respectively the coefficients of thermal expansion and isothermal compressibility of the isotropic liquid at T_{NI}. (Although ΔV is usually small, its contribution to H is not negligible since H itself is rather small.) However, the theoretical values of H are found to be much too high, usually by a factor of about 2 or 3.[71] Large discrepancies are also found in the specific heat c_V and the isothermal compressibility β in the nematic phase. The failure of the theory in this respect is evidently because the molecular field method neglects completely the effect of short range order. A naive way of accounting for this discrepancy is to assume that the molecules form small clusters of 2–3 molecules each, and that the single particle potential (2.3.8) applies to each cluster rather than to the individual molecules.[46,71] The theory is then able to give reasonable values for H, c_V and β.[71]

2.3.4. The odd–even effect: Marcelja's calculations

From chemical studies it has long been known that the end-chains of the molecules play a significant part in the stability of the mesophase. The nematic–isotropic transition temperature[76] and a number of other properties (e.g. the order parameter,[54,77] the excess specific heat,[14] the transition entropy,[14] the splay elastic constants,[78] etc.) show a pronounced alternation as the homologous series is ascended, i.e., as the number of carbon atoms in the end-chain is increased. This is often referred to as the *odd–even* effect. Qualitatively, the origin of the effect can be understood from a consideration of the molecular structure (fig. 2.3.5). In the even members of this series the disposition of the end-groups is such as to enhance the molecular anisotropy and hence the molecular order, whereas in the odd members it has the opposite effect. As the chains get longer their flexibility increases and the odd–even effect becomes less pronounced. A quantitative calculation of the contribution of the end-chains to the ordering has been made by Marcelja.[73]

The possible conformations of the alkyl chain and the corresponding internal energies can be worked out accurately as has been shown by

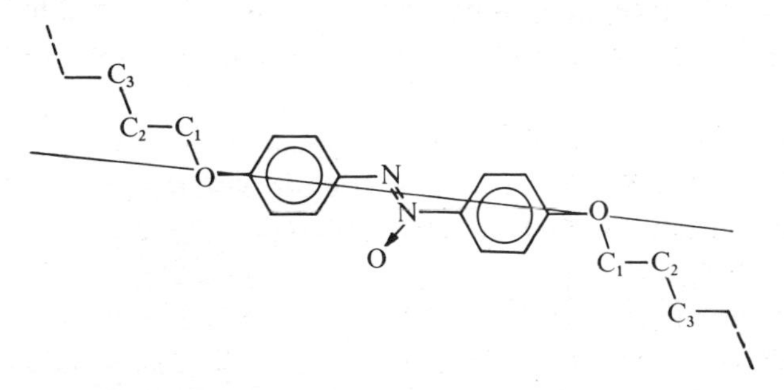

Fig. 2.3.5. Structure of *p, p'*-di-n-alkoxyazoxybenzenes. The addition of an even-numbered carbon atom in the preferred *trans* conformation is along the major molecular axis. This is not the case for an odd-numbered carbon atom.

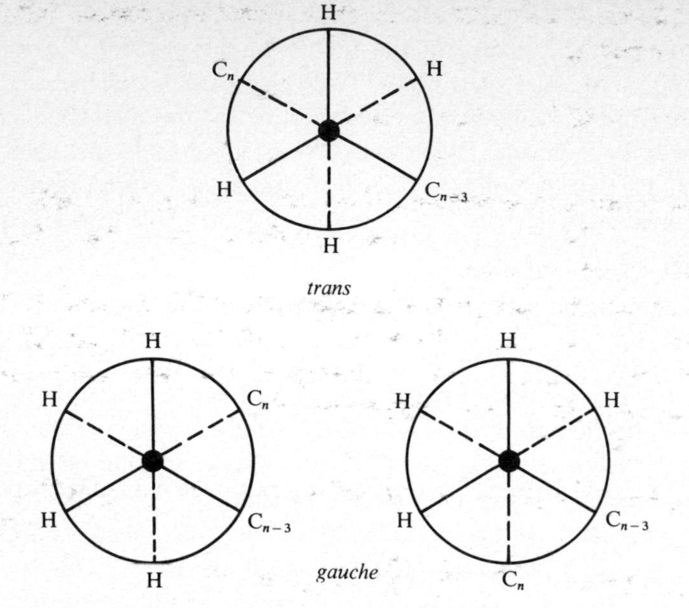

Fig. 2.3.6. The *trans* and the two symmetrical *gauche* states as viewed along the C–C bond.

Flory.[79] The valence angle θ between successive bonds is constant, while the energy as a function of the azimuthal angle φ has three minima defining the three possible states for each successive bond, viz, the extended *trans* state t and the two symmetrical *gauche* states g^+, g^- (fig. 2.3.6). Assuming nearest-neighbour interaction between chain segments, the internal energy of a conformation of a single chain may be written as

$$E_1 = E_0 + \sum_{l=4}^{n} E(\xi_l, \xi_{l-1})$$

where n is the number of carbon atoms in the end-chain, $\xi = t, g^+$ or g^-, and E_0 is the energy of the first three groups of the chain attached to the rigid part of the molecule. Next, the dispersion energy of a chain is assumed to be the sum of the interactions of each segment in a molecular field and is expressed in the Maier–Saupe form

$$E_2 = -\chi_b \sum_{l=1}^{n} \tfrac{1}{2}(3\cos^2\theta_l - 1)$$

where χ_b is a product of an interaction constant and the order parameter of the chain and θ_l the angle between the lth segment and the direction of the field. The order parameter of the chain is defined as the average over the chain segments

$$s_b = \langle (1.88/n) \sum_{l=1}^{n} \tfrac{1}{2}(3\cos^2\theta_l - 1)\rangle$$

where the factor 1.88 is introduced so that the value of s_b will range between 1 for the all-*trans* (perfectly ordered) state and 0 for the completely disordered state.

In the nematic liquid crystal, the rigid parts of the molecule and the end-chains interact. The order parameter of the rigid (aromatic or aliphatic) parts, denoted by the subscript a, and that of the chains, denoted by subscript b, may then be written as

$$s_a = (f_a)^{-1} \int \tfrac{1}{2}(3\cos^2\theta - 1)\exp[\tfrac{1}{2}\chi_a(3\cos^2\theta - 1)/k_BT]\,\mathrm{d}(\cos\theta), \tag{2.3.21}$$

$$s_b = 1.88(f_b)^{-1} \sum_{\substack{\text{all} \\ \text{conf.}}} \sum_{\substack{\text{initial} \\ \text{orient.}}} \sum_{l=1}^{n} \tfrac{1}{2}(3\cos^2\theta_l - 1)$$

$$\times \exp\left[\frac{1.88}{n}\chi_b \sum_{l=1}^{n} \tfrac{1}{2}(3\cos^2\theta_l - 1)/k_BT\right], \tag{2.3.22}$$

where f_a and f_b are the partition functions defined as

$$f_a = \int \exp[\tfrac{1}{2}\chi_a(3\cos^2\theta - 1)/k_BT]\,\mathrm{d}(\cos\theta), \tag{2.3.23}$$

$$f_b = \sum_{\substack{\text{all} \\ \text{conf.}}} \sum_{\substack{\text{initial} \\ \text{orient.}}} \sum_{l=1}^{n} \exp\left[\frac{1.88}{n}\chi_b \sum_{l=1}^{n} \tfrac{1}{2}(3\cos^2\theta_l - 1)/k_BT\right], \tag{2.3.24}$$

$$\chi_a = x_a A_{aa} s_a + x_b A_{ab} s_b,$$

$$\chi_b = x_b A_{bb} s_b + x_a A_{ba} s_a,$$

x_a and x_b are the volume fractions of the aromatic (aliphatic) rings and the end-chains respectively, and the A's interaction constants. The second summation in (2.3.22) and (2.3.24) denotes the summation over all possible orientations of the initial segment of the alkyl chain. We have ignored here the explicit dependence of the order parameters on volume.

In his calculations, Marcelja used the known values of $E(t, g^{\pm})$, $E(g^{\pm}, g^{\pm})$ and $E(g^{\pm}, g^{\mp})$ for polyethylene and estimated A_{bb} from its freezing entropy. The constant A_{aa} describing the interactions of the rigid parts can be derived, using (2.3.14), from the transition temperature of the first member of the homologous series that has no end-chain. The remaining two constants A_{ab} and A_{ba} are not independent, since the total interaction energy per mole between the rigid parts and the chains

$$Ns_a x_b A_{ab} s_b = Nn s_b x_a A_{ba} s_a,$$

where N is the Avogadro number. This gives $A_{ab}/A_{ba} = nx_a/x_b$ and consequently there is just a single adjustable parameter in the calculations.

In evaluating the internal energy it should be noted that even when $\chi_b = 0$ (non-interacting chains), there will be a certain degree of chain ordering, say s_b^0, because of the ordering of the rigid parts to which they are connected. By switching on the interactions A_{bb} and A_{ab}, the chain order increases from s_b^0 to s_b. The internal energy due to order is then

$$U = -\tfrac{1}{2}\chi_a A_{aa} s_a^2 - \tfrac{1}{2}\chi_b A_{ab} s_a s_b^0$$
$$- n\chi_a A_{ba} s_a (s_b - s_b^0) - \tfrac{1}{2} n\chi_b A_{bb} (s_b^2 - (s_b^0)^2)$$
$$+ E_{\mathrm{cnf}}(\chi_b) - E_{\mathrm{cnf}}(0), \qquad (2.3.25)$$

where E_{cnf} is the conformational energy of the end-chain, and the excess entropy is given by

$$TS = -s_a\chi_a - ns_b\chi_b + E_{\mathrm{cnf}}(\chi_b) - E_{\mathrm{cnf}}(0)$$
$$+ k_B T \ln f_a + k_B T \ln [f_b(\chi_b)/f_b(0)]. \qquad (2.3.26)$$

The transition is determined by the condition

$$F = U - TS = 0.$$

Some numerical calculations are presented in figs. 2.3.7–2.3.9. The agreement is good. Another important consequence of the theory is that depending on the strength of the interaction constant A_{ab} there is a rising trend in T_{NI} versus n for some homologous series of compounds and a decreasing trend for others.

2.3.5. Theory of dielectric anisotropy

As an example of an application of the mean field method we shall consider the theory of the dielectric anisotropy of the nematic phase.[81,82] The low frequency dielectric anisotropy of a molecule is determined by two factors: (i) the polarizability anisotropy α_a which for the elongated molecules of nematogenic compounds always makes a positive contribution (i.e., a larger contribution for the measuring field parallel to the long

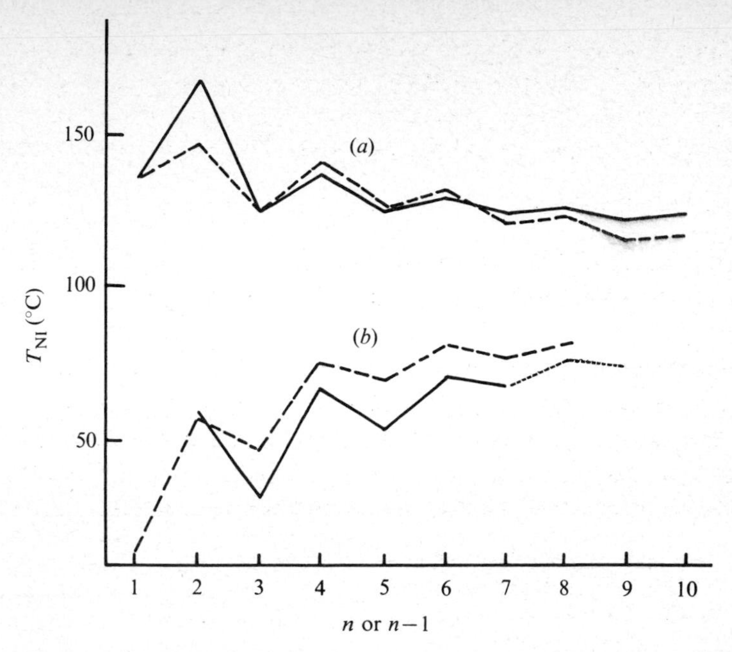

Fig. 2.3.7. Nematic–isotropic temperatures of homologous series plotted against n, the number of carbon atoms in the alkyl chain, or $n-1$. (a) p,p'-di-n-alkoxyazoxybenzenes[14] (n) and (b) p,p'-di-n-alkylazoxybenzenes[80] ($n-1$). The calculated values are shown as a dashed line. The dotted line shows the smectic–isotropic transition temperatures for the series (b). (After Marcelja.[73])

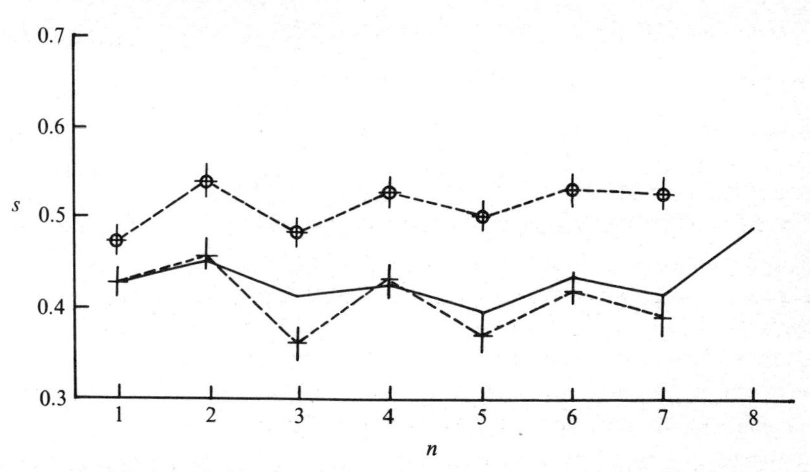

Fig. 2.3.8. Relative order parameters determined by ^{13}C nuclear magnetic resonance at T_{NI} (+) and $T_{NI}-5$ °C (⌖) plotted against the number of carbon atoms in the alkyl chain for p,p'-di-n-alkoxyazoxybenzenes. The values are normalized to 0.43 for $n=1$ at T_{NI}. The calculated values are shown as the full line. (After Pines, Ruben and Allison.[77])

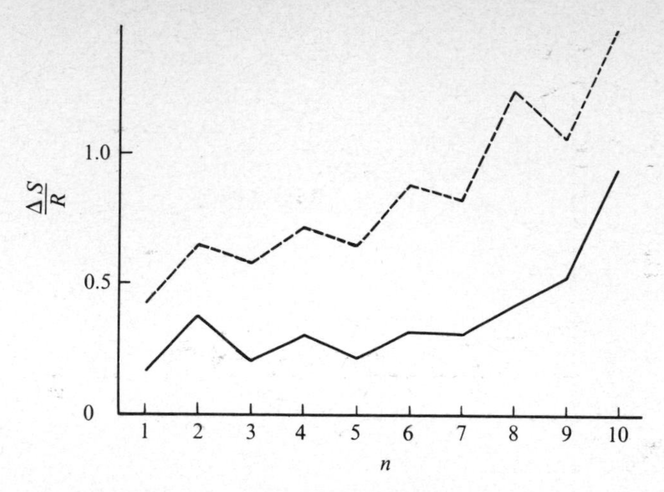

Fig. 2.3.9. Entropy of the nematic–isotropic transition plotted against the number of carbon atoms in the alkyl chain for *p,p'*-di-n-alkoxyazoxybenzenes.[14] The calculated values, shown as the dashed line, are higher for reasons discussed in § 2.3.3. (After Marcelja.[73])

molecular axis) and (ii) the dipole orientation effect. The sign of the latter contribution is positive if the net permanent dipole moment of the molecule makes only a small angle with its long axis and is negative if the angle is large. In the last case the sign of the net anisotropy depends on the relative magnitudes of the two contributions. Thus different nematic materials can exhibit widely different dielectric properties (fig. 2.3.10).

In contrast to diamagnetism where the magnetic interactions between molecules can be neglected, the polarization field in the medium becomes important when discussing dielectric anisotropy (see, for example, Bottcher[84]). Maier and Meier[81] took this into account by applying the Onsager theory.[85] The effective induced dipole moments per molecule along and perpendicular to the unique axis of the nematic liquid crystal are then given by expressions essentially similar to (2.3.2) and (2.3.3) but with appropriate correction factors for the polarization field:

$$\bar{m}_1 = (\bar{\alpha} + \tfrac{2}{3}\alpha_a s)FhE_1, \tag{2.3.27}$$

$$\bar{m}_2 = (\bar{\alpha} - \tfrac{1}{3}\alpha_a s)FhE_2, \tag{2.3.28}$$

where $h = 3\bar{\varepsilon}/(2\bar{\varepsilon}+1)$ is the cavity field factor, $\bar{\varepsilon}$ the mean dielectric constant, $F = 1/(1-\bar{\alpha}f)$ the reaction field factor, $\bar{\alpha}$ the mean polarizability, α_a the polarizability anisotropy (the direction of maximum polarizability is assumed to be the long molecular axis), $f = 4\pi N\rho(2\bar{\varepsilon}-2)/3M(2\bar{\varepsilon}+1)$, ρ the density, M the molecular weight, s the

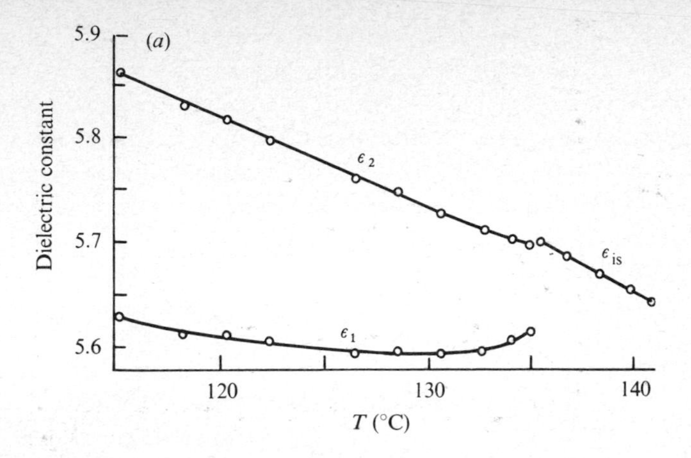

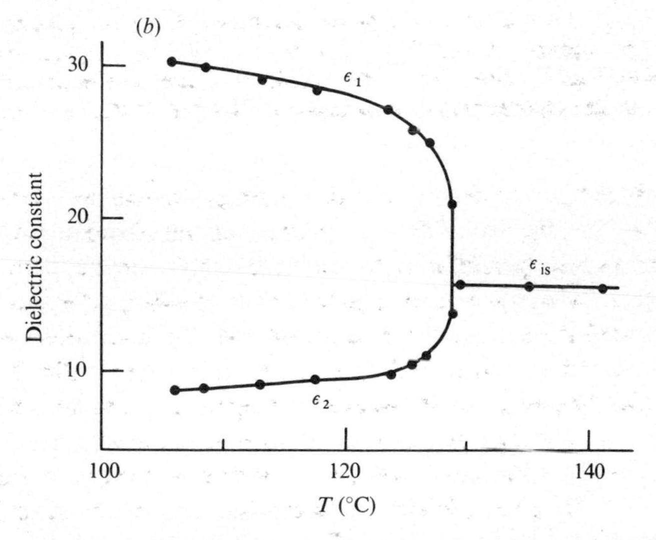

Fig. 2.3.10. Principal dielectric constants of (*a*) PAA (after Maier and Meier[(82)]) and (*b*) n(4′-ethoxybenzylidene) 4-aminobenzonitrile (after Schadt[(83)]). ε_1 and ε_2 refer to the values along and perpendicular to the optic axis of the nematic medium and ε_{is} is the isotropic value.

order parameter defined by (2.3.1), and E the applied field. Strictly speaking F and h have to be corrected for the anisotropy of the dielectric constant but we shall ignore these corrections.

To calculate the effective permanent dipole moment, we choose XYZ as the space-fixed coordinate system, Z being parallel to the unique axis of the medium, and $\xi\eta\zeta$ as the molecule-fixed coordinate system, the ξ axis coinciding with the long axis of the molecule. Let ν be the Eulerian angle between the ξ axis and the line of intersection of the XY and $\xi\eta$ planes, and ν' the angle between this line and X. Suppose that the permanent dipole moment μ is inclined at an angle β with respect to the long molecular axis. In dielectric measurements, low fields are usually employed so that the potential energy of the dipoles due to the external field is small. We can therefore write the effective dipole moments along the field directions as

$$\bar{\mu}_1 = \frac{\int_0^\pi \int_0^{2\pi} \int_0^{2\pi} [1+(\mu_z hE_1/k_B T)]\mu_z \psi(\theta) \sin\theta \, d\theta \, d\nu \, d\nu'}{\int_0^\pi \int_0^{2\pi} \int_0^{2\pi} [1+(\mu_z hE_1/k_B T)]\psi(\theta) \sin\theta \, d\theta \, d\nu \, d\nu'} \quad (2.3.29)$$

$$\bar{\mu}_2 = \frac{\int_0^\pi \int_0^{2\pi} \int_0^{2\pi} [1+(\mu_x hE_2/k_B T)]\mu_x \psi(\theta) \sin\theta \, d\theta \, d\nu \, d\nu'}{\int_0^\pi \int_0^{2\pi} \int_0^{2\pi} [1+(\mu_x hE_2/k_B T)]\psi(\theta) \sin\theta \, d\theta \, d\nu \, d\nu'} \quad (2.3.30)$$

where

$$\mu_z = F\mu[\cos\beta \cos\theta + \sin\beta \sin\nu \sin\theta]$$

$$\mu_x = F\mu[\cos\beta \sin\nu' \sin\theta + \sin\beta(\cos\nu \cos\nu' - \sin\nu \sin\nu' \cos\theta)]$$

and $\psi(\theta)\,d\theta$ is the probability of a molecule having an orientation between θ and $\theta+d\theta$. *If* $\psi(\theta)$ is given by the Maier–Saupe distribution, $\exp(-u_i/k_B T)$ where u_i is defined by (2.3.8), the integrals reduce to

$$\bar{\mu}_1 = \frac{\mu^2}{3k_B T}[1-(1-3\cos^2\beta)s]hF^2E_1 \quad (2.3.31)$$

$$\bar{\mu}_2 = \frac{\mu^2}{3k_B T}[1+\tfrac{1}{2}(1-3\cos^2\beta)s]hF^2E_2. \quad (2.3.32)$$

Since

$$\frac{\varepsilon_1 - 1}{4\pi}E_1 = \frac{N\rho}{M}(\bar{m}_1 + \bar{\mu}_1)$$

$$\varepsilon_1 = 1 + 4\pi\frac{N\rho hF}{M}\left\{\bar{\alpha} + \tfrac{2}{3}\alpha_a s + F\frac{\mu^2}{3k_B T}[1-(1-3\cos^2\beta)s]\right\} \quad (2.3.33)$$

and similarly

$$\varepsilon_2 = 1 + 4\pi\frac{N\rho hF}{M}\left\{\bar{\alpha} - \tfrac{1}{3}\alpha_a s + \frac{F\mu^2}{3k_B T}[1+\tfrac{1}{2}(1-3\cos^2\beta)s]\right\}. \quad (2.3.34)$$

Therefore

$$\varepsilon_a = \varepsilon_1 - \varepsilon_2 = \frac{4\pi N\rho}{M} hF\left[\alpha_a - \frac{F\mu^2}{2k_B T}(1 - 3\cos^2\beta)\right]s. \qquad (2.3.35)$$

Clearly if β is small, the two terms in the square brackets of (2.3.35) add to give rise to a strong positive ε_a, whereas if β is sufficiently large ε_a may be negative. In PAA, for example, μ and β are estimated to be 2.2 debyes and 62.5° respectively from Kerr constant measurements in dilute solutions and dielectric measurements in the isotropic phase. Substituting for the other known parameters for this compound and using s from theory, it turns out that ε_a should be weakly negative. The absolute value of ε_a as well as its temperature variation are in fair agreement with the experimental data.[(82)]

The dielectric constants are of course frequency dependent. The dipole orientation part of the polarization parallel to the preferred direction (1-direction) may be expected to be characterized by a relatively long relaxation time. This arises because of the strong hindering of the rotation of the longitudinal component of the dipole moment about a transverse axis. On the other hand the orientational polarization along the 2-direction will have a much faster relaxation, comparable to the Debye relaxation in normal liquids, as this involves rotation about the long axis of the molecule. If there are additional dipoles in parts of the molecule, with their own internal degrees of rotation, the corresponding relaxation times will again be similar to that in a liquid. The expected form of the dispersion curves for a compound like *p*-azoxyanisole is illustrated in fig. 2.3.11. The trends in the curves have been confirmed experimentally.[(86,87)] Meier and Saupe[(88)] have discussed the mechanism of dipole orientation in PAA and have shown that the relaxation time for polarization parallel to the 1-direction should increase from its value in the isotropic phase by a 'retardation factor' which may amount to several orders of magnitude depending on the strength of the nematic potential. However, the Maier–Saupe potential yields a retardation factor smaller than the experimental value. This is not surprising since short range order may be expected to play a dominant part in the relaxation process and the mean field theory neglects this completely. No theory of the relaxation mechanism has yet been proposed taking into account near-neighbour correlations.

2.3.6. Relationship between elasticity and orientational order

As remarked in chapter 1, a uniformly oriented film of nematic liquid crystal may be prepared by prior treatment of the surfaces with which it is

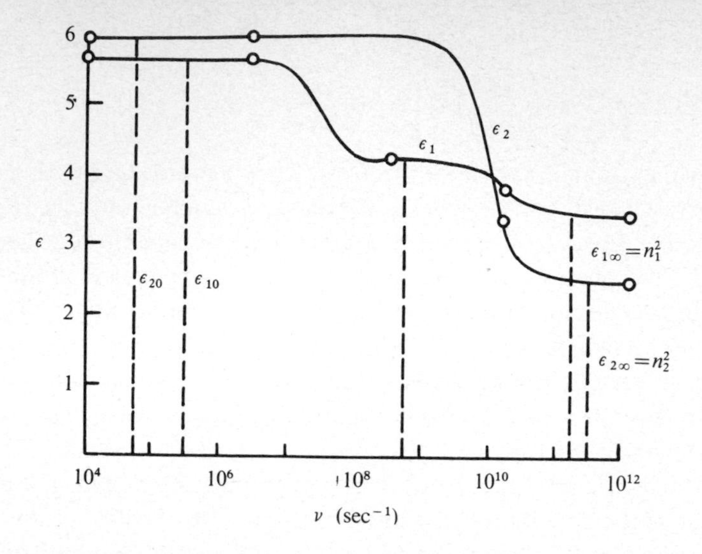

Fig. 2.3.11. Expected form of the dispersion of the principal dielectric constants of 4,4′-di-n-alkoxyazoxybenzenes. The suffix 0 refers to the static values and the suffix ∞ to the optical values. ε_1 shows the low frequency relaxation and both ε_1 and ε_2 show the normal Debye high frequency relaxation. (After Maier and Meier.[86])

in contact. If the preferred orientation imposed by the surfaces is perturbed, let us say by a magnetic field, a *curvature strain* will be introduced in the medium. The theory of such a deformation will be discussed at length in § 3.2; for the present it will suffice to state some of the important results. The free energy per unit volume of the deformed medium relative to the state of uniform orientation is

$$\delta F=\tfrac{1}{2}k_{11}\left(\frac{\partial n_x}{\partial x}\right)^2+\tfrac{1}{2}k_{22}\left(\frac{\partial n_x}{\partial y}\right)^2+\tfrac{1}{2}k_{33}\left(\frac{\partial n_x}{\partial z}\right)^2, \tag{2.3.36}$$

where **n** is a unit vector called the director representing the preferred molecular direction, $\partial n_x/\partial x$, $\partial n_x/\partial y$ and $\partial n_x/\partial z$ are respectively the *splay*, *twist* and *bend* components of curvature at any point, the curvature being defined with respect to a right-handed cartesian coordinate system XYZ with Z parallel to the preferred direction at the origin. The three principal types of deformations are illustrated schematically in fig. 2.3.12.

At the molecular level, it is obvious that curvature elasticity is a consequence of the orientational order in the liquid crystal. A quantitative relationship between them was established by Saupe[89] using the mean field theory. From (2.3.6) it is seen that the dipole–dipole part of the

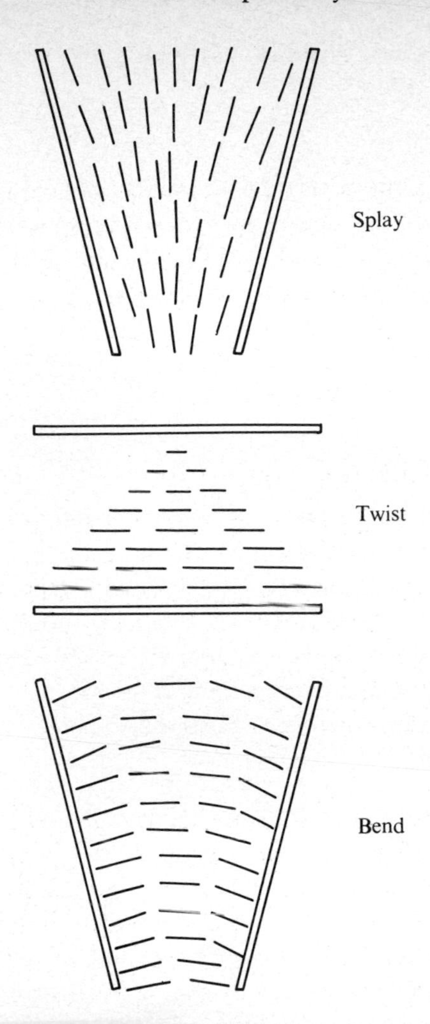

Fig. 2.3.12. The three principal types of deformation in a nematic liquid crystal.

dispersion energy of interaction of a molecule i in the average field due to its neighbours k is given by

$$u_i = -\sum_k \frac{1}{R_{ik}^6} \sum_{\mu\nu\neq 00} \frac{1}{E_{\mu\nu}-E_{00}} \left| \sum_{X,Y,Z} \left[\left(3\frac{X_{ik}^2}{R_{ik}^2} - 1 \right) x_{0\mu}^{(i)} x_{0\nu}^{(k)} \right. \right.$$

$$\left. \left. + 3\frac{X_{ik} Y_{ik}}{R_{ik}^2} \left(y_{0\mu}^{(i)} x_{0\nu}^{(k)} + x_{0\mu}^{(i)} y_{0\nu}^{(k)} \right) \right] \right|^2. \tag{2.3.37}$$

The internal energy per mole due to orientational order in the mean field approximation is then

$$U=\tfrac{1}{2}NAV^{-2}s^2,$$

where V is the molar volume. Now let us suppose that the director at the site of the kth molecule is turned through an angle α_k with respect to that at i. If we define a new system of axes $X'Y'Z'$ at the site of the kth molecule, Z' making an angle α_k with respect to the Z axis,

$$x'^{(k)}_{0\mu}=x^{(k)}_{0\mu}\cos\alpha_k-z^{(k)}_{0\mu}\sin\alpha_k,$$

$$y'^{(k)}_{0\mu}=y^{(k)}_{0\mu},$$

$$z'^{(k)}_{0\mu}=x^{(k)}_{0\mu}\sin\alpha_k+z^{(k)}_{0\mu}\cos\alpha_k.$$

Since we are dealing with small deformations, we assume that the director continues to have cylindrical symmetry about the Z' axis at the site of the kth molecule and that the order parameter s is unchanged in magnitude. Substituting for $x^{(k)}_{0\mu}$ etc. in (2.3.37) in terms of $x'^{(k)}_{0\mu}$, defining the Eulerian angles between the molecular coordinate system $\xi^{(k)}\eta^{(k)}\zeta^{(k)}$ and the $X'Y'Z'$ system and taking appropriate averages, the extra internal energy per mole of the system in the deformed state turns out to be

$$\delta U=s^2\sum_{\mu\nu\neq 00}\frac{\delta_{0\mu}\delta_{0\nu}}{E_{\mu\nu}-E_{00}}\sum_k\frac{1}{R^6_{ik}}\sin^2\alpha_k$$

$$\times\left[\left(3\frac{Z^2_{ik}}{R^2_{ik}}-1\right)^2-9\frac{Z^2_{ik}X^2_{ik}}{R^4_{ik}}\right]. \qquad (2.3.38)$$

Now for a pure bend distortion $\partial\alpha/\partial X=\partial\alpha/\partial Y=0$ and we may set $\sin\alpha_k\simeq\alpha_k\simeq R_{ik}\partial\alpha/\partial Z$. Assuming a spherically symmetric distribution, X_{ik}/R_{ik} etc. may be taken to be a constant (independent of temperature), and therefore the second sum in (2.3.38) is proportional to $V^{-4/3}$. The free energy of deformation per *unit volume* can then be expressed as

$$\delta F\simeq\delta U/V=\tfrac{1}{2}CV^{-7/3}s^2(\partial\alpha/\partial Z)^2$$

where C is a constant. But by definition

$$\delta F=\tfrac{1}{2}k_{33}(\partial\alpha/\partial Z)^2,$$

so that

$$k_{33}=CV^{-7/3}s^2. \qquad (2.3.39)$$

Similar expressions are obtained for the splay and twist elastic constants. The temperature dependence of the elastic constants of 'simple' nematics is represented well by (2.3.39).[90,91]

To calculate the absolute values of the elastic constants, one has to evaluate the second summation in the right-hand side of (2.3.38). Saupe[89] evaluated the mean value of this sum for an isotropic distribution of molecules and found $k_{11}:k_{22}:k_{33}$ to be $17:-7:11$, whereas for PAA at 120 °C it is 1.6:1:3.2. Saupe and Nehring[92] have attributed this discrepancy partly to the neglect of k_{13} in the expression for the free energy of deformation (see (3.2.8)). When this coefficient is included, $k'_{11}:k_{22}:k'_{33}$ turns out to be 5:11:5, where $k'_{11}=k_{11}-k_{13}$ and $k'_{33}=k_{33}-k_{13}$. Attempts have also been made to calculate the magnitudes of the elastic constants by assuming specific lattice models and a more general form of the intermolecular potential.[78,93,94]

A corollary to (2.3.39) is that though the order parameter as well as the elastic constants decrease rapidly with rise of temperature, the intensity of light scattering should be practically constant throughout the nematic range. Light scattering from nematic liquid crystals is governed by the long range orientational fluctuations as described by the continuum theory. It is shown in § 3.9 that the differential scattering cross-section may be written approximately as

$$\frac{\mathrm{d}\sigma}{\mathrm{d}\Omega}\sim\left(\frac{\pi\varepsilon_{\mathrm{a}}}{\lambda^2}\right)^2\frac{k_{\mathrm{B}}T}{k_{\mathrm{eff}}q^2},$$

where Ω is the irradiated volume of the liquid crystal, ε_{a} the dielectric anisotropy at optical frequencies, λ the vacuum wavelength, k_{eff} an effective average elastic constant and q the magnitude of the scattering vector. Now the variation of $k_{\mathrm{B}}T$ over the nematic range can be neglected. Since ε_{a} varies approximately as s (see (2.3.4)) and k_{eff} approximately as s^2, σ should be nearly temperature independent. Experimentally, this result was first established by Chatelain[95] in PAA and PAP. Some recent measurements on 4-methoxybenzylidene-4′-butylaniline (MBBA) are presented in fig. 2.3.13.

An important source of error in these calculations is the neglect of short range order. In particular, the theory fails for the bend and twist elastic constants when smectic-like short range order is present in the nematic liquid crystal. Under such circumstances these two constants exhibit a critical divergence as the temperature approaches the smectic–nematic point and the light scattering also shows a marked temperature dependence. The present treatment is then inadequate. This aspect of the problem will be discussed in chapter 5.

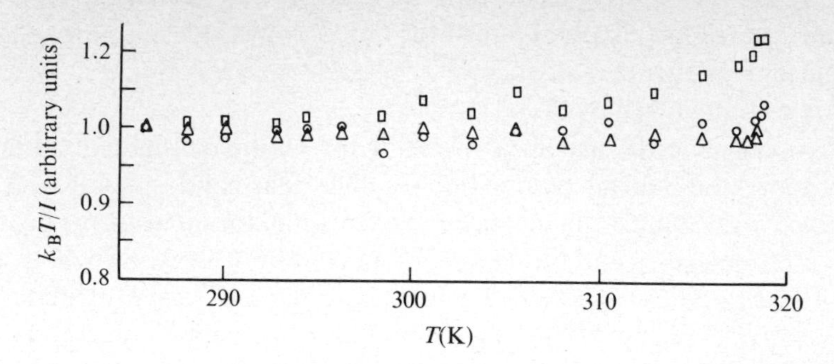

Fig. 2.3.13. The temperature dependence of the intensity (I) of light scattering from 4-methoxybenzylidene-4′-butylaniline (MBBA) in the nematic phase. The values (plotted as $k_B T/I$) are normalized to unity at 286 K. The squares, triangles and circles correspond to three different experimental configurations. For one of the configurations (squares) the data were not fully corrected for the effect of the temperature dependence of the refractive indices on the variation of one of the wave-vector components, which probably explains the slight increase near the transition temperature in this case. (After Haller and Litster.[91])

2.4. Short range order effects in the isotropic phase

2.4.1. The Landau–de Gennes model

We have noted in § 2.1.2 that though the long range order vanishes abruptly at T_{NI}, certain anomalous effects in the isotropic phase reveal that an appreciable degree of nematic-like short range order persists above the transition point. The most direct evidence of this is the very high value of the magnetic birefringence, which in the neighbourhood of T_{NI} may be as much as 100 times greater than in ordinary organic liquids[96,97] (fig. 2.4.1). Similar anomalies are seen in the flow birefringence,[98] Kerr effect[99] and nuclear spin lattice relaxation,[100] confirming the existence of strong orientational correlations between the molecules. Foex[101] observed many years ago that the magnetic birefringence exhibits a $(T-T^*)^{-1}$ dependence and drew attention to the fact that the behaviour is closely analogous to that of a ferromagnet above the Curie temperature. More recently, de Gennes[102] has proposed a phenomenological description of these pretransition effects on the basis of the Landau theory of phase transitions.[103]

Consider an expansion of the excess free energy of any ordered system in powers of a scalar order parameter s in the following form:

$$F = \tfrac{1}{2}As^2 - \tfrac{1}{3}Bs^3 + \tfrac{1}{4}Cs^4 + \cdots \qquad (2.4.1)$$

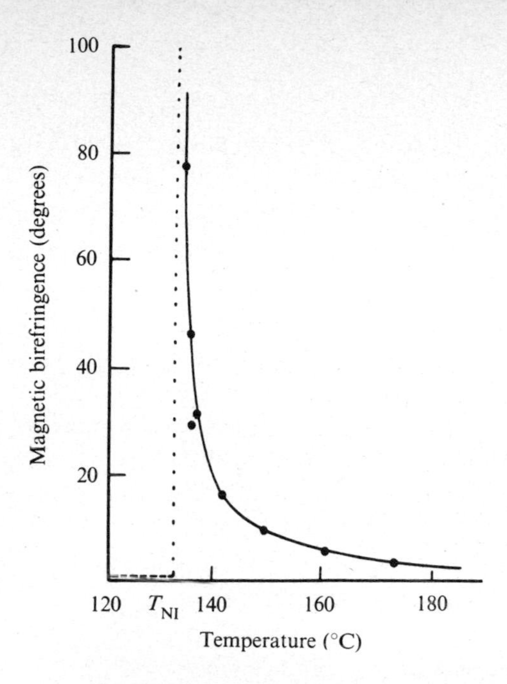

Fig. 2.4.1. Magnetic birefringence in the isotropic phase of PAA. The horizontal dashed line gives the value for nitrobenzene at 22.5 °C. (After Zadoc-Kahn.[96])

where $B>0$ and $C>0$. We observe that such an expression predicts a discontinuous transition, for putting $F=0$ and $\partial F/\partial s=0$ at the transition point T_c we get

$$s_c = 2B/3C \tag{2.4.2}$$

neglecting higher order terms. A plot of F versus s results in a family of curves similar to that shown in fig. 2.3.2.

If $B=0$, the transition is continuous and A vanishes at the transition point.[103] This is because in the disordered phase $s=0$ corresponds to a minimum of F only if $A>0$, while in the ordered phase $s\neq 0$ corresponds to a stable minimum only if $A<0$. Thus, since A is positive on one side of the transition point and negative on the other, it must vanish at the transition point itself. In the vicinity of the transition, we may therefore write

$$A = a(T-T^*) \tag{2.4.3}$$

where T^* is the second order transition temperature (e.g., the Curie temperature in the case of a ferromagnet).

If $B>0$, T^* lies below T_c and

$$(T_c-T^*)=2B^2/9aC. \tag{2.4.4}$$

For a 'weak' first order transition, B is small and $2B^2/9aC$ may be expected to be a very small quantity.

In principle, a free energy expansion of this type should be valid for nematic liquid crystals, with s denoting the usual orientational order parameter defined by (2.3.1). The term of order s^3 is not precluded by symmetry, for the states s and $-s$ represent two entirely different kinds of molecular arrangement which are not symmetry related and do not have equal free energies. In the former case, the molecules are more nearly parallel to the unique axis, while in the latter they are more nearly perpendicular to it. However, in the nematic phase s is usually quite large (greater than about 0.4) so that very many more terms have to be included in the expansion in order to draw any valid conclusions. Consequently, the model is conveniently applied only to the weakly ordered isotropic phase. We shall discuss some of these applications.

2.4.2. Magnetic and electric birefringence

The free energy per mole of the isotropic phase in the presence of an external field (magnetic or electric) may be written as

$$F=\tfrac{1}{2}a(T-T^*)s^2-\tfrac{1}{3}Bs^3+\tfrac{1}{4}Cs^4\cdots+NW(s), \tag{2.4.5}$$

where $W(s)$ is the average orientational potential energy of a molecule due to the external field and N the Avogadro number. If the external field is magnetic

$$W(s)=-\tfrac{1}{3}\chi_a H^2 s \tag{2.4.6}$$

where $\chi_a=\chi_\parallel-\chi_\perp$ is the anisotropy of diamagnetic susceptibility of the molecule. The magnetically induced order is extremely weak ($\sim 10^{-5}$) so that we may neglect cubes and higher powers of s. The condition $\partial F/\partial s=0$ then leads to the result

$$s=\tfrac{1}{3}N\chi_a H^2/a(T-T^*) \tag{2.4.7}$$

and the magnetic birefringence

$$\Delta n_M=C/a(T-T^*), \tag{2.4.8}$$

where, assuming a Lorenz–Lorentz type of relationship for the refractive index n in the absence of a field,

$$C=2\pi N^2\chi_a\eta_a H^2(n^2+2)^2/27Vn, \tag{2.4.9}$$

η_a is the anisotropy of optical polarizability of the molecule, and V the molar volume.[104] The magnetic birefringence varies essentially as $(T-T^*)^{-1}$ since the dependence of C on temperature is relatively small. Experimentally[105] this is found to be the case, with $T_{NI}-T^* \simeq 1$ K (fig. 2.4.2). As we shall see later a $(T-T^*)^{-1}$ law implies a classical mean field.

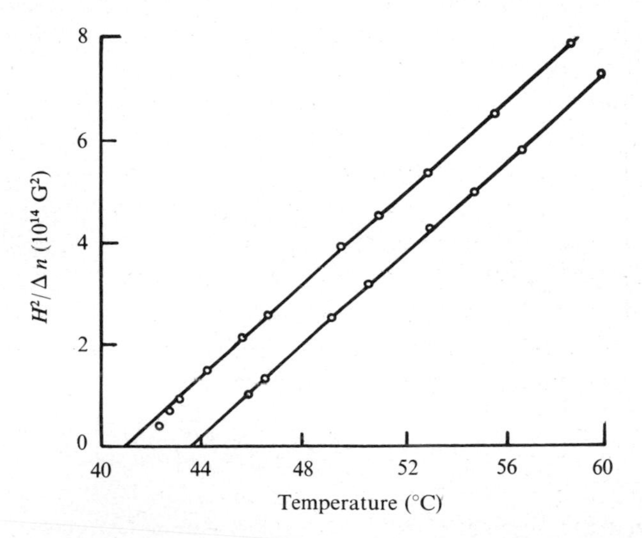

Fig. 2.4.2. Reciprocal of the Cotton–Mouton coefficient in the isotropic phase of MBBA for two samples, one slightly impure and having a lower transition point T_{NI}. Both yield the same value of $T_{NI}-T^* = 1$ °C. (After Stinson and Litster.[105])

The behaviour is slightly more complicated in the case of electric birefringence because, as explained in § 2.3.5, the orientational energy in an electric field E arises from the anisotropy of low frequency polarizability α_a as also from the net permanent dipole moment μ. An interesting example is that of PAA in which the Kerr constant actually changes sign at about T_c+5 K[99] (fig. 2.4.3). The sign reversal is easily understood.[106] The average orientational energy of the molecule due to the induced dipole moment is

$$W_1(s) = -\tfrac{1}{3}\alpha_a F h^2 E^2 s, \qquad (2.4.10)$$

and that due to the permanent dipole moment is

$$W_2(s) = -(F^2 h^2 \mu^2 E^2/6k_B T)(3\cos^2\beta - 1)s, \qquad (2.4.11)$$

where the symbols have the same meanings as in (2.3.35). In obtaining $W_2(s)$, the distribution function is supposed to involve only even powers

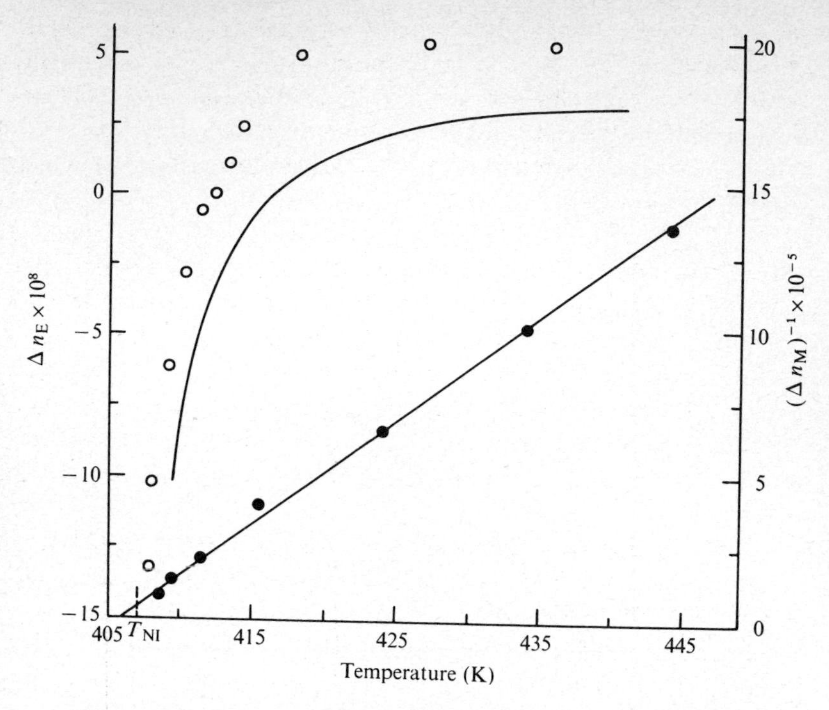

Fig. 2.4.3. The electric birefringence (Tsvetkov and Ryumtsev,[99] open circles) and the reciprocal of the magnetic birefringence (Zadoc-Kahn,[96] full circles) in the isotropic phase of PAA versus temperature. The lines represent the theoretical variations.[106]

of cos θ. This is clearly valid in the present case in view of the assumed form of the free energy expression. Putting $W(s)=W_1+W_2$,

$$F=\tfrac{1}{2}\,a(T-T^*)s^2-\tfrac{1}{3}NFh^2E^2[\alpha_a-(F\mu^2/2k_BT)(1-3\cos^2\beta)]s. \tag{2.4.12}$$

Proceeding as before

$$s=\frac{NFh^2E^2[\alpha_a-(F\mu^2/2k_BT)(1-3\cos^2\beta)]}{3a(T-T^*)} \tag{2.4.13}$$

and

$$\Delta n_E=\frac{2\pi N^2\eta_a(n^2+2)^2Fh^2E^2[\alpha_a-(F\mu^2/2k_BT)(1-3\cos^2\beta)]}{27nVa(T-T^*)}. \tag{2.4.14}$$

Δn_E can be positive or negative depending on the sign of the quantity in the square brackets of (2.4.14). The polarizability anisotropy α_a is always positive for the long molecules under consideration, but the sign of the dipole contribution depends on the angle β. If β is small, Δn_E will be strongly positive, whereas if β is sufficiently large Δn_E may be negative. Moreover, the second term in the square brackets of (2.4.14) is proportional to T^{-1}, so that in principle there can occur a reversal of sign of Δn_E with temperature. Using the values of a and T^* derived from the magnetic birefringence of PAA and substituting for μ, β, etc. (see § 2.3.5) it is found[106] that there is in fact a reversal of sign of Δn_E, though it occurs at $T_{NI}+9$ K (fig. 2.4.3). Since there is a competition between the polarizability and the permanent dipole contributions, even a small error in any of the parameters will cause an appreciable shift in the temperature at which $\Delta n_E = 0$. Taking this into consideration, the agreement may be regarded as satisfactory.

If $\beta = 0$, Δn_E given by (2.4.14) varies essentially as $(T - T^*)^{-1}$. In such materials, the electric and magnetic birefringence may be expected to exhibit the same type of behaviour over a wide temperature range. Recent measurements[107] on pure samples of hexylcyanobiphenyl, a nematogen of strong positive dielectric anisotropy with the dipole (—C≡N) pointing almost exactly along the major molecular axis, confirm this prediction (fig. 2.4.4).

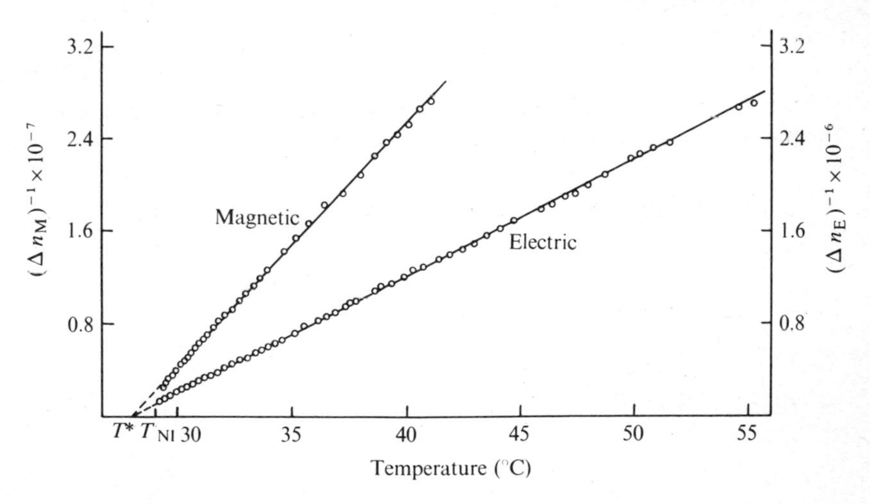

Fig. 2.4.4. Reciprocals of the magnetic and electric birefringence in the isotropic phase of 4-hexyl-4'-cyanobiphenyl versus temperature. Both give the same value of T^*($T^* = 28$ °C, $T_{NI} - T^* = 1.1$ °C). (After reference 107.)

2.4.3. Light scattering

Although in the absence of an externally applied field the equilibrium value of s in the isotropic phase is zero, there can occur fluctuations in the order parameter about the zero value.[102] This gives rise to an anomalous scattering of light.

We write the free energy expression (2.4.1) in a more general form in terms of the tensor order parameter (2.3.5), with $i=j=3$ corresponding to the long molecular axis:

$$F=F_0+\tfrac{1}{2}As_{\alpha\beta}s_{\beta\alpha}-\tfrac{1}{3}Bs_{\alpha\beta}s_{\beta\gamma}s_{\gamma\alpha}+\cdots \tag{2.4.15}$$

$$=F_0+\tfrac{1}{2}a(T-T^*)(s_{xx}^2+s_{yy}^2+s_{zz}^2+2s_{xy}^2+2s_{yz}^2+2s_{zx}^2) \tag{2.4.16}$$

neglecting higher terms. For a given scattering wave-vector $\mathbf{q}$, the differential scattering cross-section per unit volume is given by (see § 3.9)

$$\frac{d\sigma}{d\Omega}=\pi^2\lambda^{-4}\langle\delta\varepsilon_{\alpha\beta}^2\rangle_{\mathbf{q}} \tag{2.4.17}$$

where

$$\delta\varepsilon_{\alpha\beta}=\varepsilon_{\alpha\beta}-\bar{\varepsilon}=\tfrac{2}{3}\Delta\varepsilon s_{\alpha\beta}, \tag{2.4.18}$$

$\bar{\varepsilon}$ is the mean dielectric constant and $\Delta\varepsilon$ is the dielectric anisotropy when the molecules are all exactly parallel to one another (see (2.3.2)). Both $\bar{\varepsilon}$ and $\Delta\varepsilon$ refer, of course, to optical frequencies. If the incident light is linearly polarized along z and the scattered light polarized along x

$$\left(\frac{d\sigma}{d\Omega}\right)_{zx}=\frac{4\pi^2}{9\lambda^4}(\Delta\varepsilon)^2\langle s_{zx}^2\rangle_{\mathbf{q}}. \tag{2.4.19}$$

From the equipartition theorem

$$\tfrac{1}{2}a(T-T^*)\langle 2s_{zx}^2\rangle_{\mathbf{q}}=\tfrac{1}{2}k_BT,$$

so that

$$\left(\frac{d\sigma}{d\Omega}\right)_{zx}=\frac{4\pi^2}{18\lambda^4}\frac{(\Delta\varepsilon)^2k_BT}{a(T-T^*)}. \tag{2.4.20}$$

If the incident and scattered radiations are both polarized along z,

$$a(T-T^*)\langle s_{zz}^2\rangle_{\mathbf{q}}=\tfrac{2}{3}k_BT$$

(where we have made use of the condition $s_{xx}+s_{yy}+s_{zz}=0$) and

$$\left(\frac{d\sigma}{d\Omega}\right)_{zz}=\frac{8\pi^2}{27\lambda^4}\frac{(\Delta\varepsilon)^2k_BT}{a(T-T^*)}. \tag{2.4.21}$$

Thus the intensity of scattering should vary essentially as $(T-T^*)^{-1}$ (fig. 2.4.5) and the ratio of the scattered light polarized along z to that polarized along x should be 4/3. Both these predictions have been verified quantitatively for MBBA.[108]

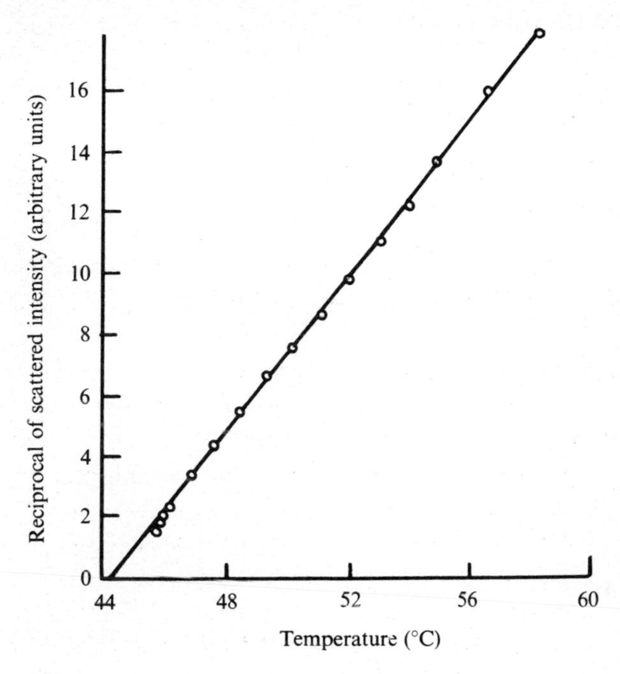

Fig. 2.4.5. Reciprocal of the intensity of light scattering in the isotropic phase of MBBA versus temperature. (After Stinson and Litster.[105])

If the order parameter varies gradually from point to point, the free energy expression should include gradient terms as well, which can be written in the form

$$\tfrac{1}{2}L_1\,\partial_\alpha s_{\beta\gamma}\,\partial_\alpha s_{\beta\gamma}+\tfrac{1}{2}L_2\,\partial_\alpha s_{\alpha\gamma}\,\partial_\beta s_{\beta\gamma} \tag{2.4.22}$$

where $\partial_\alpha=\partial/\partial x_\alpha$, repeated indices being subject to the usual summation convention, and L_1, L_2 are constants. To elucidate the consequences of these additional terms let us suppose that $s_{zz}=s$, $s_{xx}=s_{yy}=-\tfrac{1}{2}s$, $s_{xy}=s_{yz}=s_{zx}=0$. Also, let s be a function of z only. The free energy is then

$$\begin{aligned}F&=F_0+\tfrac{3}{4}a(T-T^*)s^2+\frac{3L_1+2L_2}{4}\left(\frac{ds}{dz}\right)^2\\&=F_0+\tfrac{3}{4}a(T-T^*)\left[s^2+\xi^2\left(\frac{ds}{dz}\right)^2\right],\end{aligned} \tag{2.4.23}$$

where

$$\xi = \left(\frac{L_1 + \frac{2}{3}L_2}{a(T-T^*)}\right)^{1/2} \tag{2.4.24}$$

may be called the coherence length. The spatial correlation function has the form

$$\langle s(0)s(R)\rangle = \text{const. } k_B T \exp(-R/\xi) \quad (R \gtrsim \xi),$$

and the scattering cross-section (2.4.21) for a given scattering wave-vector $\mathbf{q}$ is modified to [102]

$$\left(\frac{d\sigma}{d\Omega}\right)_{zz} = \frac{8\pi^2}{27\lambda^4}(\Delta\varepsilon)^2 \frac{k_B T}{a(T-T^*)(1+\xi^2 q^2)}. \tag{2.4.25}$$

Experimentally it is found that the angular variation of the intensity of scattering is rather small [109,110] proving that the coherence length is much smaller than the wavelength of light ($q\xi < 0.1$), but by careful measurements Stinson and Litster[110] have established that $\xi \propto (T-T^*)^{-1/2}$, in accordance with (2.4.24).

2.4.4. Flow birefringence

To discuss flow birefringence, we have to make use of some results of the continuum theory developed in chapter 3. In analogy with (3.1.38) we write for the isotropic phase[102]

$$\begin{bmatrix} t_{\alpha\beta} \\ \varphi_{\alpha\beta} \end{bmatrix} = \begin{bmatrix} \eta & \mu \\ \mu' & \nu \end{bmatrix} \begin{bmatrix} d_{\alpha\beta} \\ R_{\alpha\beta} \end{bmatrix}, \tag{2.4.26}$$

where $t_{\alpha\beta}$, $\varphi_{\alpha\beta}$ are treated as 'forces' and $d_{\alpha\beta}$, $R_{\alpha\beta}$ as 'fluxes'; $t_{\alpha\beta}$ is the viscous stress tensor;

$$\varphi_{\alpha\beta} = -\frac{\partial F}{\partial s_{\alpha\beta}}$$

$$= -A s_{\alpha\beta} \text{ from (2.4.15);}$$

$$d_{\alpha\beta} = \frac{1}{2}\left(\frac{\partial v_\alpha}{\partial x_\beta} + \frac{\partial v_\beta}{\partial x_\alpha}\right), \; v \text{ being the velocity;}$$

$$R_{\alpha\beta} = \frac{ds_{\alpha\beta}}{dt}.$$

For an incompressible fluid $d_{\alpha\alpha} = 0$. All four tensors are symmetric and traceless. Further from Onsager's relations, $\mu = \mu'$.

Now consider shear flow along x with a velocity gradient $\mathrm{d}v/\mathrm{d}z$. The flow induces a birefringence proportional to the velocity gradient with the principal axes of the index ellipsoid inclined at 45° to the x, z axes. In the steady state $R_{\alpha\beta} = 0$, $\varphi_{xz} = \frac{1}{2}\mu\,\mathrm{d}v/\mathrm{d}z$. Therefore

$$s_{xz} = -\frac{\mu}{2a(T-T^*)}\frac{\mathrm{d}v}{\mathrm{d}z}. \qquad (2.4.27)$$

No other components of s exist, and

$$s_{\alpha\beta} = \begin{bmatrix} 0 & 0 & s_{xz} \\ 0 & 0 & 0 \\ s_{xz} & 0 & 0 \end{bmatrix}.$$

Transforming to the x', z' axes which are inclined at 45° to x, z,

$$s'_{\alpha\beta} = \begin{bmatrix} s_{xz} & 0 & 0 \\ 0 & 0 & 0 \\ 0 & 0 & -s_{xz} \end{bmatrix}. \qquad (2.4.28)$$

Therefore $x'yz'$ represent the principal axes of the order parameter tensor. The difference between the dielectric constants (at optical frequencies) for polarizations along the x' and z' axes is

$$\delta\varepsilon = \tfrac{2}{3}\Delta\varepsilon\,(s_{11} - s_{33}) = \tfrac{4}{3}\Delta\varepsilon s_{xz}$$

where, as in (2.4.18), $\Delta\varepsilon$ is the dielectric anisotropy when all the molecules are exactly parallel to one another. Putting $\delta\varepsilon = 2\bar{n}\,\delta n$ and $\Delta\varepsilon = 2\bar{n}\Delta n$, the flow birefringence

$$\delta n = -\frac{2}{3}\frac{\Delta n\mu}{a(T-T^*)}\frac{\mathrm{d}v}{\mathrm{d}z} \qquad (2.4.29)$$

where Δn is the birefringence when the molecules are all perfectly parallel. The flow birefringence may therefore be expected to show an anomalous increase as the temperature approaches the transition. This was observed by Tolstoi and Fedotov[98] many years ago in PAA. The recent experiments of Martinoty, Candau and Debeauvais[11] on MBBA have confirmed the temperature dependence predicted by de Gennes's equation (2.4.29) (see fig. 2.4.6).

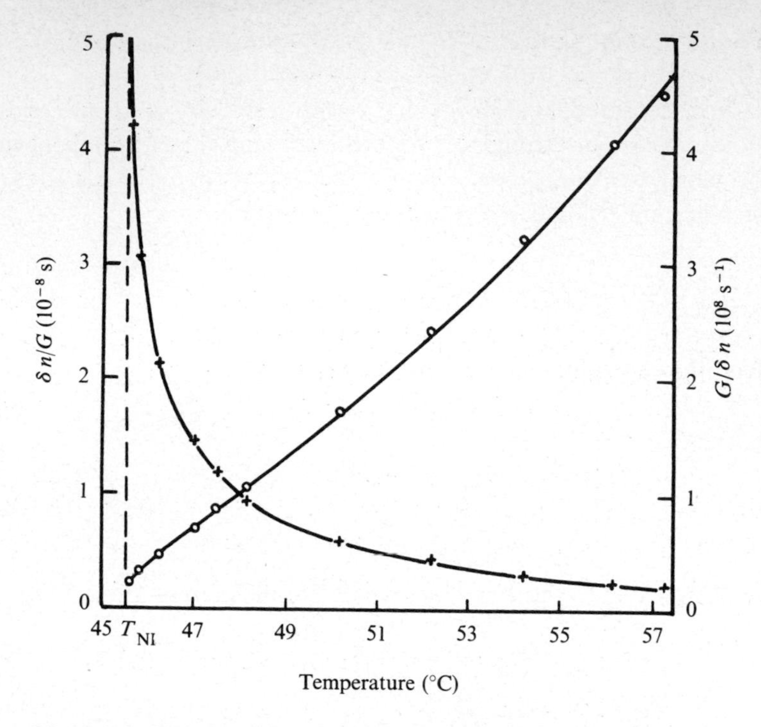

Fig. 2.4.6. Flow birefringence in the isotropic phase of MBBA. Crosses represent the birefringence (δn) per unit shear rate (G) and circles the reciprocal of this quantity. (After Martinoty, Candau and Debeauvais.[111])

2.4.5. Comparison with the Maier–Saupe theory

It is of interest to examine the relationship between the phenomenological model that we have just discussed and the molecular statistical theory of Maier and Saupe. The free energy of the weakly ordered isotropic phase in the presence of an external magnetic field is, according to the mean field theory,

$$F=N\left\{\frac{1}{2}\frac{As(s+1)}{V^2}\right.$$
$$\left.-k_BT\log\int_0^{\pi/2}\exp\left[\frac{3}{2k_BT}\left(\frac{As}{V^2}+\frac{1}{3}\chi_aH^2\right)\cos^2\theta\right]\sin\theta\,d\theta\right\}. \tag{2.4.30}$$

Expanding and integrating

$$F = Nk_B T\left[\frac{As^2}{2k_B T^2 V^2}(T-T^*) - 0.0762\frac{A^3 s^3}{8k_B^3 T^3 V^6}\right.$$

$$\left. + 0.0122\frac{A^4 s^4}{16k_B^4 T^4 V^8}\right] - \tfrac{1}{3}N\chi_a H^2 s, \qquad (2.4.31)$$

where $T^* = A/5k_B V_c^2$. This expression is identical in form to the free energy expansion (2.4.1) of the Landau model. However, it does not yield a satisfactory value of T^*. Since $(A/k_B T_{NI} V_c^2) = 4.54$, $T^*/T_{NI} = 0.908$. For PAA, $T_{NI} = 408$ K so that $T_{NI} - T^* \sim 40$ K, whereas empirically $T_{NI} - T^* \sim 1$ K.

Clearly near-neighbour correlations have to be allowed for in the molecular statistical approach to give a better description of the pretransition effects. A step in this direction has been taken recently[115–117] which we shall proceed to consider in the next section.

2.5. Near-neighbour correlations: Bethe's method

2.5.1. The Krieger–James approximation

The theory is based on a method developed originally by Bethe[112] for treating order–disorder effects in binary alloys. We suppose that every molecule is surrounded by z neighbours ($z \geqslant 3$) and that no two of the z neighbours are nearest neighbours of each other. (This implies that in writing down the interactions between the central molecule and its z neighbours, we neglect the interaction between any two of the z neighbours.) Let the pair potential between the central molecule 0 and one of its neighbours j be $E(\theta_{0j})$, where θ_{0j} is a function of the usual spherical coordinates θ_0, φ_0, θ_j, φ_j, and let every outer shell molecule j be coupled with the remaining (external) molecules of the uniaxial medium by an interaction potential $V(\theta_j)$. The relative weight for a given configuration of a cluster of $z+1$ molecules is then

$$\prod_{j=1}^{z} f(\theta_{0j})g(\theta_j), \qquad (2.5.1)$$

where

$$f(\theta_{0j}) = \exp[-E(\theta_{0j})/k_B T],$$

$$g(\theta_j) = \exp[-V(\theta_j)/k_B T].$$

The relative probability of the central molecule assuming a certain orientation θ_0, φ_0 is

$$\int \cdots \int \prod_{j=1}^{z} f(\theta_{0j})g(\theta_j)\,\mathrm{d}(\cos\theta_j)\,\mathrm{d}\varphi_j, \tag{2.5.2}$$

while that for an outer shell molecule, say 1, to assume an orientation θ_1, φ_1 is

$$\iint f(\theta_{01})g(\theta_1)\,\mathrm{d}(\cos\theta_0)\,\mathrm{d}\varphi_0$$
$$\times \int \cdots \int \prod_{j=2}^{z} f(\theta_{0j})g(\theta_j)\,\mathrm{d}(\cos\theta_j)\,\mathrm{d}\varphi_j. \tag{2.5.3}$$

If we postulate that these two probabilities are identical when 0 and 1 have the same orientation, we obtain the following consistency relation due to Chang[113]:

$$\left[\iint f(\theta_{0j})g(\theta_j)\,\mathrm{d}(\cos\theta_j)\,\mathrm{d}\varphi_j\right]^z = \iint f(\theta_{01})g(\theta_1)$$
$$\times \left[\iint f(\theta_{0j})g(\theta_j)\,\mathrm{d}(\cos\theta_j)\,\mathrm{d}\varphi_j\right]^{z-1} \mathrm{d}(\cos\theta_0)\,\mathrm{d}\varphi_0 \tag{2.5.4}$$

which has to be satisfied for all values of θ, φ.

This condition was expressed in a slightly different form by Krieger and James.[114] The relative probability that the central molecule 0 and one of its neighbours, say 1, are oriented along θ_0, φ_0 and θ_1, φ_1 respectively is

$$\psi(\theta_0, \varphi_0; \theta_1, \varphi_1)$$
$$= f(\theta_{01})g(\theta_1) \prod_{j=2}^{z} \iint f(\theta_{0j})g(\theta_j)\,\mathrm{d}(\cos\theta_j)\,\mathrm{d}\varphi_j. \tag{2.5.5}$$

Krieger and James postulated that this probability should be the same irrespective of which molecule is regarded as the central one, i.e.,

$$\psi(\theta_0, \varphi_0; \theta_1, \varphi_1) = \psi(\theta_1, \varphi_1; \theta_0, \varphi_0),$$

so that

$$\frac{g(\theta_0)}{[\iint f(\theta_{0j})g(\theta_j)\,\mathrm{d}(\cos\theta_j)\,\mathrm{d}\varphi_j]^{z-1}} = \rho \text{ (constant)}, \tag{2.5.6}$$

which again has to be satisfied for all values of θ, φ.

Methods involving less stringent assumptions, which have the advantage that the relevant consistency relations can be solved exactly, have been put forward, as we shall see later, but first we shall discuss the application of the Krieger–James method.[115–117]

The isotropic phase. We consider the isotropic phase of a nematic liquid crystal in which a weak long range orientational order ($\sim 10^{-4}$ or less) has been induced by an external field (magnetic or electric) acting along the z axis of a space-fixed cartesian coordinate system xyz. Let $W(\theta)$ be the orientational contribution to the potential energy of a molecule due to the field. The relative weight for a given configuration of the cluster of $z+1$ molecules is now

$$\prod_{j=1}^{z} f(\theta_{0j})g(\theta_j)h(\theta_0)h(\theta_j), \tag{2.5.7}$$

where

$$h(\theta_0) = \exp[-W(\theta_0)/k_B T], \text{ etc.}$$

Therefore

$$\psi(\theta_0, \varphi_0; \theta_1, \varphi_1) = f(\theta_{01})h(\theta_0)g(\theta_1)h(\theta_1)$$

$$\times \prod_{j=2}^{z} \iint f(\theta_{0j})g(\theta_j)h(\theta_j)\, \mathrm{d}(\cos\theta_j)\, \mathrm{d}\varphi_j. \tag{2.5.8}$$

We shall suppose that

$$E(\theta_{0j}) = -B^* P_2(\cos\theta_{0j}), \tag{2.5.9}$$

$$V(\theta_j) = -BP_2(\cos\theta_j), \tag{2.5.10}$$

where $P_2(\cos\theta) = \frac{1}{2}(3\cos^2\theta - 1)$ is the Legendre polynomial of the second order. The assumption here is that the potential energy is independent of φ, i.e., that the distribution function is cylindrically symmetric. We also ignore the volume dependence of the potential function. Since we are dealing with a very weakly ordered system, we may resort to the infinitesimal approximation and expand $f(\theta_{0j})$ as

$$f(\theta_{0j}) = \tfrac{1}{2}D \sum_{q=0}^{\infty} (2q+1)c_q P_q(\cos\theta_{0j}), \tag{2.5.11}$$

where

$$c_n = \frac{\int P_n(\cos\theta_{0j})f(\theta_{0j})\, \mathrm{d}(\cos\theta_{0j})}{\int f(\theta_{0j})\, \mathrm{d}(\cos\theta_{0j})} \tag{2.5.12}$$

is a measure of the short range order in the absence of an external field, and D the denominator of (2.5.12). In view of the form of the interaction (2.5.9), q can take only even values in (2.5.11). Similarly we write

$$g(\theta_j)h(\theta_j) = \sum_{q=0}^{\infty} a_q P_q(\cos\theta_j). \tag{2.5.13}$$

From (2.5.11), (2.5.13), (2.5.6) and (2.5.8) it follows that

$$\frac{\sum_{q=0}^{\infty} a_q P_q(\cos\theta_0)}{h(\theta_0)\left[\sum_{l=0}^{\infty} a_l c_l P_l(\cos\theta_0)\right]^{z-1}} = \frac{\sum_{q=0}^{\infty} a_q P_q(\cos\theta_1)}{h(\theta_1)\left[\sum_{l=0}^{\infty} a_l c_l P_l(\cos\theta_1)\right]^{z-1}}$$

$$= \rho \text{ (constant)}, \tag{2.5.14}$$

where l can take only even values. The consistency relation then reduces to the form

$$\sum_{q=0}^{\infty} a_q P_q(\cos\theta) = \rho h(\theta) \sum_{m=0}^{\infty} A'_m P_m(\cos\theta) \tag{2.5.15}$$

where

$$A'_m = (m+\tfrac{1}{2}) \int P_m(\cos\theta)\left[\sum_{m=0}^{\infty} a_l c_l P_l(\cos\theta)\right]^{z-1} \mathrm{d}(\cos\theta).$$

From (2.5.13) and (2.5.15), it can be seen that for the weakly ordered isotropic phase

$$1 + (B/k_{\mathrm{B}}T)P_2(\cos\theta) = A'_0 P_0(\cos\theta) + A'_2 P_2(\cos\theta). \tag{2.5.16}$$

Remembering that $P_0(\cos\theta) = 1$, and equating the coefficients of $P_0(\cos\theta)$ and $P_2(\cos\theta)$ respectively,

$$B/k_{\mathrm{B}}T = A'_2/A'_0 = (z-1)a_2 c_2, \tag{2.5.17}$$

which is the solution of the consistency relation in the limit $a_2 \ll 1$.

The long range order parameter s is given by

$$s = \langle P_2(\cos\theta_0)\rangle$$

$$= \frac{\int\cdots\int P_2(\cos\theta_0)\psi(\theta_0, \varphi_0; \theta_j, \varphi_j)\,\mathrm{d}(\cos\theta_0)\,\mathrm{d}\varphi_0\,\mathrm{d}(\cos\theta_j)\,\mathrm{d}\varphi_j}{\int\cdots\int \psi(\theta_0, \varphi_0; \theta_j, \varphi_j)\,\mathrm{d}(\cos\theta_0)\,\mathrm{d}\varphi_0\,\mathrm{d}(\cos\theta_j)\,\mathrm{d}\varphi_j}. \tag{2.5.18}$$

Applying the consistency relation and using (2.5.12) and (2.5.13),

$$s = \tfrac{1}{5}a_2(1 + c_2). \tag{2.5.19}$$

Therefore,

$$B = \frac{5k_{\mathrm{B}}Ts(z-1)c_2}{1 + c_2}. \tag{2.5.20}$$

Since c_2 depends only on B^*, all the properties can be deduced in terms of this single parameter. If the externally applied field is magnetic,

$$W(\theta) = -\tfrac{1}{3}\chi_{\mathrm{a}} H^2 P_2(\cos\theta),$$

and hence from (2.5.13)

$$a_2 = \frac{B}{k_B T} + \frac{\chi_a H^2}{3k_B T}. \tag{2.5.21}$$

Using (2.5.17), (2.5.20) and (2.5.21)

$$s = \frac{\chi_a H^2}{15k_B T} \frac{1+c_2}{1-(z-1)c_2}. \tag{2.5.22}$$

Expressing c_2 given by (2.5.12) as a series and substituting in the denominator of (2.5.22)

$$s = s_{\text{mag}} = \frac{\chi_a H^2}{15k_B} \frac{1+c_2}{T-T^*} \tag{2.5.23}$$

where

$$T^* = \frac{(z-1)B^*}{5k_B}\left[1 + \frac{B^*}{7k_B T} - \frac{B^{*2}}{35k_B^2 T^2} + \cdots\right].$$

From (2.5.22) and (2.5.23) it is clear that when $T = T^*$

$$c_2 = 1/(z-1) \tag{2.5.24}$$

which at once determines the value of B^*. Similarly it can be shown that under the action of an electric field

$$s_{\text{elec}} = \frac{Fh^2E^2}{15k_B}\left[\alpha_a - \frac{F\mu^2}{2k_B T}(1 - 3\cos^2\beta)\right]\frac{1+c_2}{T-T^*}, \tag{2.5.25}$$

where the symbols have already been defined in § 2.3.5. Equations (2.5.23) and (2.5.25) are formally similar to (2.4.7) and (2.4.13) of the Landau–de Gennes model, except that T^* exhibits a very slight temperature dependence. We shall show presently that this value of T^* is an improvement on that predicted by the Maier–Saupe theory.

In the absence of an external field, the internal energy of the isotropic liquid is

$$U = -\tfrac{1}{2}NzB^*c_2 \tag{2.5.26}$$

and the specific heat

$$c_V = \frac{9NzB^{*2}}{8k_B T^2}(\langle\cos^4\theta\rangle - \langle\cos^2\theta\rangle^2); \tag{2.5.27}$$

c_V turns out to be approximately of the right magnitude but varies rather too slowly with temperature.[115]

The ordered phase In this case, the infinitesimal approximation is certainly not valid and the solving of the Krieger–James consistency relation is not so straightforward. Assuming $E(\theta_{0j})$ and $V(\theta_j)$ to be given by (2.5.9) and (2.5.10) respectively, those pairs of values of B^*/k_BT and B/k_BT which give a constant ratio (2.5.6) for every $P_2(\cos\theta_0)$ represent solutions at that temperature. It turns out in actual practice that with the potentials chosen the ratio is not constant with angle. When the best pairs of values of B^*/k_BT and B/k_BT are substituted into Chang's relation (2.5.4) there are discrepancies, which in the neighbourhood of $\theta_0 \simeq \pi/2$ amount to 2–3 per cent. However, there is a marked improvement in the accuracy if we take

$$V(\theta_j) = -BP_2(\cos\theta_j) - CP_4(\cos\theta_j) \tag{2.5.28}$$

(which is the form of the potential assumed in recent extensions of the Maier–Saupe theory[72]). The values of B and C can be derived in terms of B^* at every temperature by successive iterations so that, as before, the properties of the system can be deduced in terms of a single parameter. It is found that Chang's relation is now satisfied to better than 0.1 per cent at $T \simeq T_{NI}$ for $z = 8$.[117] In principle, greater precision can be achieved by including higher order terms, but the computation becomes too tedious and has not yet been attempted. In the isotropic phase, i.e., in the infinitesimal approximation, the contribution of the P_4 term will of course be negligible.

The long range order parameter s is given by (2.5.18) as before, while the internal energy of the system is

$$U = -\tfrac{1}{2}NzB^*\langle P_2(\cos\theta_{0j})\rangle, \tag{2.5.29}$$

where

$$\langle P_2(\cos\theta_{0j})\rangle = \frac{\int\cdots\int P_2(\cos\theta_{0j})\psi(\theta_0,\varphi_0;\theta_j,\varphi_j)\,\mathrm{d}(\cos\theta_0)\,\mathrm{d}\varphi_0\,\mathrm{d}(\cos\theta_j)\,\mathrm{d}\varphi_j}{\int\cdots\int \psi(\theta_0,\varphi_0;\theta_j,\varphi_j)\,\mathrm{d}(\cos\theta_0)\,\mathrm{d}\varphi_o\,\mathrm{d}(\cos\theta_j)\,\mathrm{d}\varphi_j}. \tag{2.5.30}$$

The plot of U versus $1/T$ (at constant volume) shows the characteristic sigmoid shape (fig. 2.5.1). The curves for the ordered and disordered phases meet at the second order transition point T^* which, as we have seen earlier, is also the temperature at which the short range order parameter $\langle P_2(\cos\theta_{0j})\rangle$ in the isotropic phase $= 1/(z-1)$. The first order transition point T_{NI} is the temperature at which the shaded areas are equal, i.e., when the Helmholtz free energies of the ordered and disordered phases are the same. The present calculation (using (2.5.9) and

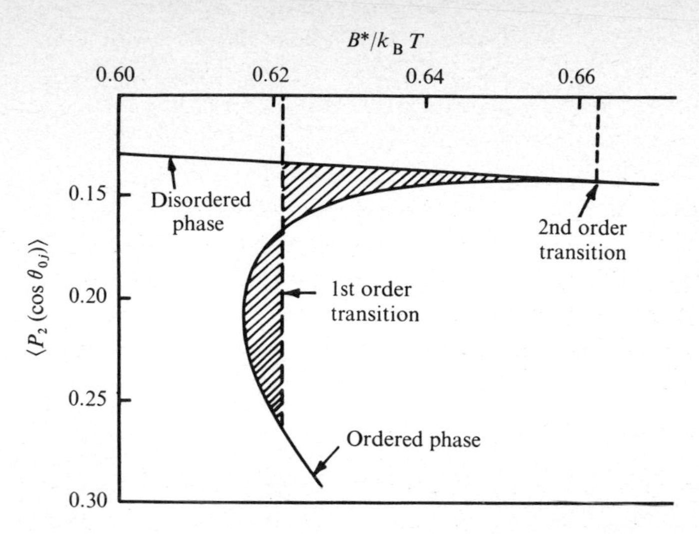

Fig. 2.5.1. Plot of $\langle P_2(\cos\theta_{0j})\rangle(=-2U/NzB^*)$ versus B^*/k_BT for $z=8$.[117] At the first order transition temperature the shaded areas are equal so that the Helmholtz free energy of the ordered and disordered phases are the same. At the second order transition temperature $\langle P_2(\cos\theta_{0j})\rangle = c_2 = 1/(z-1)$.

(2.5.28)) gives $(T_{NI}-T^*)/T_{NI}=0.062$ for $z=8$, which is an improvement over the mean field value of 0.092, though still much higher than the experimental value of 0.003.

The curves for s, $\langle P_4(\cos\theta)\rangle$ and $\langle P_2(\cos\theta_{0j})\rangle$ as functions of temperature are shown in fig. 2.5.2. All three parameters change discontinuously at T_{NI}; the long range order drops to zero at the transition, while the short range order persists even in the isotropic phase. At the transition, $s=0.400$ which is slightly lower than the Maier–Saupe value of 0.4292.

2.5.2. Other approximations

Ypma and Vertogen[118] made the weaker assumption that the *average* value of $P_2(\cos\theta)$ is the same for the central and outer shell molecules. The consistency relation is then expressible as

$$\iint P_2(\cos\theta_0)\left[\iiint f(\theta_{0j})g(\theta_j)\,\mathrm{d}(\cos\theta_j)\,\mathrm{d}\varphi_j\right]^z \mathrm{d}(\cos\theta_0)\,\mathrm{d}\varphi_0$$

$$=\int\ldots\int P_2(\cos\theta_1)\left[\iiint f(\theta_{0j})g(\theta_j)\,\mathrm{d}(\cos\theta_j)\,\mathrm{d}\varphi_j\right]^{z-1}$$

$$\times f(\theta_{01})g(\theta_1)\,\mathrm{d}(\cos\theta_0)\,\mathrm{d}\varphi_0\,\mathrm{d}(\cos\theta_1)\,\mathrm{d}\varphi_1. \tag{2.5.31}$$

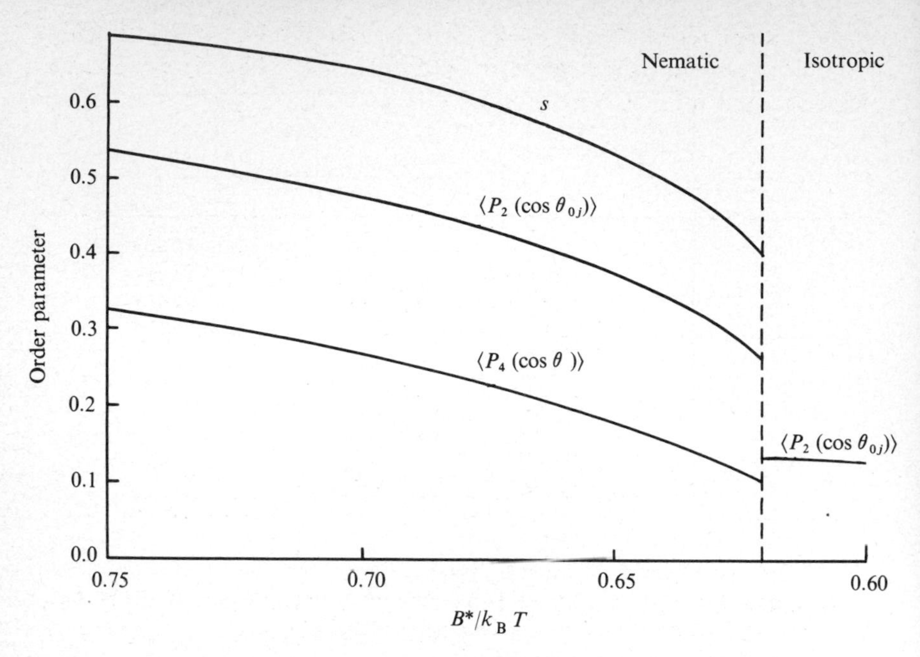

Fig. 2.5.2. Short range order parameter $\langle P_2(\cos\theta_{0j})\rangle$ and the long range order parameters s and $\langle P_4(\cos\theta)\rangle$ versus B^*/k_BT calculated in the Krieger–James approximation.[117] The long range order disappears at the transition but the short range order persists even in the isotropic phase.

Another somewhat similar approach is as follows:[117,119] the orientational order of a molecule is evaluated (i) by considering it in the mean field due to all its z neighbours, and (ii) by treating one of the z interactions exactly and replacing the rest of the interactions of this coupled pair by the effective field of $z-1$ neighbours on each of them. It is then assumed that (i) and (ii) should give the same results. This leads to the relation

$$\frac{\int P_2(\cos\theta_0)[g(\theta_0)]^z\,\mathrm{d}(\cos\theta_0)}{\int [g(\theta_0)]^z\,\mathrm{d}(\cos\theta_0)} = \frac{\int\ldots\int P_2(\cos\theta_0)f(\theta_{01})g(\theta_0)^{z-1}g(\theta_1)^{z-1}\,\mathrm{d}(\cos\theta_0)\,\mathrm{d}\varphi_0\,\mathrm{d}(\cos\theta_1)\,\mathrm{d}\varphi_1}{\int\cdots\int f(\theta_{01})g(\theta_0)^{z-1}g(\theta_1)^{z-1}\,\mathrm{d}(\cos\theta_0)\,\mathrm{d}\varphi_0\,\mathrm{d}(\cos\theta_1)\,\mathrm{d}\varphi_1}. \tag{2.5.32}$$

Some calculations based on (2.5.31) and (2.5.32), taking $E(\theta_{0j})$ and $V(\theta_j)$ to be given by (2.5.9) and (2.5.10), are presented in table 2.5.1. It is seen that the two methods yield similar results. An interesting fact first pointed out by Ypma and Vertogen is that $(T_{NI}-T^*)/T_{NI}$ approaches the experimental value with decreasing z.

TABLE 2.5.1

	z	Equation (2.5.31)[118]	Equation (2.5.32)[117]	Maier–Saupe approx.
s_c	4	0.413	0.420	0.429
	8	0.424	0.426	
$(T_{NI}-T^*)/T_{NI}$	4	0.0288	0.0288	0.0918
	8	0.0569	0.0567	

Again (2.5.10) is found to be inadequate; the solutions of (2.5.31) and (2.5.32) satisfy Chang's relation only to about 3 per cent at $\theta \simeq \pi/2$. However, it turns out that when $V(\theta_j)$ is taken to be of the form (2.5.28) and the additional condition is imposed that the average $P_4(\cos\theta)$ is the same for the central and outer shell molecules, the solutions do fulfil (2.5.4) to better than 0.1 per cent at all angles for $z=8$[117] and yield results that are practically identical to those of the Krieger–James method. It may be emphasized that thermodynamic equilibrium of the system requires that (2.5.4) be satisfied accurately.

In the infinitesimal approximation, i.e., in the isotropic phase, the differences between these various approaches become insignificant and in effect the theory reduces in all cases to that presented in § 2.5.1.

2.5.3. Antiferroelectric short range order

The vast majority of nematogens are polar compounds but the absence of ferroelectricity in the nematic phase shows that there is equal probability of the dipoles pointing in either direction. Because of this it is generally assumed that the permanent dipolar contribution to the orientational order is negligibly small. However, a simple calculation shows that the interaction between neighbouring dipoles is by no means trivial compared with dispersion forces, particularly in strongly polar materials. We shall now consider a model which takes into account the influence of permanent dipoles and is at the same time consistent with the non-polar character of the medium.[116,117]

Suppose as before that the molecules are cylindrically symmetric so that the dipole moment is along the major molecular axis. Now if a dipole is fixed at O as shown in fig. 2.5.3, I and II represent situations of

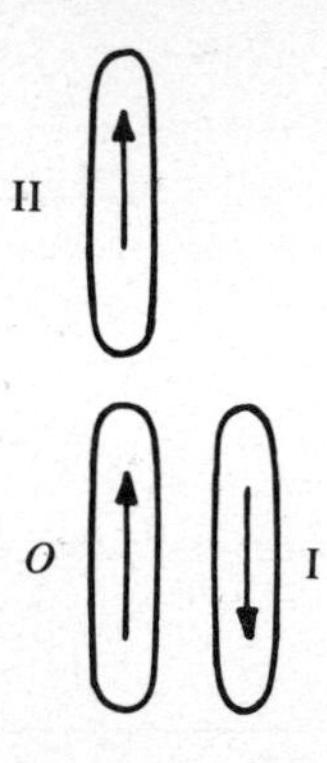

Fig. 2.5.3. Preferred orientation of neighbouring dipoles in the end-on and broadside-on positions. However, because of the anisotropic shape of the molecules, situation I is much more important than II in the nematic structure and there results a net antiparallel correlation between neighbouring dipoles.

minimum energy for a neighbouring dipole in the broad-side-on and end-on positions respectively. Evidently, by virtue of the anisotropic shape of the molecule, situation I will be much more important, i.e., there will be a greater tendency for the nearest neighbours to assume an antiparallel orientation. However, the absence of long range translational order in the nematic fluid precludes the possibility of antiferroelectric long range order. To express this in a mathematically tractable form we shall resort again to the Bethe approximation. We modify the pair potential (2.5.9) to

$$E(\theta_{0j}) = A^* P_1(\cos\theta_{0j}) - B^* P_2(\cos\theta_{0j}), \tag{2.5.33}$$

which favours an antiparallel arrangement of the permanent dipoles, but let the interaction between j and the rest of the medium continue to be (2.5.28) as before. Here P_1 is the Legendre polynomial of the first order. Fig. 2.5.4 gives the curves for the long range parameter s and the short range parameters $\langle P_1(\cos\theta_{0j})\rangle$ and $\langle P_2(\cos\theta_{0j})\rangle$ calculated on the basis of the modified potential, taking $A^*/B^* = 0.5$. $\langle P_1(\cos\theta_{0j})\rangle$ is negative signifying antiparallel order. $(T_{NI} - T^*)/T_{NI}$ is now 0.059 for $z = 8$.

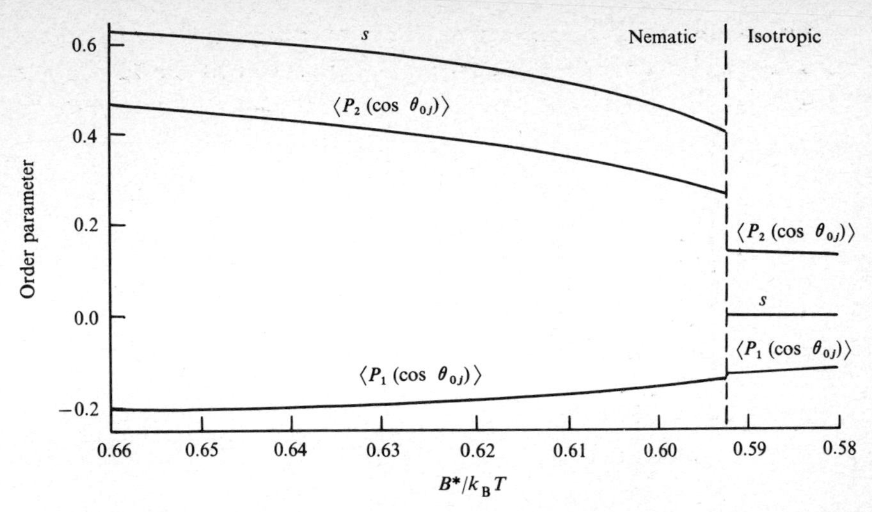

Fig. 2.5.4. Short range order parameters $\langle P_1(\cos\theta_{0j})\rangle$, $\langle P_2(\cos\theta_{0j})\rangle$ and the long range order parameter s versus B^*/k_BT. The negative value of $\langle P_1(\cos\theta_{0j})\rangle$ signifies antiparallel ordering. The curves are calculated for $A^*/B^*=0.5$.[117]

The first important question that needs to be considered is whether such an antiparallel correlation leads to the right sign and magnitude of the dielectric anisotropy. Proceeding in the usual manner, the principal dielectric constants turn out to be

$$\varepsilon_1=1+\frac{4\pi N\rho hF}{M}\left[\bar{\alpha}+\tfrac{2}{3}\alpha_a s+\frac{F\mu^2}{3k_BT}(2s+1)+\frac{F\mu^2}{k_BT}\langle\cos\theta_0\cos\theta_j\rangle\right] \tag{2.5.34}$$

and

$$\varepsilon_2=1+\frac{4\pi N\rho hF}{M}\left[\bar{\alpha}-\tfrac{1}{3}\alpha_a s+\frac{F\mu^2}{k_BT}\langle\sin^2\theta_0\cos^2\varphi_0\rangle+\frac{F\mu^2}{k_BT}\langle\cos\varphi_0\cos\varphi_j\sin\theta_0\sin\theta_j\rangle\right], \tag{2.5.35}$$

where the symbols F, h, etc., have the same meanings as in § 2.3.5. Using the theoretically derived s of fig. 2.5.4, ε_1, ε_2 as well as $\bar{\varepsilon}=\frac{1}{3}(\varepsilon_1+2\varepsilon_2)$ calculated from these equations are presented in fig. 2.5.5 for a strongly polar (nitrile) compound of the type studied by Schadt[83] (see fig.

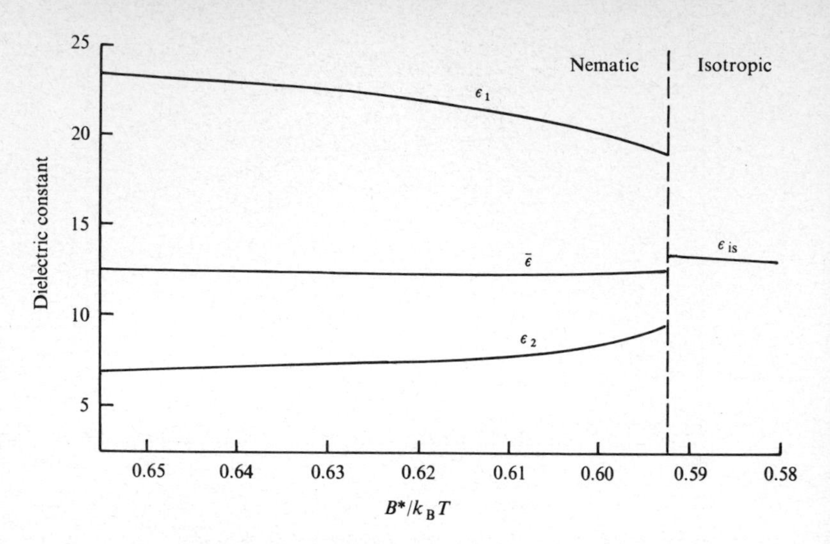

Fig. 2.5.5. Theoretical variation of the dielectric constants ε_1, ε_2 and $\bar{\varepsilon}=\frac{1}{3}(\varepsilon_1+2\varepsilon_2)$; $(A^*/B^*=0.5)$.[117] The theory predicts that $\bar{\varepsilon}$ in the nematic phase should be slightly lower than the extrapolated value of ε_{is}, the dielectric constant in the isotropic phase. This is found to be the case experimentally (see fig. 2.5.6).

2.3.10(*b*)). The parameters used in the calculations are: $\mu=5$ debyes along the major molecular axis, $\alpha=28\times10^{-24}\ \text{cm}^3$ and $\alpha_a=15\times10^{-24}\ \text{cm}^3$ (evaluated approximately by assuming addivity of bond polarizabilities extrapolated to low frequency). Since thermal expansion is ignored, the rate of variation with temperature is reduced especially near the transition, but apart from that it is clear from a comparison with fig. 2.3.10(*b*) that the dielectric anisotropy ε_a is of the right order of magnitude.

An interesting consequence of the theory is that the mean dielectric constant $\bar{\varepsilon}$ should *increase* by a few per cent on going from the nematic to the isotropic phase because of the diminution in $\langle P_1(\cos\theta_{0j})\rangle$. This is found to be the case experimentally in a number of strongly positive materials[83,107] (fig. 2.5.6). (A similar increase is seen in some negatively anisotropic materials also, e.g., PAA[82] (see fig. 2.3.10); this can probably be explained as due to an antiparallel correlation between the longitudinal components of the dipole moments. As far as the transverse components are concerned there will not be on the average any orientational correlation for position I of fig. 2.5.3 because of the cylindrically

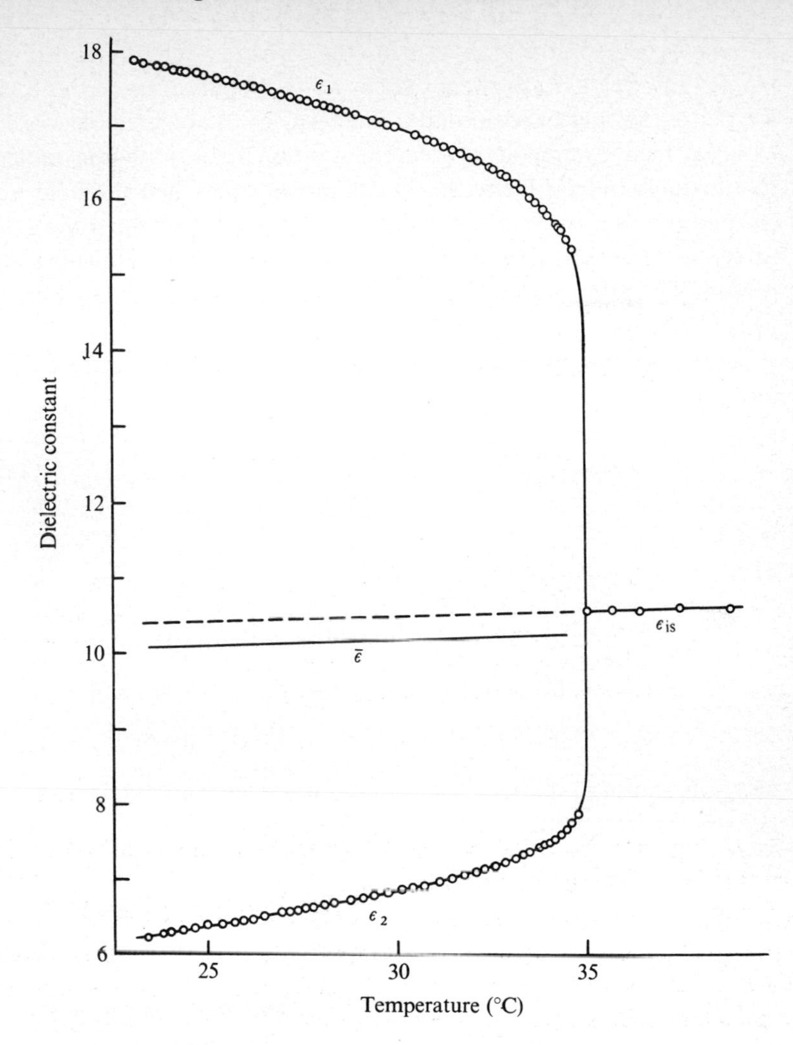

Fig. 2.5.6 Principal dielectric constants of 4-n-pentyl-4-cyanobiphenyl (5CB); $\bar{\varepsilon} = \frac{1}{3}(\varepsilon_1 + 2\varepsilon_2)$ is calculated from the measured values of ε_1 and ε_2. The dashed line denotes the extrapolated value of ε_{is}. (After reference 107.)

symmetric distribution about the optic axis, but there will be an antiparallel correlation for position II which is however likely to be so weak that it can probably be neglected.)

The theory of the Kerr effect in the isotropic phase is now somewhat more complicated because one has to take into account the field-induced long range order parameter $\langle P_1(\cos\theta)\rangle$ in addition to the usual long range parameter s. Moreover, in contrast to dielectric measurements which employ low fields, Kerr effect studies require fields of the order of a few kV per cm so that the orientational energy of the polar molecule in the external field cannot be neglected in comparison with the molecular interactions. Indeed Helfrich[120] has shown that when the field is very strong there is actually a slight shift of the transition temperature T_{NI} in materials of very strong positive dielectric anisotropy. However, in the present discussion we shall ignore this small change of T_{NI} ($\sim 1\times 10^{-3}$ K per kV cm^{-1}).

The relative weight for a given configuration of the cluster of $z+1$ molecules is now

$$\prod_{j=1}^{z} f(\theta_{0j})g(\theta_j)\chi(\theta_0)\chi(\theta_j) \tag{2.5.36}$$

where

$$f(\theta_{0j})=\exp[-E(\theta_{0j})/k_{\mathrm{B}}T]$$
$$g(\theta_j)=\exp[\{AP_1(\cos\theta_j)+BP_2(\cos\theta_j)\}/k_{\mathrm{B}}T]$$
$$\chi(\theta_0)=\exp[\{Fh\mu EP_1(\cos\theta_0)+\tfrac{1}{3}\alpha_a Fh^2E^2P_2(\cos\theta_0)\}/k_{\mathrm{B}}T]$$
$$\chi(\theta_j)=\exp[\{Fh\mu EP_1(\cos\theta_j)+\tfrac{1}{3}\alpha_a Fh^2E^2P_2(\cos\theta_j)\}/k_{\mathrm{B}}T]$$

and A is another interaction parameter to be determined. Proceeding as before, the consistency relation takes the form

$$\frac{1+a_1P_1(\cos\theta)+a_2P_2(\cos\theta)}{\chi(\theta)[1+a_1c_1P_1(\cos\theta)+a_2c_2P_2(\cos\theta)]^{z-1}}=\text{constant} \tag{2.5.37}$$

and the electrically induced long range order can be shown to be

$$s_{\text{elec}}=\left[\frac{\alpha_a Fh^2E^2}{3k_{\mathrm{B}}T}+\frac{(Fh\mu E)^2}{3k_{\mathrm{B}}^2T^2}\frac{1-(z-1)c_1^2}{\{1-(z-1)c_1\}^2}\right]\frac{1+c_2}{5\{1-(z-1)c_2\}}$$
$$+\frac{1}{15}\frac{c_1(Fh\mu E)^2}{k_{\mathrm{B}}^2T^2\{1-(z-1)c_1\}^2}, \tag{2.5.38}$$

where c_1 and c_2 are defined by (2.5.12).

If the applied field is magnetic, $F=h=1$, $\mu=0$, and replacing α_a by χ_a, (2.5.38) reduces to (2.5.22). Since there is no spontaneous $\langle P_1(\cos\theta)\rangle$ type of long range order in the nematic phase, the second order transition point T^* is still determined by the condition $c_2=1/(z-1)$. Thus both s_{mag} and s_{elec} vary as $(T-T^*)^{-1}$ to a good approximation. This is in accord with experiment[107] (see fig. 2.4.4). However if the dipolar interactions are extremely large, s_{elec} may exhibit a slightly slower variation at temperatures well above T^*.[116] This type of behaviour seems to have been observed in some compounds[121] but further studies are necessary to establish this point.

The model may also have implications for the phenomenon of curvature electricity discussed in § 3.11. If the molecules possess 'shape polarity' in addition to a strong dipole moment along the major axis, then the possibility exists that a splay deformation will polarize the material, and conversely that an electric field will produce a splay in the structure (see fig. 3.11.1). Qualitatively it can be seen that the magnitude of the effect will be diminished because of the tendency of the neighbouring dipoles to be antiparallel.

Direct X-ray evidence for such antiparallel local ordering in the nematic and isotropic phases of 4′-n-pentyl- and 4′-n-heptyl-4-cyanobiphenyls (5CB and 7CB), both of which are strongly polar compounds, have been reported very recently by Leadbetter, Richardson and Colling.[122] They have found that the meridional reflexions correspond to a repeat distance along the preferred axis of about 1.4 times the molecular length, which they have interpreted as due to an overlapping head-to-tail arrangement of the neighbouring molecules (fig. 2.5.7). Similar conclusions have been drawn by Lydon and Coakley,[123] who have observed that the layer spacing in the smectic *A* phase of the octyl compound 8CB is again far in excess of the molecular length. In MBBA, on the other hand, the repeat distance is approximately equal to the molecular length. These studies appear to lend strong support to the model of antiparallel correlation that we have just discussed.

Thus the application of the Bethe approximation has been useful in bringing out some qualitative features of short range order in nematics, but it certainly cannot be regarded as adequate from a quantitative point of view. A more rigorous molecular statistical description of these effects still remains to be developed.

2.6. The nematic liquid crystal free surface

We now turn our attention briefly to the nematic liquid crystal surface. In the case of simple liquids, the structural features of the liquid–vapour transition zone has been investigated in considerable detail. It has been suggested by Croxton and Ferrier[124] that under certain circumstances relatively ordered states may develop in the vicinity of the surface in the form of stable density oscillations in the transition profile. We shall now discuss the formal extension of these ideas to the nematic liquid crystal and show that the possibility exists of orientationally ordered states

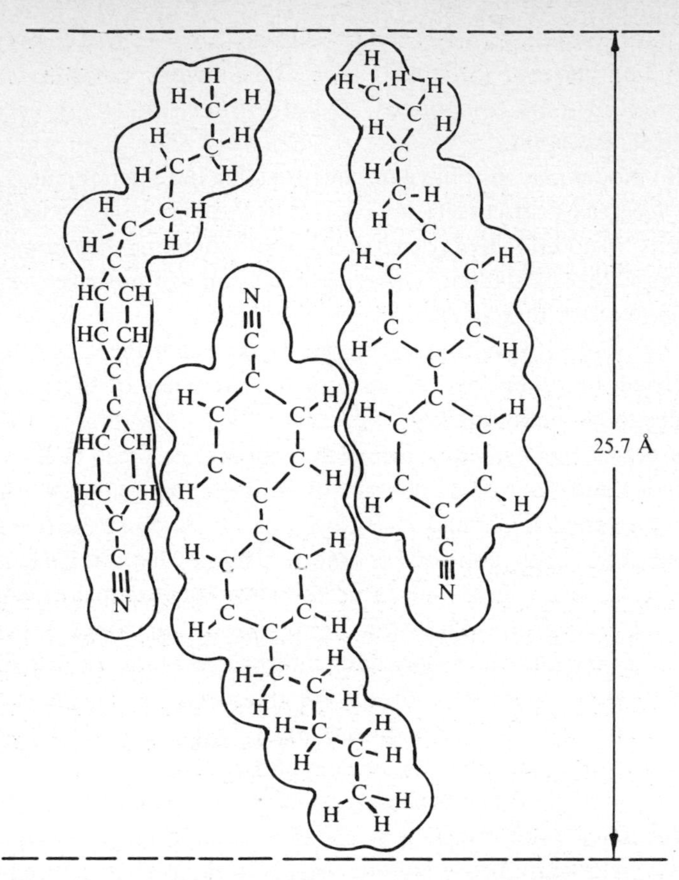

Fig. 2.5.7. Schematic diagram of antiparallel local structure in 5CB resulting in a repeat distance along the nematic axis of about 1.4 times the molecular length. (Proposed by Leadbetter, Richardson and Colling.[122])

developing near the surface.[125] This has important consequences particularly in relation to the temperature dependence of the surface tension γ. The gradient of the surface tension versus temperature characteristic is directly related to the surface excess entropy per unit area as follows:[44]

$$\frac{d\gamma}{dT} = -(S_\sigma - S_\beta), \tag{2.6.1}$$

where σ and β refer to the surface and bulk states respectively. Thus if ordered states develop near the nematic surfaces S_σ may be less than S_β and γ may actually show a positive slope. Such a trend was observed by Ferguson and Kennedy[126] many years ago and has been confirmed by recent measurements carried out under equilibrium conditions.[127]

To study the orientational contributions to the $\gamma(T)$ characteristic, we begin with a pair potential of the form proposed by McMillan:[128]

$$u_{12}(r_{12}, \cos\theta_{12}) = -u_0 \exp[-(r_{12}/r_0)^2] P_2(\cos\theta_{12}), \qquad (2.6.2)$$

where r_{12} is the separation of the centres of mass of the molecules, θ_{12} is the relative orientation of their long axes, and r_0 a constant of the order of a molecular length. In the bulk liquid, we assume a single particle potential in the mean field approximation

$$u_1(\cos\theta_1) = -u_o s P_2(\cos\theta_1), \qquad (2.6.3)$$

where s is the usual orientational order parameter. To allow for spatial delocalization of the interaction between the molecule and its environment in the vicinity of the surface, we modify (2.6.3) to

$$u_1(z_1, \cos\theta_1) = -u_0 s(z) P_2(\cos\theta_1). \qquad (2.6.4)$$

An oscillatory density profile is unlikely to develop in nematic systems (though it may conceivably arise in smectics). We shall therefore assume a simple profile of the form shown in fig. 2.6.1 for the single particle distribution function $g_{(1)}(z)$ which describes the spatial distribution of the centres of gravity of the molecules. The function $g_{(1)}(z)$ is related to the single particle potential (of mean force) $\psi(z)$ as

$$g_{(1)}(z) = \exp(-\psi(z)/k_B T)$$

so that $\psi(z)$ serves to decouple the interaction across the transition zone.

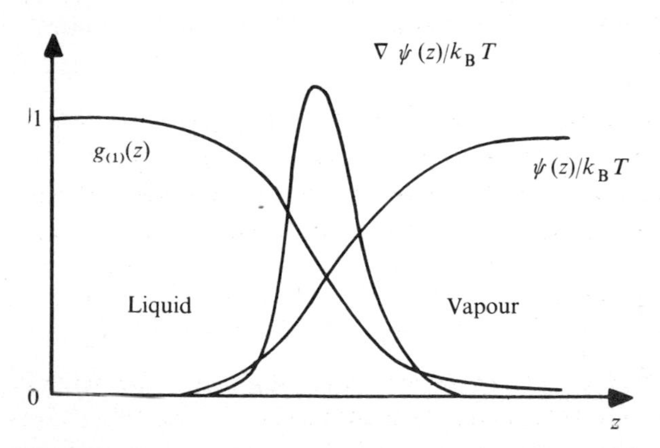

Fig. 2.6.1. Schematic variation of the single particle distribution of molecular centres $g_{(1)}(z)$, the potential of mean force $\psi(z)$ and its gradient $\nabla\psi(z)/k_B T$ in the vicinity of the nematic liquid crystal surface.

A convenient decoupling function is

$$\exp(-\psi(z)/k_{\mathrm{B}}T)-1$$

which we take to describe the spatial modification of the single particle potential as the molecule passes through the collective field in the transition zone. The surface field may then be taken as $-\nabla\psi(z)$ (fig. 2.6.1). We also propose an associated local orientational potential

$$1-\exp(-c\nabla\psi(z)/k_{\mathrm{B}}T)$$

where c is a constant governing the strength and range of the orienting torque. We therefore arrive at the following single particle potential

$$u_1(z_1, \cos\theta_1) = -u_0P_2(\cos\theta_1)[s-\xi_0\{\exp(-\psi(z)/k_{\mathrm{B}}T)-1\} + \xi_1\{1-\exp(-c\nabla\psi(z)/k_{\mathrm{B}}T)\}] \tag{2.6.5}$$

where ξ_0 and ξ_1 are coefficients to be determined and the quantity in the square brackets represents an effective order parameter. Thus the structural delocalization across the transition zone diminishes the local order, while the surface field enhances it. Now the single particle potential is expressible in terms of the pair potential (2.6.2) as

$$u_1(z_1, \cos\theta_1) = \frac{\iint u_{12}(r_{12}, \cos\theta_{12})g_{(1)}(z_2, \cos\theta_2)\,\mathrm{d}z\,\mathrm{d}\hat{z}}{\iint g_{(1)}(z_2, \cos\theta_2)\,\mathrm{d}z\,\mathrm{d}\hat{z}} \tag{2.6.6}$$

where the integrals range over all positions z and orientations $\hat{z}$ of molecule 2, and

$$g_{(1)}(z_2, \cos\theta_2) = \exp[-u_1(z_2, \cos\theta_2)/k_{\mathrm{B}}T].$$

From a comparison of (2.6.5) and (2.6.6) we may formally write

$$s = \langle P_2(\cos\theta_2)\rangle_{g(1)(2)}, \tag{2.6.7}$$

$$\xi_0 = \langle[\exp(-\psi(z_2)/k_{\mathrm{B}}T)-1]P_2(\cos\theta_2)\rangle_{g(1)(2)}, \tag{2.6.8}$$

$$\xi_1 = \langle[1-\exp(-c\nabla\psi(z_2)/k_{\mathrm{B}}T)]P_2(\cos\theta_2)\rangle_{g(1)(2)}, \tag{2.6.9}$$

and these must be determined self consistently. As $z\to-\infty$, $s(z)\to s$, the asymptotic bulk value. Using the standard statistical thermodynamic relations it can be shown that the surface excess entropy per unit area

$(=-\mathrm{d}\gamma/\mathrm{d}T)$ developed at the liquid surface:

$$-\frac{\mathrm{d}\gamma}{\mathrm{d}T}=-\frac{u_0\rho}{T}\int_{-\infty}^{\infty} g_{(1)}(z)[s^2-\xi_0^2\{\exp(-\psi(z)/k_\mathrm{B}T)-1\}^2$$

$$+\xi_1^2\{1-\exp(-c\nabla\psi(z)/k_\mathrm{B}T)\}^2]\,\mathrm{d}z+\frac{u_0\rho}{T}\int_{-\infty}^{0} s^2\,\mathrm{d}z$$

$$+\rho k_\mathrm{B}\int_{-\infty}^{\infty} g_{(1)}(z)\ln\Xi(z,s,\xi_0,\xi_1)\,\mathrm{d}z$$

$$-\rho k_\mathrm{B}\int_{-\infty}^{0}\ln\Xi(z,s)\,\mathrm{d}z, \qquad (2.6.10)$$

where

$$\Xi(z,s)=\int_0^1 \exp\{(u_0 s/k_\mathrm{B}T)P_2(\cos\theta)\}\,\mathrm{d}(\cos\theta),$$

$$\Xi(z,s,\xi_0,\xi_1)=\int_0^1 \exp[(u_0/k_\mathrm{B}T)\{s-\xi_0(\exp(-\psi(z)/k_\mathrm{B}T)-1)$$

$$+\xi_1(1-\exp(-c\nabla\psi(z)/k_\mathrm{B}T))\}P_2(\cos\theta)]\,\mathrm{d}(\cos\theta).$$

The expression for the surface excess free energy per unit area may also be written down in a similar manner.

The surface tension will of course be dependent on the angle which the molecules make with the surface[129] and it is known that this angle varies from substance to substance. For example, light scattering studies[130] indicate that in MBBA the angle is about 75° and changes with temperature, while in PAA it is nearly zero and temperature independent. The theory as formulated above does not explicitly take into account the orientation of the molecules relative to the surface, but merely defines an effective order parameter $s(z)$ to describe the net degree of local order irrespective of the orientation. The information regarding the orientation is assumed to be contained indirectly in $g_{(1)}(z)$. For example, one would anticipate different forms of the transition profile depending on whether the molecules are aligned parallel or perpendicular to the surface, and if there is any thermal variation of the orientation, this would be incorporated in $g_{(1)}(z, T)$, which in any case already exhibits a pronounced spatial relaxation with increasing temperature.

In the absence of explicit knowledge of $g_{(1)}(z)$, only the formal aspects of (2.6.10) can be examined. Some possibilities are indicated in figs. 2.6.2 and 2.6.3. Fig. 2.6.2 shows the schematic variation of the effective order

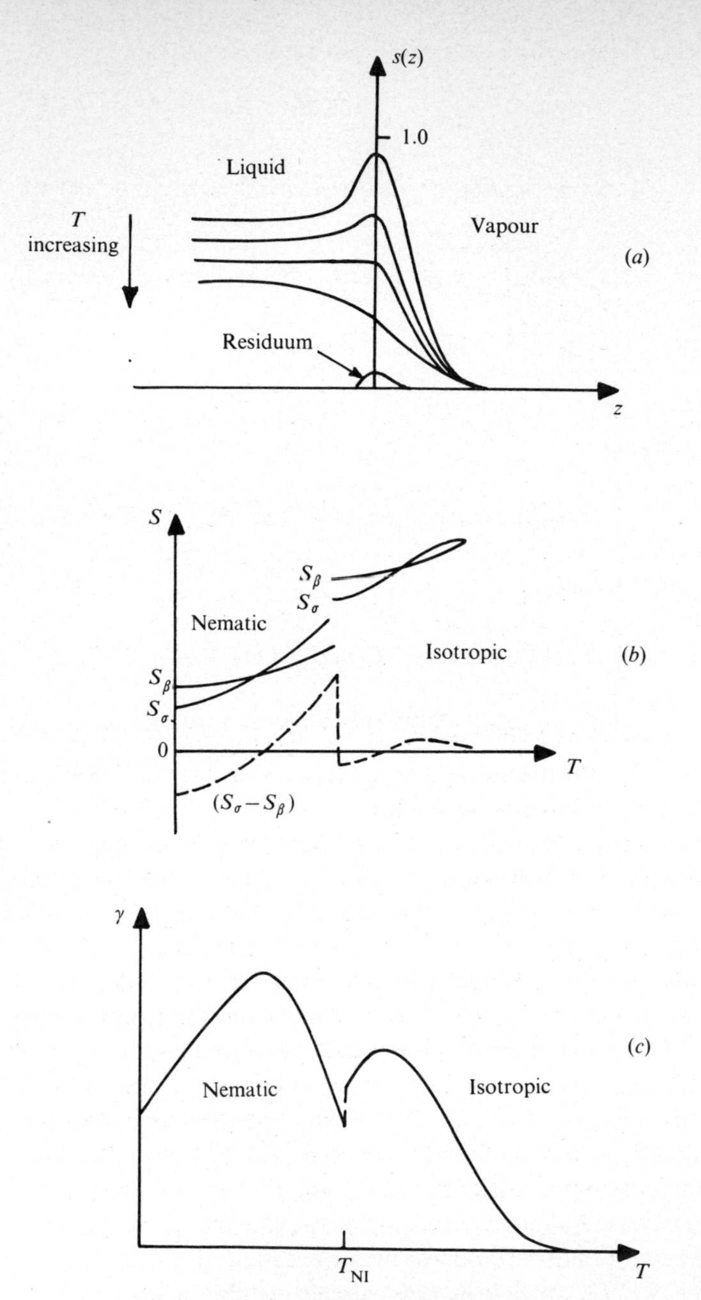

Fig. 2.6.2. Schematic variation of (*a*) the effective order parameter $s(z)$ in the vicinity of the nematic liquid crystal surface, (*b*) the bulk and surface entropy curves with temperature on the basis of $s(z, T)$, and (*c*) the $\gamma(T)$ characteristic.[125] (For explanation, see text.)

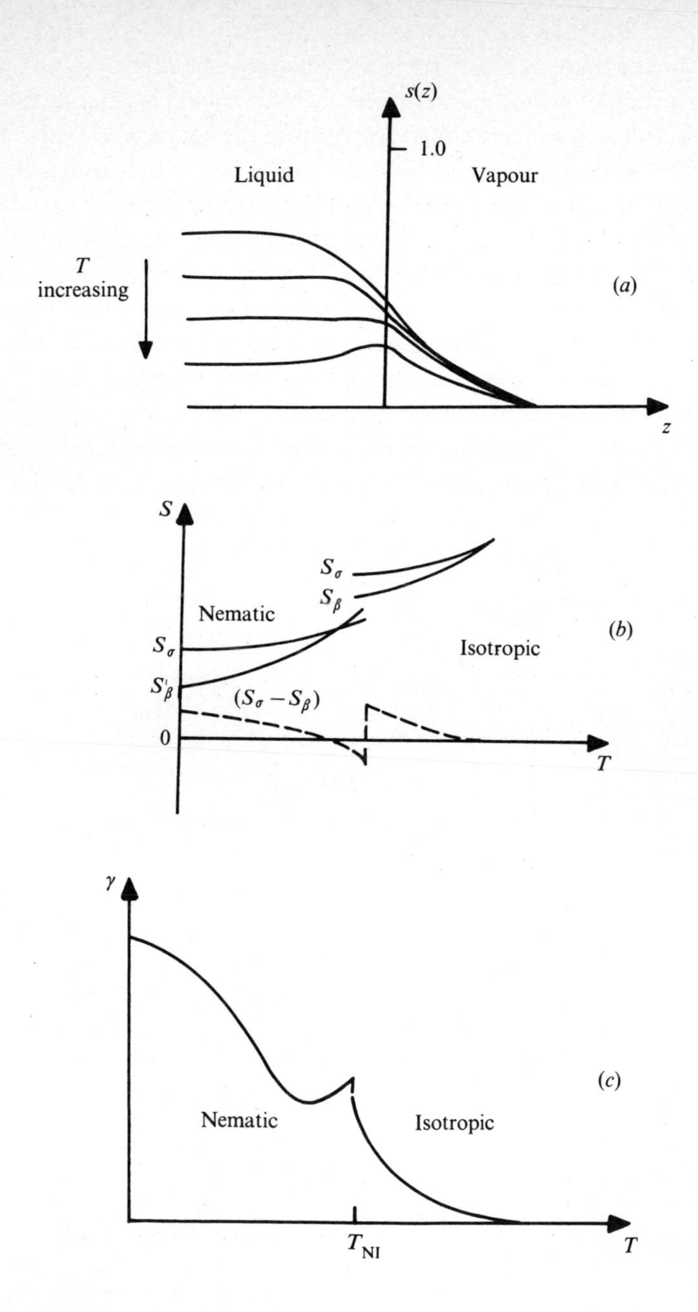

Fig. 2.6.3. See legend of fig. 2.6.2.

parameter $s(z)$ in the vicinity of the nematic liquid crystal surface. At low temperatures there is a net surface enhancement of the orientational order attributed to surface field effects overcoming spatial delocalizational disorder. With increasing temperature the balance is reversed, and progressive spatial and surface field relaxation results in a surface depression of the local order parameter. With the catastrophic bulk variation of s at the transition temperature, a weak surface residuum may persist for a short thermal range beyond the transition. Fig. 2.6.3 illustrates another possible situation. While at low temperatures there is a net depression of the surface value of $s(z)$, the bulk value of the order parameter becomes sufficiently low just prior to the transition for an effective surface enhancement to develop. Since positive slopes in the $\gamma(T)$ characteristic depend only upon the *relative* values of the bulk and surface entropies per unit area, an increase in γ may be anticipated immediately before T_{NI}.

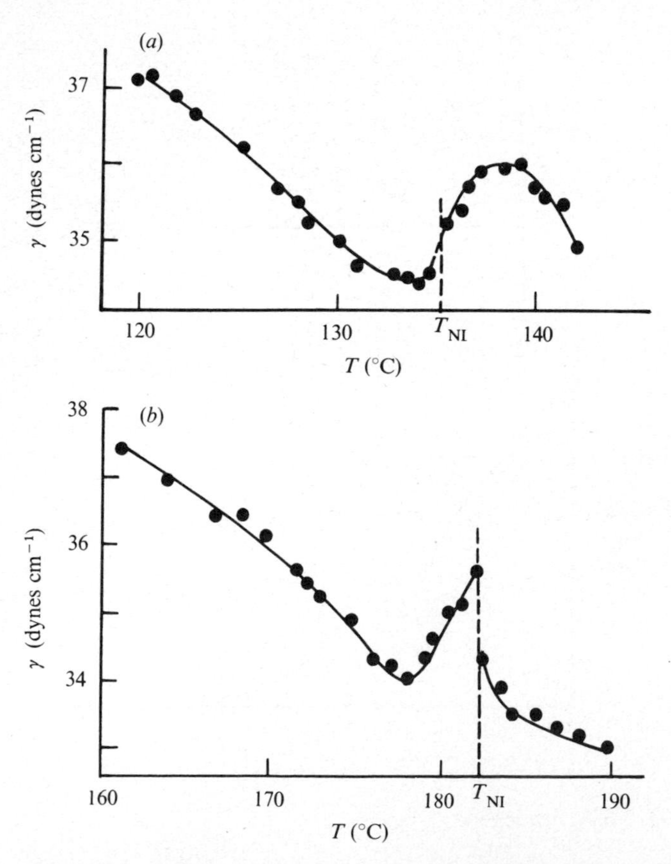

Fig. 2.6.4. Surface tension–temperature characteristic curves for (*a*) PAA and (*b*) *p*-anisaldazine. (After reference 127.)

Fig. 2.6.4 presents a recent determination of the temperature variation of the surface tension of PAA and *p*-anisaldazine under equilibrium conditions using the pendant drop method.[127] The drop was completely enclosed in a thermostatically controlled chamber filled with inert atmosphere and maintained in equilibrium with its saturated vapour. The observed features can be seen to be in qualitative agreement with the theory.

3

Continuum theory of the nematic state

3.1. The Ericksen–Leslie theory

In this chapter we shall discuss the continuum theory of nematic liquid crystals and some of its applications. Many of the most important physical phenomena exhibited by the nematic phase, such as its unusual flow properties or its response to electric and magnetic fields, can be studied by regarding the liquid crystal as a continuous medium. The foundations of the continuum model were laid in the late 1920s by Oseen[1] and Zöcher[2] who developed a static theory which proved to be quite successful. The subject lay dormant for nearly thirty years afterwards until Frank[3] re-examined Oseen's treatment and presented it as a theory of curvature elasticity. Dynamical theories were put forward by Anzelius[4] and Oseen,[1] but the formulation of general conservation laws and constitutive equations describing the mechanical behaviour of the nematic state is due to Ericksen[5,6] and Leslie[7]. Other continuum theories have been proposed,[8] but it turns out that the Ericksen–Leslie approach is the one that is most widely used in discussing the nematic state.

The nematic liquid crystal differs from a normal liquid in that it is composed of rod-like molecules with the long axes of neighbouring molecules aligned approximately parallel to one another. To allow for this anisotropic structure, we introduce a vector $\mathbf{n}$ to represent the direction of preferred orientation of the molecules in the neighbourhood of any point. This vector is called the *director*. Its orientation can change continuously and in a systematic manner from point to point in the medium (except at singularities). Thus external forces and fields acting on the liquid crystal can result in a translational motion of the fluid as also in an orientational motion of the director.

3.1.1. Conservation laws and the entropy inequality

We begin by writing down the conservation or balance laws (Ericksen[6]). We shall employ the cartesian tensor notation, repeated tensor indices being subject to the usual summation convention. The comma denotes partial differentiation with respect to spatial coordinates and the superposed dot a material time derivative. For example,

$$T_{,i} = \partial T/\partial x_i, \qquad v_{i,j} = \partial v_i/\partial x_j$$

and

$$\dot{T} = \mathrm{d}T/\mathrm{d}t.$$

We shall consider the medium to be incompressible ($v_{i,i} = 0$, where v_i is the linear velocity) and at constant temperature ($\dot{T} = T_{,i} = 0$). We shall assume further that the director is of constant magnitude. This implies that the external forces and fields responsible for elastic deformation, viscous flow etc., are very much weaker than the molecular interactions giving rise to the spontaneous alignment of the neighbouring molecules. It is indeed a valid assumption in all the static and dynamic phenomena discussed in this chapter. We may therefore conveniently choose $\mathbf{n}$ to be a dimensionless unit vector ($n_i n_i = 1$).

Let the material volume be V bounded by a surface A. The conservation laws take the following form:

Conservation of mass

$$\frac{\mathrm{d}}{\mathrm{d}t}\int_V \rho \,\mathrm{d}V = 0, \tag{3.1.1}$$

where ρ is the density.

Conservation of linear momentum

$$\frac{\mathrm{d}}{\mathrm{d}t}\int_V \rho v_i \,\mathrm{d}V = \int_V f_i \,\mathrm{d}V + \int_A t_{ji} \,\mathrm{d}A_j, \tag{3.1.2}$$

where f_i is the body force per unit volume and t_{ji} the stress tensor.

Conservation of energy

$$\frac{\mathrm{d}}{\mathrm{d}t}\int_V (\tfrac{1}{2}\rho v_i v_i + U + \tfrac{1}{2}\rho_1 \dot{n}_i \dot{n}_i)\,\mathrm{d}V$$

$$= \int_V (f_i v_i + G_i \dot{n}_i)\,\mathrm{d}V + \int_A (t_i v_i + s_i \dot{n}_i)\,\mathrm{d}A, \tag{3.1.3}$$

where ρ_1 is a material constant having the dimensions of moment of inertia per unit volume (ML^{-1}), U the internal energy per unit volume, G_i the external director body force (which has the dimensions of torque per unit volume since n_i has been chosen to be dimensionless), $t_i = t_{ji}\nu_j$ the surface force per unit area acting across the plane whose unit normal is ν_j, and $s_i = \pi_{ji}\nu_j$ the director surface force (which has the dimensions of torque per unit area). We assume here that there are no heat sources or sinks.

Conservation of angular momentum

$$\frac{\mathrm{d}}{\mathrm{d}t}\int_V (\rho e_{ijk}x_j v_k + \rho_1 e_{ijk} n_j \dot{n}_k)\,\mathrm{d}V$$

$$= \int_V (e_{ijk}x_j f_k + e_{ijk} n_j G_k)\,\mathrm{d}V + \int_A (e_{ijk}x_j t_k + e_{ijk} n_j s_k)\,\mathrm{d}A, \tag{3.1.4}$$

or in vector notation,

$$\frac{\mathrm{d}}{\mathrm{d}t}\int_V [\rho(\mathbf{r}\times\mathbf{v}) + \rho_1(\mathbf{n}\times\dot{\mathbf{n}})]\,\mathrm{d}V$$

$$= \int_V [(\mathbf{r}\times\mathbf{f}) + (\mathbf{n}\times\mathbf{G})]\,\mathrm{d}V + \int_A [(\mathbf{r}\times\mathbf{t}) + (\mathbf{n}\times\mathbf{s})]\,\mathrm{d}A.$$

Finally, we have *Oseen's equation*:

$$\int_V \rho_1 \ddot{n}_i\,\mathrm{d}V = \int_V (G_i + g_i)\,\mathrm{d}V + \int_A \pi_{ji}\,\mathrm{d}A_j, \tag{3.1.5}$$

where g_i is the intrinsic director body force, which has the dimensions of torque per unit volume and whose existence is independent of G_i. Converting surface integrals into volume integrals and simplifying, (3.1.1)–(3.1.5) lead to the following differential equations:

$$\dot{\rho} = 0, \tag{3.1.6}$$

$$\rho\dot{v}_i = f_i + t_{ji,j}, \tag{3.1.7}$$

$$\dot{U} = t_{ji}d_{ij} + \pi_{ji}N_{ij} - g_i N_i, \tag{3.1.8}$$

$$\rho_1\ddot{n}_i = G_i + g_i + \pi_{ji,j}, \tag{3.1.9}$$

where

$$t_{ji} - \pi_{kj}n_{i,k} + g_j n_i = t_{ij} - \pi_{ki}n_{j,k} + g_i n_j, \tag{3.1.10}$$

$$N_i = \dot{n}_i - w_{ik}n_k,$$

$$N_{ij} = \dot{n}_{i,j} - w_{ik}n_{k,j},$$

$$2d_{ij} = v_{i,j} + v_{j,i},$$

$$2w_{ij} = v_{i,j} - v_{j,i}.$$

N_i may be interpreted as the angular velocity of the director relative to that of the fluid. It should be emphasized that the stress tensor t_{ji} is asymmetric. When $n_i = 0$, (3.1.6)–(3.1.10) reduce to the familiar equations of hydrodynamics for a normal fluid.

In conjunction with the balance equations we make use of the inequality

$$\frac{\mathrm{d}}{\mathrm{d}t}\int_V S\,\mathrm{d}V \geqslant 0,$$

where S is the entropy per unit volume. Defining the Helmholtz free energy function per unit volume

$$F = U - TS,$$

we obtain for a system in isothermal equilibrium

$$t_{ji}d_{ij} + \pi_{ji}N_{ij} - g_iN_i - \dot{F} \geqslant 0. \qquad (3.1.11)$$

3.1.2. Constitutive equations

In order to develop the theory, it is necessary to set up constitutive equations for the quantities F, t_{ji}, π_{ji} and g_i (Leslie[7]). We assume that these quantities are single-valued functions of

$$n_i, \quad n_{i,j}, \quad \dot{n}_i \quad \text{and} \quad v_{i,j}. \qquad (3.1.12)$$

We now invoke the fundamental principle of classical physics that material properties are indifferent to the frame of reference or the observer.[9] Hence the constitutive equations should be invariant under proper orthogonal transformations. It is seen that $\dot{n}_i$ and $v_{i,j}$ do not transform as tensors. The parameters (3.1.12) must therefore be replaced by

$$n_i, \quad n_{i,j}, \quad N_i, \quad \text{and} \quad d_{ij}.$$

Thus $\dot{F}$ may be expanded as

$$\dot{F} = \frac{\partial F}{\partial n_i}\dot{n}_i + \frac{\partial F}{\partial n_{i,j}}\frac{\mathrm{d}}{\mathrm{d}t}(n_{i,j}) + \frac{\partial F}{\partial N_i}\dot{N}_i + \frac{\partial F}{\partial d_{ij}}\frac{\mathrm{d}}{\mathrm{d}t}(d_{ij}).$$

But

$$\dot{n}_i = N_i + w_{ij}n_j = N_i - w_{ji}n_j,$$

$$\frac{\mathrm{d}}{\mathrm{d}t}(n_{i,j}) = N_{ij} - w_{ki}n_{k,j} - (d_{kj} + w_{kj})n_{i,k},$$

and

$$\frac{\partial F}{\partial n_{i,j}}\frac{\mathrm{d}}{\mathrm{d}t}(n_{i,j})=\frac{\partial F}{\partial n_{i,j}}N_{ij}-\frac{\partial F}{\partial n_{i,k}}n_{j,k}w_{ji}-\frac{\partial F}{\partial n_{k,j}}n_{k,i}d_{ij}-\frac{\partial F}{\partial n_{k,i}}n_{k,j}w_{ji}.$$

Therefore

$$\dot{F}=\frac{\partial F}{\partial n_i}N_i+\frac{\partial F}{\partial n_i}n_jw_{ij}+\frac{\partial F}{\partial n_{i,j}}N_{ij}+\frac{\partial F}{\partial N_i}\dot{N}_i+\frac{\partial F}{\partial d_{ij}}\dot{d}_{ij}-\frac{\partial F}{\partial n_{i,k}}n_{j,k}w_{ji}$$
$$-\frac{\partial F}{\partial n_{k,j}}n_{k,i}d_{ij}-\frac{\partial F}{\partial n_{k,i}}n_{k,j}w_{ji}.$$

Hence (3.1.11) becomes

$$\left(t_{ji}+\frac{\partial F}{\partial n_{k,j}}n_{k,i}\right)d_{ij}+\left(\pi_{ji}-\frac{\partial F}{\partial n_{i,j}}\right)N_{ij}-\left(g_i+\frac{\partial F}{\partial n_i}\right)N_i$$
$$+\left(n_j\frac{\partial F}{\partial n_i}+n_{j,k}\frac{\partial F}{\partial n_{i,k}}+n_{k,j}\frac{\partial F}{\partial n_{k,i}}\right)w_{ji}$$
$$-\left(\frac{\partial F}{\partial N_i}\right)\dot{N}_i-\left(\frac{\partial F}{\partial d_{ij}}\right)\frac{\mathrm{d}}{\mathrm{d}t}(d_{ij})\geqslant 0. \tag{3.1.13}$$

In view of the constitutive assumptions, it is clear that w_{ji}, N_{ij}, $\dot{N}_i$ and $\dot{d}_{ij}$ can be varied arbitrarily and independently of all other quantities and hence their coefficients must vanish, i.e.,

$$\frac{\partial F}{\partial N_i}=0, \qquad \frac{\partial F}{\partial d_{ij}}+\frac{\partial F}{\partial d_{ji}}=0,$$

or

$$F=F(n_i, n_{i,j}); \tag{3.1.14}$$

$$\left(n_i\frac{\partial F}{\partial n_j}+n_{i,k}\frac{\partial F}{\partial n_{j,k}}+n_{k,i}\frac{\partial F}{\partial n_{k,j}}\right)$$
$$-\left(n_j\frac{\partial F}{\partial n_i}+n_{j,k}\frac{\partial F}{\partial n_{i,k}}+n_{k,j}\frac{\partial F}{\partial n_{k,i}}\right)=0; \tag{3.1.15}$$

$$\pi_{ji}-\frac{\partial F}{\partial n_{i,j}}=0; \tag{3.1.16}$$

and (3.1.13) reduces to

$$\left(t_{ji}+\frac{\partial F}{\partial n_{k,j}}n_{k,i}\right)d_{ij}-\left(g_i+\frac{\partial F}{\partial n_i}\right)N_i\geqslant 0. \tag{3.1.17}$$

Let us write the stress and the intrinsic director body force as

$$t_{ji} = t^0_{ji} + t'_{ji},$$
$$g_i = g^0_i + g'_i,$$

where the superscript 0 denotes the isothermal static deformation value and the prime the hydrodynamic part. Equations (3.1.14) and (3.1.16) prove that π_{ji} does not depend on d_{ij} or N_i so that $\pi_{ji} = \pi^0_{ji}$. Substituting in (3.1.17)

$$\left(t^0_{ji} + \frac{\partial F}{\partial n_{k,j}} n_{k,i}\right) d_{ij} - \left(g^0_i + \frac{\partial F}{\partial n_i}\right) N_i + (t'_{ji} d_{ij} - g'_i N_i) \geqslant 0. \tag{3.1.18}$$

Since d_{ij} and N_i can be chosen arbitrarily and independently of the static parts t^0_{ji} and g^0_i,

$$t^0_{ji} = -\frac{\partial F}{\partial n_{k,j}} n_{k,i}, \tag{3.1.19}$$

$$g^0_i = -\frac{\partial F}{\partial n_i}, \tag{3.1.20}$$

$$t'_{ji} d_{ij} - g'_i N_i \geqslant 0. \tag{3.1.21}$$

Further, using (3.1.10) and (3.1.15)

$$t'_{ji} + g'_j n_i = t'_{ij} + g'_i n_j. \tag{3.1.22}$$

The incompressibility condition implies that the stress is indeterminate to an arbitrary pressure. Similarly there is a certain degree of indeterminacy in g^0_i and π_{ji} for if we replace

$$g^0_i \text{ by } \gamma n_i - \beta_j n_{i,j} + g^0_i$$

and

$$\pi_{ij} \text{ by } \beta_i n_j + \pi_{ij},$$

(3.1.8) continues to be satisfied because

$$n_i \dot{n}_i = n_i N_i = n_i N_{ij} + n_{i,j} N_i = 0.$$

Thus (3.1.19), (3.1.20) and (3.1.16) become

$$t^0_{ji} = -p\delta_{ji} - \frac{\partial F}{\partial n_{k,j}} n_{k,i}, \tag{3.1.23}$$

$$g^0_i = \gamma n_i - \beta_j n_{i,j} - \frac{\partial F}{\partial n_i}, \tag{3.1.24}$$

$$\pi_{ji} = \beta_j n_i + \frac{\partial F}{\partial n_{i,j}}, \tag{3.1.25}$$

where p, γ and β_i are arbitrary constants, while the hydrodynamic components fulfil the inequality (3.1.21).

For an isothermal static deformation, (3.1.9) becomes an equation of equilibrium:

$$g_i^0 + \pi_{ji,j} + G_i = 0. \tag{3.1.26}$$

Substituting for g_i^0 and π_{ji} from (3.1.24) and (3.1.25),

$$\left(\frac{\partial F}{\partial n_{i,j}}\right)_{,j} - \frac{\partial F}{\partial n_i} + G_i + \gamma n_i = 0. \tag{3.1.27}$$

3.1.3. Coefficients of viscosity

We next consider the nature of t'_{ji} and g'_i, the hydrodynamic components of the stress tensor and the intrinsic director body force. We assume that they are linear functions of n_i, N_i and d_{ij} and omit higher order terms:[5,7]

$$\left.\begin{aligned} t'_{ji} &= A^0_{ji} + A^1_{jik} N_k + A^2_{jikm} d_{km}, \\ g'_i &= B^0_i + B^1_{ij} N_j + B^2_{ijk} d_{jk}, \end{aligned}\right\} \tag{3.1.28}$$

where A and B are functions of n_i (at constant T and ρ). They can therefore be expanded as

$$\begin{aligned} A^0_{ji} &= \alpha_0 \delta_{ji} + \alpha_1 n_i n_j, \\ A^1_{jik} &= \alpha_2 \delta_{ji} n_k + \alpha_3 \delta_{jk} n_i + \alpha_4 \delta_{ik} n_j + \alpha_5 n_i n_j n_k, \\ A^2_{jikm} &= \alpha_6 \delta_{ji} n_k n_m + \alpha_7 n_j n_m \delta_{ik} + \alpha_8 n_i n_j \delta_{km} + \alpha_9 \delta_{jk} n_i n_m \\ &\quad + \alpha_{10} \delta_{jm} n_i n_k + \alpha_{11} \delta_{im} n_j n_k + \alpha_{12} \delta_{ji} \delta_{km} \\ &\quad + \alpha_{13} \delta_{jk} \delta_{im} + \alpha_{14} n_i n_j n_k n_m, \\ B^0_i &= \gamma_0 n_i, \\ B^1_{ij} &= \gamma_1 \delta_{ij} + \gamma_2 n_i n_j, \\ B^2_{ijk} &= \gamma_3 \delta_{ij} n_k + \gamma_4 \delta_{ik} n_j + \gamma_5 \delta_{jk} n_i + \gamma_6 n_i n_j n_k. \end{aligned}$$

Substituting in (3.1.28) and remembering that $N_i n_i = d_{ii} = 0$,

$$\begin{aligned} t'_{ji} &= (\alpha_0 + \alpha_6 d_{km} n_k n_m) \delta_{ji} + (\alpha_1 + \alpha_{14} d_{km} n_k n_m) n_i n_j \\ &\quad + \alpha_{13} d_{ji} + \alpha_{15} d_{kj} n_i n_k + \alpha_{16} d_{ki} n_j n_k + \alpha_3 n_i N_j + \alpha_4 n_j N_i \end{aligned} \tag{3.1.29}$$

where

$$\alpha_{15} = \alpha_9 + \alpha_{10}, \qquad \alpha_{16} = \alpha_7 + \alpha_{11},$$

and

$$g'_i = (\gamma_0 + \gamma_6 d_{kj} n_k n_j) n_i + \gamma_9 d_{ik} n_k + \gamma_1 N_i \qquad (3.1.30)$$

where

$$\gamma_9 = \gamma_3 + \gamma_4.$$

But in view of (3.1.22)

$$\gamma_9 = \alpha_{16} - \alpha_{15} \qquad \gamma_1 = \alpha_4 - \alpha_3. \qquad (3.1.31)$$

In addition, t'_{ji} and g'_i must satisfy the entropy inequality (3.1.21). Substituting (3.1.29) and (3.1.30) in (3.1.21) we get

$$\begin{aligned}&[(\alpha_0 + \alpha_6 d_{km} n_k n_m)\delta_{ij}]d_{ij} + [\alpha_1 n_i n_j]d_{ij} - (\gamma_0 + \gamma_6 d_{kj} n_k n_j) n_i N_i \\ &\quad + [(\alpha_{14} d_{km} n_k n_m d_{ij} n_i n_j + \alpha_{15} d_{jk} d_{ij} n_i n_k + \alpha_{16} d_{ik} d_{ij} n_j n_k) \\ &\quad + \alpha_{13} d_{ij} d_{ji} + (-\gamma_9 d_{ik} n_k N_i - \gamma_1 N_i N_i)] \geqslant 0,\end{aligned}$$

i.e.,

$$\alpha_1 n_i n_j d_{ij} + (\text{quadratic in } d_{ij}, N_i) \geqslant 0.$$

As d_{ij} can be chosen arbitrarily, $\alpha_1 = 0$. Also the coefficient of δ_{ji} $(= -p$ say) in t_{ji} and that of $n_i (= \gamma$ say) in g'_i are arbitrary since $d_{ii} = 0$ and $n_i N_i = 0$. Putting:

$$\begin{aligned}\mu_1 &= \alpha_{14} & \mu_4 &= \alpha_{13}\\ \mu_2 &= \alpha_4 & \mu_5 &= \alpha_{16}\\ \mu_3 &= \alpha_3 & \mu_6 &= \alpha_{15}\\ \lambda_1 &= \gamma_1 & \lambda_2 &= \gamma_9\end{aligned}$$

we obtain

$$\begin{aligned}t'_{ji} &= \mu_1 n_k n_m d_{km} n_i n_j + \mu_2 n_j N_i + \mu_3 n_i N_j \\ &\quad + \mu_4 d_{ji} + \mu_5 n_j n_k d_{ki} + \mu_6 n_i n_k d_{kj}, \qquad (3.1.32)\end{aligned}$$

$$g'_i = \lambda_1 N_i + \lambda_2 n_j d_{ji}, \qquad (3.1.33)$$

where, making use of (3.1.31),

$$\lambda_1 = \mu_2 - \mu_3, \qquad \lambda_2 = \mu_5 - \mu_6. \qquad (3.1.34)$$

We have omitted the terms of the form $p\delta_{ij}$ in (3.1.32) and γn_i in (3.1.33) as they do not contribute to the hydrodynamic effects and can be combined with the corresponding terms in (3.1.23) and (3.1.24) respectively.

Substituting for t'_{ji} and g'_i it is found that the left-hand side of (3.1.21) is a positive and definite quantity if the following inequalities are satisfied:[7]

$$\left.\begin{aligned} &\mu_4 \geqslant 0,\\ &2\mu_1+3\mu_4+2\mu_5+2\mu_6 \geqslant 0,\\ &2\mu_4+\mu_5+\mu_6 \geqslant 0,\\ &-4\lambda_1(2\mu_4+\mu_5+\mu_6) \geqslant (\mu_2+\mu_3-\lambda_2)^2. \end{aligned}\right\} \tag{3.1.35}$$

Finally, from (3.1.23), (3.1.24), (3.1.32) and (3.1.33),

$$\begin{aligned} t_{ji} &= t^0_{ji}+t'_{ji}\\ &= -p\delta_{ji} - \frac{\partial F}{\partial n_{k,j}} n_{k,i} + \mu_1 n_k n_m d_{km} n_i n_j\\ &\quad + \mu_2 n_j N_i + \mu_3 n_i N_j + \mu_4 d_{ji}\\ &\quad + \mu_5 n_j n_k d_{ki} + \mu_6 n_i n_k d_{kj}, \end{aligned} \tag{3.1.36}$$

$$\begin{aligned} g_i &= g^0_i + g'_i\\ &= \gamma n_i - \beta_j n_{i,j} - \frac{\partial F}{\partial n_i} + \lambda_1 N_i + \lambda_2 n_j d_{ji}. \end{aligned} \tag{3.1.37}$$

The μ's represent the six coefficients of viscosity of a nematic liquid crystal. However, the number of independent coefficients reduces to 5 if we assume Onsager's reciprocal relations.

3.1.4. Parodi's relation

From (3.1.21) we observe that the rate of entropy production per unit volume

$$T\dot{S} = t'_{ji} d_{ij} - g'_i N_i$$

where t'_{ji} is given by (3.1.32) and g'_i by (3.1.33). Since t'_{ji} is an asymmetric tensor it can be resolved into a symmetric component Y_{ij} and an antisymmetric component Z_{ij}, where

$$\begin{aligned} Y_{ij} &= \mu_1 d_{kp} n_k n_p n_i n_j + \mu_4 d_{ij} + \tfrac{1}{2}(\mu_2+\mu_3)(n_i N_j + N_i n_j)\\ &\quad + \tfrac{1}{2}(\mu_5+\mu_6)(d_{ki} n_k n_j + d_{kj} n_k n_i),\\ Z_{ji} &= \tfrac{1}{2}(\mu_2-\mu_3)(N_j n_i - n_j N_i) + \tfrac{1}{2}(\mu_5-\mu_6)(d_{ki} n_j n_k - d_{kj} n_i n_k). \end{aligned}$$

As $Z_{ji} = -Z_{ij}$ and $d_{ij} = d_{ji}$, it follows that

$$Z_{ji} d_{ij} = 0,$$

and therefore

$$T\dot{S} = Y_{ji}d_{ij} - g_i'N_i.$$

Thus the entropy production can be separated into two parts, one due to the linear motion of the fluid and the other due to the orientational motion of the director. Now $\dot{\mathbf{n}} = \mathbf{\Omega}\times\mathbf{n}$, where $\mathbf{\Omega}$ is the angular velocity of the director. It is then easily shown that

$$g_i'N_i = J_{ji}(\Omega_{ij} - w_{ij}),$$

where $J_{ji} = n_j g_i' - n_i g_j'$ is the torque exerted by the director. Consequently

$$T\dot{S} = Y_{ji}d_{ij} + J_{ij}(\Omega_{ij} - w_{ij}).$$

Y_{ji}, J_{ij} may be regarded as 'fluxes' and d_{ij}, $(\Omega_{ij} - w_{ij})$ as 'forces'. The relation between the fluxes and forces is

$$\begin{bmatrix} Y_{ji} \\ J_{ji} \end{bmatrix} = \begin{bmatrix} D^1_{jirs} & D^2_{jirs} \\ D^3_{jirs} & D^4_{jirs} \end{bmatrix} \begin{bmatrix} d_{rs} \\ \Omega_{rs} - w_{rs} \end{bmatrix}, \tag{3.1.38}$$

where

$$D^1_{jirs} = \mu_1 n_i n_j n_r n_s + \mu_4 \delta_{ir}\delta_{js} + \tfrac{1}{2}(\mu_5 + \mu_6)(\delta_{ir}n_s n_j + \delta_{js}n_r n_i),$$

$$D^2_{jirs} = \tfrac{1}{2}(\mu_2 + \mu_3)(n_i n_s \delta_{rj} - n_r n_j \delta_{is}),$$

$$D^3_{jirs} = \tfrac{1}{2}(\mu_6 - \mu_5)(n_i n_s \delta_{rj} - n_r n_j \delta_{is}),$$

$$D^4_{jirs} = \tfrac{1}{2}(\mu_2 - \mu_3)(n_i n_s \delta_{rj} + n_r n_j \delta_{is}).$$

From Onsager's reciprocal relations in irreversible processes[10] it follows that $[D]$ is symmetric, i.e.,

$$D^2_{jirs} = D^3_{jirs}$$

or

$$\mu_2 + \mu_3 = \mu_6 - \mu_5. \tag{3.1.39}$$

This is referred to as *Parodi's relation.*[11] The number of independent viscosity coefficients is therefore reduced from 6 to 5.

It is well known that Truesdell[12] is of the view that Onsager's relations do not apply to phenomena like heat conduction, viscosity and diffusion since there is no unambiguous way of selecting the fluxes and forces. It would appear therefore that there may be some doubts as to the validity of Parodi's relation. Experimental data are not precise enough to establish this point unequivocally,[13] though the relation has been tacitly assumed to be true in a number of recent discussions.

3.2. Curvature elasticity: the Oseen–Zöcher–Frank equations

We have shown that for an incompressible fluid and an isothermal deformation (see (3.1.14))

$$F=F(n_i, n_{i,j}).$$

We may therefore expand the free energy per unit volume of a deformed liquid crystal relative to that in the state of uniform orientation as

$$F=k_{ij}n_{i,j}+\tfrac{1}{2}k_{ijlm}n_{i,j}n_{l,m}.$$

We neglect higher order terms in the expansion as we are concerned only with infinitesimal deformations. Since these deformations relate to changes in the orientation of the director, $n_{i,j}$ may appropriately be called the *curvature strain tensor* (Frank[3]). In order to define the components of this tensor, let us choose a local right-handed system of cartesian coordinates with $\mathbf{n}$ parallel to z at the origin. The components of strain are then

$$\textit{Splay}: \frac{\partial n_x}{\partial x}, \quad \frac{\partial n_y}{\partial y}$$

$$\textit{Twist}: \frac{\partial n_y}{\partial x}, \quad \frac{\partial n_x}{\partial y}$$

$$\textit{Bend}: \frac{\partial n_x}{\partial z}, \quad \frac{\partial n_y}{\partial z}.$$

We ignore $\partial n_z/\partial x$, $\partial n_z/\partial y$ and $\partial n_z/\partial z$. The three types of deformations are shown in fig. 2.3.12. The curvature strain tensor may therefore be written as

$$n_{i,j}=\begin{bmatrix} \dfrac{\partial n_x}{\partial x} & \dfrac{\partial n_x}{\partial y} & \dfrac{\partial n_x}{\partial z} \\ -\dfrac{\partial n_y}{\partial x} & \dfrac{\partial n_y}{\partial y} & \dfrac{\partial n_y}{\partial z} \\ 0 & 0 & 0 \end{bmatrix} = \begin{bmatrix} a_1 & a_2 & a_3 \\ a_4 & a_5 & a_6 \\ 0 & 0 & 0 \end{bmatrix} \text{(say).}$$

Now any of the three types of deformation destroys the centre of symmetry of the liquid crystal. The strain tensor is therefore an axial second rank tensor which vanishes identically under a centro-symmetric operation. Since the free energy is a scalar, the components of k_{ij} also form an axial second rank tensor:

$$k_{ij}=\begin{bmatrix} k_{11} & k_{12} & k_{13} \\ k_{21} & k_{22} & k_{23} \\ k_{31} & k_{32} & k_{33} \end{bmatrix},$$

or using an abbreviated one-index notation

$$k_i = \begin{bmatrix} k_1 & k_2 & k_3 \\ k_4 & k_5 & k_6 \\ k_7 & k_8 & k_9 \end{bmatrix}. \tag{3.2.1}$$

In general k_i has 9 components, but the presence of symmetry in the liquid crystal reduces this number.[14] The distribution of molecules around any point is cylindrically symmetric, so that the choice of the x axis is arbitrary apart from the requirement that it should be normal to the director axis z. Therefore

$$k_i = \begin{bmatrix} k_1 & k_2 & 0 \\ -k_2 & k_1 & 0 \\ 0 & 0 & 0 \end{bmatrix}.$$

Additional symmetry operations reduce the number of moduli even further:

Enantiomorphic and polar	$k_1 \neq 0, \quad k_2 \neq 0$
Enantiomorphic and non-polar	$k_1 = 0, \quad k_2 \neq 0$
Non-enantiomorphic and polar	$k_1 \neq 0, \quad k_2 = 0$
Non-enantiomorphic and non-polar	$k_1 = 0, \quad k_2 = 0.$

The tensor k_{ijlm} (or k_{ij} in the abbreviated notation of (3.2.1)) has 81 components in general, but as a_7, a_8 and a_9 are zero there remain only 36. The presence of cylindrical symmetry reduces this number to 18 with only 5 independent constants:

$$k_{ij} = \begin{bmatrix} k_{11} & k_{12} & 0 & -k_{12} & k_{15} & 0 \\ k_{12} & k_{22} & 0 & k_{24} & k_{12} & 0 \\ 0 & 0 & k_{33} & 0 & 0 & 0 \\ -k_{12} & k_{24} & 0 & k_{22} & -k_{12} & 0 \\ k_{15} & k_{12} & 0 & -k_{12} & k_{11} & 0 \\ 0 & 0 & 0 & 0 & 0 & k_{33} \end{bmatrix},$$

where $k_{15} = k_{11} - k_{22} - k_{24}$. In the absence of polarity or enantiomorphy, $k_{12} = 0$.

The free energy of deformation may therefore be written in the form

$$\begin{aligned} F = {} & k_1(a_1 + a_5) + k_2(a_2 + a_4) + \tfrac{1}{2}k_{11}(a_1 + a_5)^2 \\ & + \tfrac{1}{2}k_{22}(a_2 + a_4)^2 + \tfrac{1}{2}k_{33}(a_3^2 + a_6^2) \\ & + k_{12}(a_1 + a_5)(a_2 + a_4) - (k_{22} + k_{24})(a_1 a_5 + a_2 a_4). \end{aligned} \tag{3.2.2}$$

As the free energy must be positive definite, it can be shown[15] by standard algebraical methods[16] that

$$k_{11} \geqslant 0, \qquad k_{22} \geqslant 0, \qquad k_{33} \geqslant 0,$$

$$|k_{24}| \leqslant k_{22}, \qquad |k_{11} - k_{22} - k_{24}| \leqslant k_{11}.$$

In tensor notation (3.2.2) reads as

$$F = \tfrac{1}{2}k_{11}(s + n_{i,i})^2 + \tfrac{1}{2}k_{22}(q + n_i e_{ijk} n_{k,j})^2 + \tfrac{1}{2}k_{33} n_i n_j n_{k,i} n_{k,j} + \tfrac{1}{2}(k_{22} + k_{24})[n_{i,j} n_{j,i} - (n_{i,i})^2], \qquad (3.2.3)$$

where $s = k_1/k_{11}$ is the permanent splay and $q = k_2/k_{22}$ the permanent twist. There is no physical polarity along the direction $\mathbf{n}$ in any known nematic or cholesteric substance. The molecules themselves may be polar but the absence of ferroelectricity confirms that there is equal probability of their pointing in either direction. We shall therefore set $k_1 = k_{12} = 0$. Substituting for F in (3.1.27) we then obtain

$$k_2(e_{ijk} n_{k,j}) + \tfrac{1}{2}k_{22}[2 e_{ijk} n_{k,j} n_p e_{pqr} n_{r,q} - e_{ijk} n_j (n_p e_{pqr} n_{r,q})_{,k}] + \tfrac{1}{2}k_{33} n_k n_{j,k} n_{i,j} + G_i = 0. \qquad (3.2.4)$$

It is seen that k_{24} plays no role in (3.2.4) and can be omitted as far as equilibrium situations are concerned (Ericksen[17]). When external body torques are absent ($G_i = 0$), the solutions of (3.2.4) are

$$\left.\begin{aligned} n_1 &= \cos(qz + \psi), \\ n_2 &= \sin(qz + \psi), \end{aligned}\right\} \qquad (3.2.5)$$

where

$$\frac{\partial^2 \psi}{\partial x^2} + \frac{\partial^2 \psi}{\partial y^2} = 0. \qquad (3.2.6)$$

Equation (3.2.5) describes a cholesteric structure with a twist per unit length of k_2/k_{22}. In the absence of enantiomorphy, $k_2 = 0$ and the structure is nematic. The solutions of (3.2.6) describe the configurations around line singularities in the structure which we shall consider in some detail in a later section (§ 3.5.1).

In vector notation the free energy density may be written in the more compact form

$$F = k_2(\mathbf{n} \cdot \nabla \times \mathbf{n}) + \tfrac{1}{2}k_{11}(\nabla \cdot \mathbf{n})^2 + \tfrac{1}{2}k_{22}(\mathbf{n} \cdot \nabla \times \mathbf{n})^2 + \tfrac{1}{2}k_{33}(\mathbf{n} \times \nabla \times \mathbf{n})^2, \qquad (3.2.7)$$

with $k_2 = 0$ in the nematic case. Nehring and Saupe[18] have expressed the view that second order terms in the free energy expansion cannot be neglected and have proposed the following more general expression instead of (3.2.7):

$$F = k_2(\mathbf{n}\cdot\nabla\times\mathbf{n}) + k'_{11}(\nabla\cdot\mathbf{n})^2 + k_{22}(\mathbf{n}\cdot\nabla\times\mathbf{n})^2 + k'_{33}(\mathbf{n}\times\nabla\times\mathbf{n})^2 + 2k_{13}[\nabla\cdot(\nabla\cdot\mathbf{n})\mathbf{n}] \qquad (3.2.8)$$

where $k'_{11} = k_{11} - 2k_{13}$ and $k'_{33} = k_{33} + 2k_{13}$. As far as simple splay and bend deformations are concerned k_{11} and k_{33} have in effect been rescaled and consequently there appears to be no straightforward method of determining the absolute value of k_{13}. As the importance of these additional terms has not yet been established we shall throughout assume the first order theory to be valid.

3.3. Summary of equations of the continuum theory

In the following sections of this chapter we shall apply the continuum theory to study the behaviour of the nematic phase in various physical situations. For convenience we set out below the most important equations of the theory which we shall be referring to constantly:

$$\rho\dot{v}_i = f_i + t_{ji,j}, \qquad (3.3.1)$$

$$\rho_1\ddot{n}_i = G_i + g_i + \pi_{ji,j}, \qquad (3.3.2)$$

where ρ is the density of the fluid (assumed to be incompressible and at constant temperature), ρ_1 a material constant having the dimensions of moment of inertia per unit volume, n_i a dimensionless unit vector called the director, v_i the linear velocity, f_i the body force per unit volume, t_{ji} the stress tensor, G_i the external director body force, g_i the intrinsic director body force and π_{ji} the director surface stress.

The stress tensor t_{ji} may be separated into a static (or elastic) part and a hydrodynamic (or viscous) part:

$$t_{ji} = t^0_{ji} + t'_{ji}, \qquad (3.3.3)$$

where

$$t^0_{ji} = -p\delta_{ij} - \frac{\partial F}{\partial n_{k,j}} n_{k,i}, \qquad (3.3.4)$$

$$t'_{ji} = \mu_1 n_k n_m d_{km} n_i n_j + \mu_2 n_j N_i + \mu_3 n_i N_j + \mu_4 d_{ji} + \mu_5 n_j n_k d_{ki} + \mu_6 n_i n_k d_{kj}, \qquad (3.3.5)$$

F is the free energy per unit volume given by

$$F = \tfrac{1}{2}(k_{11} - k_{22})n_{i,i}n_{j,j} + \tfrac{1}{2}k_{22}n_{i,j}n_{i,j}$$

$$+ \tfrac{1}{2}(k_{33} - k_{22})n_i n_j n_{l,i} n_{l,j}, \tag{3.3.6}$$

$$= \tfrac{1}{2}k_{11}(\boldsymbol{\nabla} \cdot \mathbf{n})^2 + \tfrac{1}{2}k_{22}(\mathbf{n} \cdot \boldsymbol{\nabla} \times \mathbf{n})^2 + \tfrac{1}{2}k_{33}(\mathbf{n} \times \boldsymbol{\nabla} \times \mathbf{n})^2, \tag{3.3.7}$$

$$N_i = \dot{n}_i - w_{ik}n_k, \tag{3.3.8}$$

$$2d_{ij} = v_{i,j} + v_{j,i}, \tag{3.3.9}$$

$$2w_{ij} = v_{i,j} - v_{j,i}, \tag{3.3.10}$$

p is an arbitrary (indeterminate) constant, $\mu_1 \ldots \mu_6$ the coefficients of viscosity, and k_{11}, k_{22}, k_{33} the elastic constants. Similarly the intrinsic director body force g_i may be written in two parts

$$g_i = g_i^0 + g_i', \tag{3.3.11}$$

where

$$g_i^0 = \gamma n_i - \beta_j n_{i,j} - (\partial F/\partial n_i), \tag{3.3.12}$$

$$g_i' = \lambda_1 N_i + \lambda_2 n_j d_{ji}, \tag{3.3.13}$$

γ and β_j are arbitrary (indeterminate) constants, and

$$\left.\begin{aligned} \lambda_1 &= \mu_2 - \mu_3, \\ \lambda_2 &= \mu_5 - \mu_6. \end{aligned}\right\} \tag{3.3.14}$$

Also, according to Parodi's relation

$$\mu_2 + \mu_3 = \mu_6 - \mu_5. \tag{3.3.15}$$

The director surface stress

$$\pi_{ji} = \beta_j n_i + (\partial F/\partial n_{i,j}). \tag{3.3.16}$$

For static deformations,

$$\left(\frac{\partial F}{\partial n_{i,j}}\right)_{,j} - (\partial F/\partial n_i) + G_i + \gamma n_i = 0. \tag{3.3.17}$$

3.4. Distortions due to a magnetic field: static theory

3.4.1. The Freedericksz effect

The simplest method of measuring the three elastic constants of a nematic liquid crystal is by studying the deformations due to an external magnetic field (Freedericksz and Tsvetkov,[19] Zöcher[2]). The geometry has to be so chosen that the orienting effect of the field conflicts with the orientations imposed by the surfaces with which the liquid crystal is in contact.

To develop a static theory of such deformations we apply the equation of equilibrium (3.3.17) where G_i is the external director body force due to the magnetic field **H**. If $\chi_\parallel$ and $\chi_\perp$ are the principal diamagnetic susceptibilities per unit volume along and perpendicular to the director axis respectively,

$$G_i = \chi_a H_j n_j H_i, \tag{3.4.1}$$

where $\chi_a = \chi_\parallel - \chi_\perp$.

Let us consider first the case of a nematic film in which the initial undisturbed orientation of the director is throughout parallel to the glass plates. The magnetic field **H** is now applied perpendicular to the director and to the plates (fig. 3.4.1(*a*)). For this geometry, $\mathbf{n} = (\cos\theta, 0, \sin\theta)$, $\mathbf{H} = (0, 0, H)$ and $\mathbf{G} = (0, 0, \chi_a H^2 \sin\theta)$. The free energy of elastic deformation (3.3.6) reduces in this case to

$$F = \tfrac{1}{2}\{k_{11} n_{z,z}^2 + k_{22}[(1-n_z^2)n_{x,z}^2 - n_z^2 n_{z,z}^2] + k_{33} n_z^2 (n_{x,z}^2 + n_{z,z}^2)\}.$$

By straightforward substitution in (3.3.17) and simplification, we obtain the equilibrium condition

$$\frac{\mathrm{d}}{\mathrm{d}z}\left[(k_{11}\cos^2\theta + k_{33}\sin^2\theta)\left(\frac{\mathrm{d}\theta}{\mathrm{d}z}\right)^2 + \chi_a H^2 \sin^2\theta\right] = 0.$$

As is to be expected, the deformation involves the splay and bend moduli, k_{11} and k_{33} respectively, and not the twist modulus k_{22}. Because of the orienting influence of the glass surfaces, $\theta = 0$ at $z = 0$ and d, where d is the thickness of the film. Therefore θ attains a maximum value θ_m at $z = d/2$ and from symmetry considerations $\theta(z) = \theta(d - z)$. Since $\mathrm{d}\theta/\mathrm{d}z = 0$ at $z = d/2$, we get

$$(\chi_a)^{1/2} H \int_0^{d/2} \mathrm{d}z = \int_0^{\theta_m} \left[\frac{k_{11}\cos^2\theta + k_{33}\sin^2\theta}{\sin^2\theta_m - \sin^2\theta}\right]^{1/2} \mathrm{d}\theta.$$

Transforming to a new variable λ given by $\sin\lambda = \sin\theta/\sin\theta_m$,

$$\tfrac{1}{2}(\chi_a)^{1/2} H d = \int_0^{\pi/2} \left[\frac{k_{11}(1 - \sin^2\theta_m \sin^2\lambda) + k_{33}\sin^2\theta_m \sin^2\lambda}{1 - \sin^2\theta_m \sin^2\lambda}\right]^{1/2} \mathrm{d}\lambda.$$

Taking the limit $\theta_m = 0$ gives

$$H_c = \frac{\pi}{d}\left(\frac{k_{11}}{\chi_a}\right)^{1/2}. \tag{3.4.2}$$

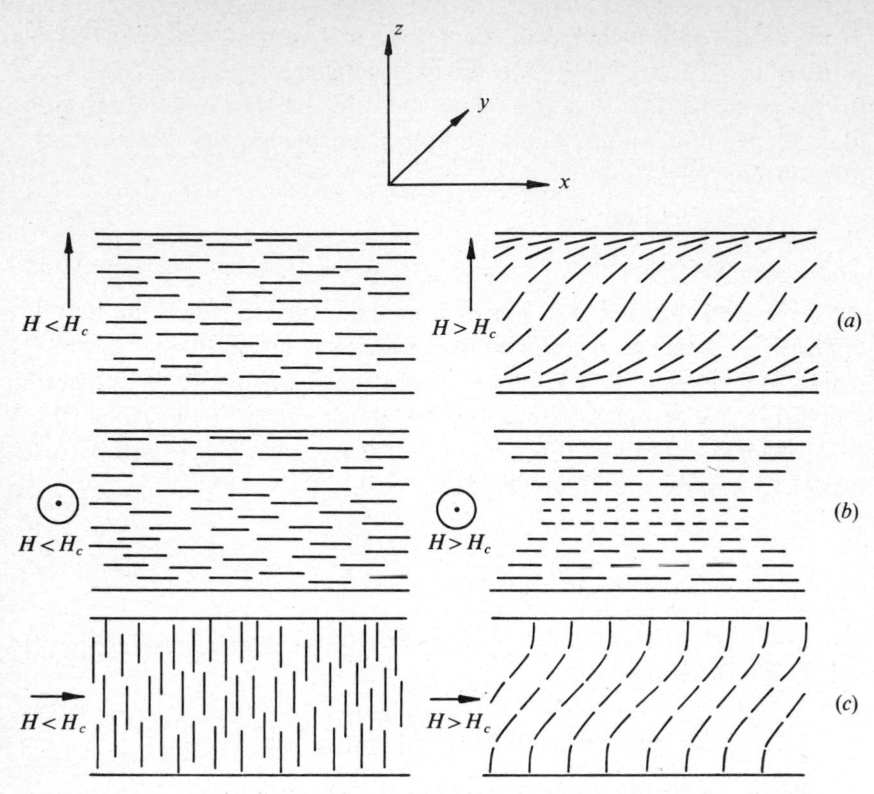

Fig. 3.4.1. The three principal types of Freedericksz deformation.

In other words, deformation occurs only above a certain critical field H_c. This is referred to as the *Freedericksz effect*. The threshold condition can be used for a direct determination of the splay modulus k_{11}.

For $H > H_c$, the deformation at any arbitrary point can be computed from the expressions[20,21]

$$\frac{H}{H_c} = \frac{2}{\pi}\int_0^{\theta_m}\left(\frac{1+q\sin^2\theta}{\sin^2\theta_m - \sin^2\theta}\right)^{1/2} d\theta$$
$$= 1 + \tfrac{1}{4}(q+1)\theta_m^2 + \ldots, \tag{3.4.3}$$

$$\frac{z}{d} = \frac{H_c}{\pi H}\int_0^{\theta}\left(\frac{1+q\sin^2\theta}{\sin^2\theta_m - \sin^2\theta}\right)^{1/2} d\theta$$
$$= \frac{1}{\pi}\arcsin(\theta/\theta_m)$$
$$-\theta(\theta_m^2 - \theta^2)^{1/2}\frac{1+3q+\ldots}{12\pi[1+\tfrac{1}{4}(q+1)\theta_m^2+\ldots]}, \tag{3.4.4}$$

where $q = (k_{33} - k_{11})/k_{11}$.

Two other important geometries are illustrated in fig. 3.4.1. For $\mathbf{n}=(\cos\theta, \sin\theta, 0)$, $\mathbf{H}=(0, H, 0)$ (fig. 3.4.1(*b*)),

$$H_c=(\pi/d)(k_{22}/\chi_a)^{1/2} \tag{3.4.5}$$

and for $\mathbf{n}=(\sin\theta, 0, \cos\theta)$, $\mathbf{H}=(H, 0, 0)$ (fig. 3.4.1(*c*)),

$$H_c=(\pi/d)(k_{33}/\chi_a)^{1/2}. \tag{3.4.6}$$

De Gennes[22] has introduced a parameter which he has called the magnetic coherence length to define the thickness of the transition layer near the boundary. Consider, for example, a nematic liquid crystal occupying the half space $z>0$. Let the wall, the xy plane at $z=0$, impose an orientation along x and let the magnetic field be applied along y (analogous to the geometry of fig. 3.4.1(*b*)). If $\varphi(=\frac{1}{2}\pi-\theta)$ is the angle made by the director with the field, the equilibrium condition is easily shown to be

$$\frac{\mathrm{d}}{\mathrm{d}z}\left[\xi\left(\frac{\mathrm{d}\varphi}{\mathrm{d}z}\right)^2+\cos^2\varphi\right]=0$$

where $\xi=(k_{22}/\chi_a)^{1/2}H^{-1}$. Integrating, subject to the boundary condition that when $z\to\infty$, $\varphi\to 0$ and $\mathrm{d}\varphi/\mathrm{d}z\to 0$,

$$\tan(\varphi/2)=\exp(-z/\xi).$$

ξ is the magnetic coherence length. It is usually of the order of a micron for a field of 10^4 gauss and increases with diminishing field. If the sample thickness d is very much greater than ξ, most of the material will be aligned in the field direction.

The experiment for the determination of k_{11} or k_{33} consists of measuring the variation of birefringence for light incident normal to the film. With linearly polarized light incident and a suitable analyzer (e.g. a combination of a $\lambda/4$ plate and a linear polarizer), the transmitted intensity shows a sudden change when the field attains the threshold value. A measurement of H_c in the geometries (*a*) and (*c*) of fig. 3.4.1 therefore gives the elastic constant k_{11} or k_{33} directly. As the field is gradually increased further, the intensity exhibits oscillations because of the change in phase retardation (fig. 3.4.2). The observed variation in the retardation is found to be in conformity with that expected from (3.4.3) and (3.4.4).[21] In principle, measurements beyond H_c using the geometry (*a*) of fig. 3.4.1 should yield both k_{11} and k_{33}.

However, the threshold for a twist deformation cannot be detected optically when viewed along the twist axis. This is because of the large birefringence (δn) of the medium for this direction of propagation (the case $\beta\ll\gamma$ in § 4.1.1). Thus with the experimental geometry of fig.

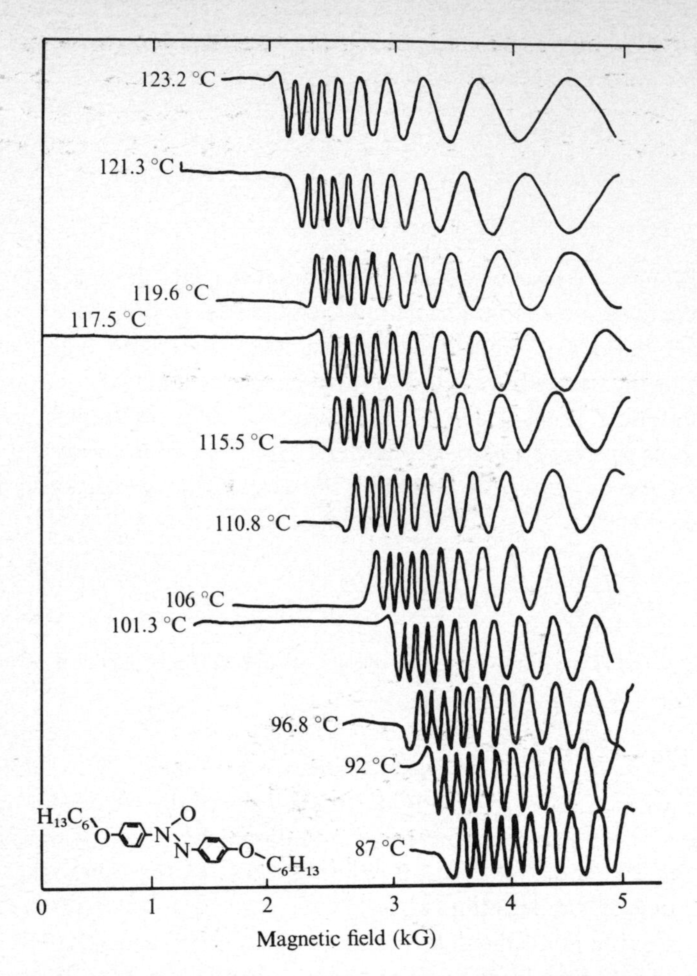

Fig. 3.4.2. Raw recorder traces of interference oscillations due to the change in the sample birefringence with deformation for hexyloxyazoxybenzene at various temperatures. Polarizer and crossed analyser inclined at 45° to the principal axes of the specimen. The sudden onset of oscillations occurs at the threshold field. The increase in the threshold for the successive traces illustrates the rapid temperature variation of the elastic constant. Sample thickness 45 μm. $T_{NI} =$ 128.5 °C. (After Gruler, Scheffer and Meier.[21])

3.4.1(*b*) in which the director is anchored parallel to the walls at either end and light is incident normal to the film, the state of polarization of the emergent beam is indistinguishable from that of the beam emerging from the untwisted nematic. For this reason Freedericksz and Tsvetkov[19] used a total internal reflexion technique by letting the light beam fall at an appropriate angle on the specimen contained between a convex lens and a prism. A simple and more direct method has been proposed recently.[23] If the ellipsoid of refractive index is viewed obliquely, say at 5 or 10° to the director (fig. 3.4.3), the effective δn is reduced to a low value and the

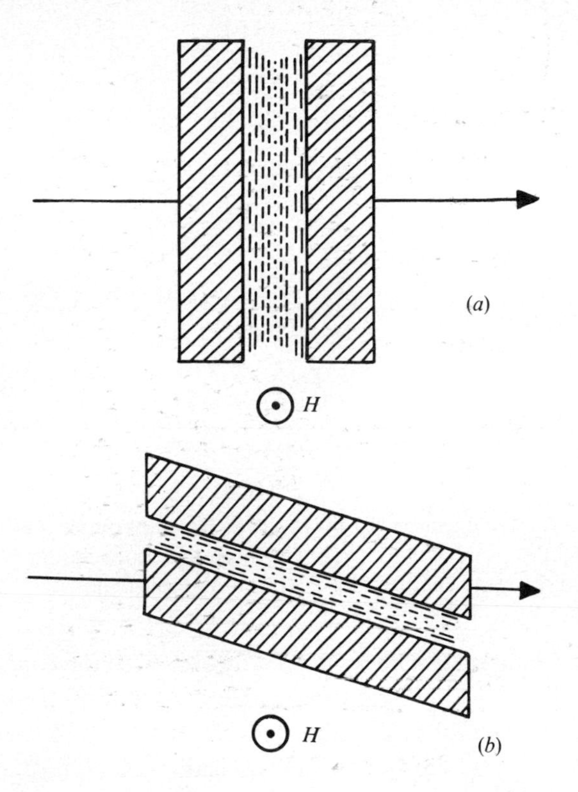

Fig. 3.4.3 (*a*) The usual experimental configuration for the optical observation of the Freedericksz effect. Light is incident normal to the film. However, for reasons discussed in the text, this arrangement is unsuitable for observing a twist deformation. (*b*) 'Oblique' configuration which enables the optical detection of a twist deformation.[23] The magnetic field is perpendicular to the plane of the paper in both cases.

deformed medium can be shown to be optically equivalent to a rotator and a retarder (the case $\beta \sim \gamma$ in § 4.1.1). Hence a twist deformation produces a change in the state of polarization of the emergent beam which can be detected by the usual optical methods. In fig. 3.4.4 we present k_{22} for two compounds determined by this technique.

When the medium is twisted, the principal axes of the equivalent retarder will evidently be rotated with respect to those of the undistorted

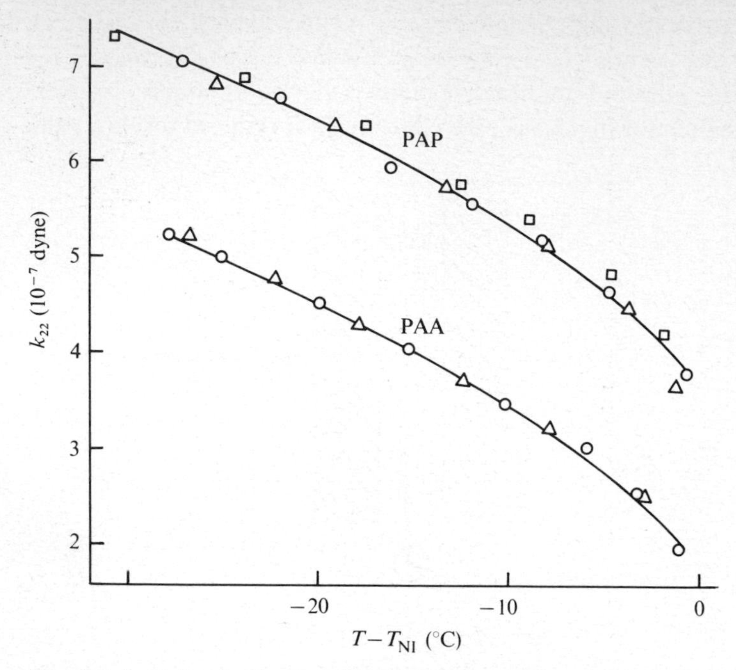

Fig. 3.4.4. Twist elastic constant (k_{22}) versus temperature for PAA and PAP. Squares, circles and triangles represent independent measurements on different samples. (P. P. Karat, unpublished.)

one. The angle of rotation can be measured by observing the concoscopic interference figures in a convergent beam. Here again it is essential that the rays make a sufficiently large angle with the twist axis to reduce the effective δn. This method has been used by Cladis[24] to determine k_{22}.

It should be emphasized that if the magnetic field is not strictly perpendicular to the initial undisturbed orientation of the director, the distortion does not set in abruptly (fig. 3.4.5) and the experimental determination of the elastic constants becomes somewhat unreliable. When the field is applied exactly at right angles, there is equal probability of the director turning through an angle φ or $-\varphi$ with respect to H. Consequently a number of disclination walls dividing the domains having different preferred orientations are formed in the specimen (see § 3.5.2). This serves as a useful criterion for checking the alignment of the field.

Orientation at the glass surfaces Errors may also be introduced by weak anchoring at the glass surfaces. The usual procedure for obtaining a planar or homogeneous structure is to rub the surface with dry lens tissue or cloth along a fixed direction.[25] Berreman has shown that geometrical

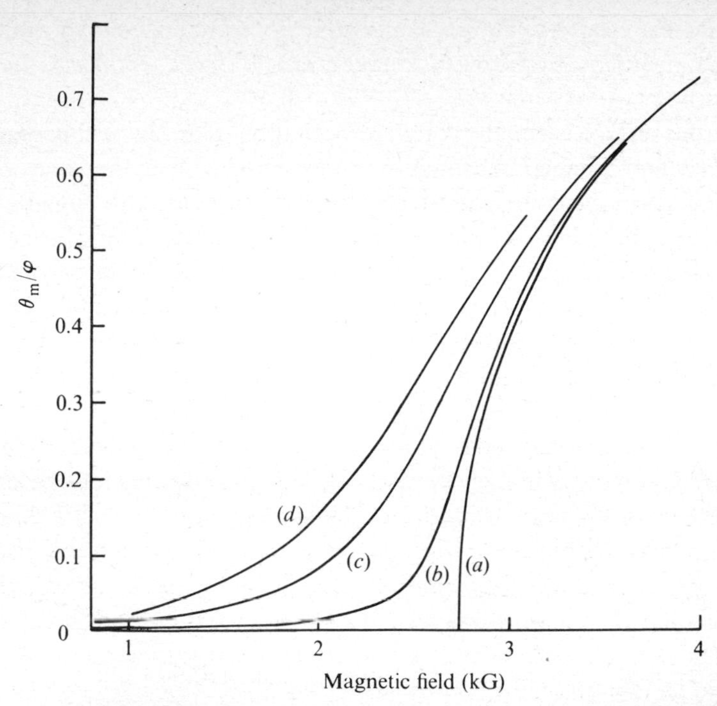

Fig. 3.4.5. Theoretical curves illustrating the relaxation of the Freedericksz threshold as the magnetic field is tilted away from the normal to the initial orientation of the director. The calculations have been made for PAA in the twist geometry for a sample of thickness 12.7 μm. (*a*) Field normal to the director, $\varphi = 90°$, (*b*) $\varphi = 89°$, (*c*) $\varphi = 85°$ and (*d*) $\varphi = 80°$. θ_m is the maximum deformation in the midplane of the sample. (Kini and Ranganath, unpublished.)

factors play an important role in producing such alignment, for rubbing tends to corrugate the surface.[25] Assuming that both ends of the molecule have equal affinity for the surface material so that they lie flat against the surface, it is obvious that more elastic energy is required for the molecules to lie across the rubbed direction than for them to lie parallel to it (fig. 3.4.6). A simple calculation shows that this extra energy

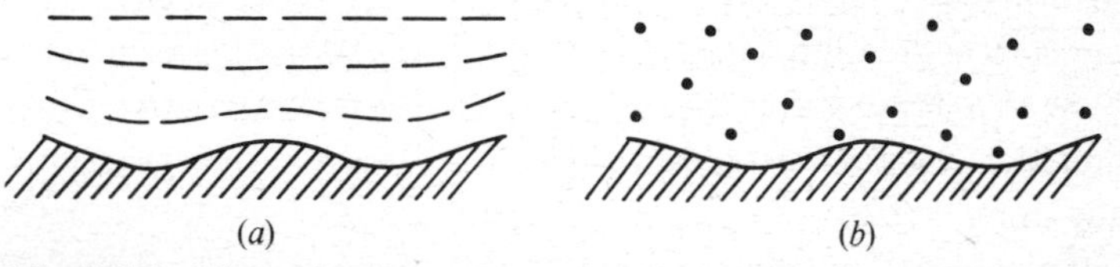

Fig. 3.4.6. The orienting effect of grooves. Extra elastic energy is required for the nematic director to lie across the grooves on the solid surface as in (*a*) rather than to lie parallel to them as in (*b*).

for the material *near the surface* is quite appreciable and almost impossible to overcome by means of a magnetic field. The anchoring is therefore firm.

If the surface is equally rough in both directions (as can happen if it is etched and cleaned thoroughly) there would be a tendency for the molecules to stand upright. Perpendicular or homeotropic alignment may also be achieved by coating the surface with surfactants in which case the detailed intermolecular forces probably play a significant part. There is evidence that homeotropic alignment is not always rigid. The consequences of weak anchoring on the Freedericksz transition have been discussed by Rapini and Papoular.[26]

Electric fields Measurements can also be made using electric fields.[21] There is complete analogy between the electric and magnetic fields as far as the threshold is concerned, but the analogy fails above the threshold. This is because dielectric susceptibilities are so large that the distortion gives rise to a non-uniformity of the field in the medium, which is practically absent in the magnetic field case. An important precaution to be taken in electric field measurements is that the sample has to be pure to avoid conduction-induced instabilities (see § 3.10).

Capacitance measurements The deformation at the threshold field in geometries (*a*) and (*c*) of fig. 3.4.1 may be detected conveniently by measuring the change in the capacitance.[27] A slight disadvantage with this technique is that relatively large areas have to be uniformly oriented, which is not easily achieved in practice. Non-uniform areas and edge effects tend to destroy the sharpness of the threshold. With the optical method, on the other hand, a well oriented area of less than 1 mm^2 is adequate.

3.4.2. The twisted nematic cell

Another configuration of much practical interest is the twisted nematic film.[28,29] The liquid crystal is sandwiched between two glass plates with the director aligned homogeneously (parallel to the walls). A twist is now imposed on the liquid crystal by turning one of the plates in its own plane about an axis normal to the film. A magnetic field above a critical strength applied along the twist axis, results in a deformation as shown schematically in fig. 4.1.4.

When $\mathbf{H}=0$, we have $\mathbf{n}=[\cos\{\varphi(z)\}, \sin\{\varphi(z)\}, 0]$, where $\varphi(0)=-\varphi_0$ and $\varphi(d)=\varphi_0$, d being the film thickness. When $\mathbf{H}=(0,0,H)$, $\mathbf{n}=[\cos\{\theta(z)\}\cos\{\varphi(z)\}, \cos\{\theta(z)\}\sin\{\varphi(z)\}, \sin\{\theta(z)\}]$ where θ is the angle made by $\mathbf{n}$ with the xy plane. From (3.3.17) we get

$$f(\theta)\frac{\mathrm{d}^2\theta}{\mathrm{d}z^2}+\frac{1}{2}\frac{\mathrm{d}f(\theta)}{\mathrm{d}\theta}\left(\frac{\mathrm{d}\theta}{\mathrm{d}z}\right)^2-\frac{1}{2}\frac{\mathrm{d}g(\theta)}{\mathrm{d}\theta}\left(\frac{\mathrm{d}\varphi}{\mathrm{d}z}\right)^2$$
$$+\chi_a H^2 \sin\theta\cos\theta=0 \qquad (3.4.7)$$

and

$$g(\theta)\frac{\mathrm{d}^2\varphi}{\mathrm{d}z^2}+\frac{\mathrm{d}g(\theta)}{\mathrm{d}\theta}\frac{\mathrm{d}\theta}{\mathrm{d}z}\frac{\mathrm{d}\varphi}{\mathrm{d}z}=0, \qquad (3.4.8)$$

where

$$f(\theta)=k_{11}\cos^2\theta+k_{33}\sin^2\theta,$$
$$g(\theta)=(k_{22}\cos^2\theta+k_{33}\sin^2\theta)\cos^2\theta.$$

After integration, (3.4.7) and (3.4.8) yield

$$f(\theta)\left(\frac{\mathrm{d}\theta}{\mathrm{d}z}\right)^2+g(\theta)\left(\frac{\mathrm{d}\varphi}{\mathrm{d}z}\right)^2+\chi_a H^2\sin^2\theta=A, \qquad (3.4.9)$$

and

$$g(\theta)\frac{\mathrm{d}\varphi}{\mathrm{d}z}=B, \qquad (3.4.10)$$

where A and B are constants. Equation (3.4.9) may be rewritten as

$$f(\theta)\left(\frac{\mathrm{d}\theta}{\mathrm{d}z}\right)^2+\frac{B^2}{g(\theta)}+\chi_a(\sin^2\theta)H^2=A. \qquad (3.4.11)$$

Because of strong anchoring at the walls, $\theta=0$ at $z=0$ and d, while θ_m, the maximum value of θ, occurs at the midplane $z=d/2$. Since $\mathrm{d}\theta/\mathrm{d}z=0$ at $z=d/2$,

$$A=\{B^2/g(\theta_m)\}+\chi_a H^2\sin^2\theta_m.$$

Substituting in (3.4.11)

$$f(\theta)\left(\frac{\mathrm{d}\theta}{\mathrm{d}z}\right)^2+B^2\left(\frac{1}{g(\theta)}-\frac{1}{g(\theta_m)}\right)+\chi_a H^2(\sin^2\theta-\sin^2\theta_m)=0$$

or

$$z=\int_0^\theta\left(\frac{f(\psi)}{\chi_a H^2(\sin^2\theta_m-\sin^2\psi)+B^2[\{1/g(\theta_m)\}-\{1/g(\psi)\}]}\right)^{1/2}\mathrm{d}\psi$$
$$=\int_0^\theta (N(\psi))^{1/2}\,\mathrm{d}\psi, \quad \text{say.}$$

Similarly from (3.4.8)

$$\varphi = -\varphi_0 + \int_0^{\theta} (N(\psi))^{1/2} \frac{B}{g(\psi)} \mathrm{d}\psi.$$

Therefore

$$d = 2\int_0^{\theta_m} (N(\psi))^{1/2}\, \mathrm{d}\psi \tag{3.4.12}$$

and

$$\varphi_0 = \int_0^{\theta_m} (N(\psi))^{1/2} \frac{B}{g(\psi)} \mathrm{d}\psi. \tag{3.4.13}$$

Transforming to a new variable given by $\sin\lambda = \sin\psi/\sin\theta_m$, (3.4.12) and (3.4.13) become

$$d = 2\int_0^{\pi/2} \left(\frac{f(\psi)}{\chi_a H^2 - B^2 M(\psi)}\right)^{1/2} \frac{\mathrm{d}\lambda}{(1-\sin^2\theta_m \sin^2\lambda)^{1/2}}$$

and

$$\varphi_0 = \int_0^{\pi/2} \left(\frac{f(\psi)}{\chi_a H^2 - B^2 M(\psi)}\right)^{1/2} \frac{B\,\mathrm{d}\lambda}{g(\psi)(1-\sin^2\theta_m \sin^2\lambda)^{1/2}},$$

where

$$M(\psi) = \frac{1}{(\sin^2\theta_m - \sin^2\psi)}\left(\frac{1}{g(\theta)} - \frac{1}{g(\theta_m)}\right)$$
$$= [k_{33} - 2k_{22} - (k_{33} - k_{22})(\sin^2\psi + \sin^2\theta_m)]/g(\theta)\, g(\theta_m).$$

Taking the limit $\theta_m \to 0$, we have $\theta \to 0$, $f(\theta) \to k_{11}$, $g(\theta) \to k_{22}$, $\mathrm{d}\varphi/\mathrm{d}z \to 2\varphi_0/d$, $B \to 2k_{22}\varphi_0/d$, $M \to (k_{33} - 2k_{22})/k_{22}^2$ and

$$d = 2\int_0^{\pi/2} \left(\frac{k_{11}}{\chi_a H^2 - (4\varphi_0^2/d^2)(k_{33} - 2k_{22})}\right)^{1/2} \mathrm{d}\lambda$$

or

$$H_c = (2/\chi_a^{1/2} d)[k_{11}(\pi/2)^2 + (k_{33} - 2k_{22})\varphi_0^2]^{1/2}, \tag{3.4.14}$$

which is the critical field for the deformation to occur.[28] Thus a measurement of H_c in this geometry gives k_{22}, if k_{11} and k_{33} are known.[30] However, the deformation cannot be detected optically by observation in polarized light at normal incidence until the field is well above the threshold value H_c (see fig. 4.1.5). A more convenient way of detecting H_c would appear to be by measuring the change in the capacitance.

A noteworthy feature of the twisted nematic is that the transmitted intensity (through a pair of polarizers) as a function of field shows a 'bilevel' behaviour. It can therefore be used as an optical shutter and has important possibilities in display technology, as was first pointed out by Schadt and Helfrich[29] (see § 4.1.1).

3.5. Disclinations

3.5.1. Schlieren textures

As remarked in chapter 1, the nematic state is named for the threads that can be seen within the fluid under a microscope (plate 3(a)). In thin films sandwiched between glass plates these threads can be seen end on. A typical example of the texture in a plane film of thickness about 10 μm between crossed polarizers – the *structures à noyaux* or *schlieren textures* – is given in plate 3(b). The black brushes originating from the points are due to 'line singularities' perpendicular to the layer. In analogy with dislocations in crystals, Frank[3] proposed the term 'disinclinations' which has since been modified to *disclinations* in current usage.

The brushes are regions where the director (or the local optical axis) is either parallel or perpendicular to the plane of polarization of the incident light. The polarization is unchanged by the material in these regions and the light is therefore extinguished by the crossed analyser. Some points have four black brushes while others have only two. The positions of the points remain unchanged on rotating the crossed polarizers but the brushes themselves rotate continuously showing that the orientation of the director changes continuously about the disclinations. The sense of rotation may either be the same as that of the polarizers (*positive* disclinations), or opposite (*negative* disclinations). The rate of rotation is about equal to that of the polarizers when the disclination has four brushes and is twice as fast when it has only two.

The strength of a disclination is defined as $s=\frac{1}{4}$(number of brushes). So far, $s=+\frac{1}{2}$, $-\frac{1}{2}$, $+1$ and -1 have been observed. Neighbouring disclinations connected by brushes are of opposite signs and the sum of the strengths of all disclinations in a sample tends to be zero. Often, particularly at temperatures close to T_{NI}, disclinations of opposite signs are seen to attract each other and coalesce. They may then disappear altogether ($s_1+s_2=0$) or form a new singularity ($s_1+s_2=s'$).

The significance of these textures was understood by Lehmann[31] and Friedel,[32] but a mathematical treatment of the actual configuration around disclinations was given by Oseen[1] and Frank.[3] A more complete discussion of the theory of schlieren textures is due to Nehring and Saupe.[33]

Consider a space-fixed cartesian coordinate system xyz with the z axis normal to the nematic layer, and let the components of the director be $(\sin\theta\cos\psi, \sin\theta\sin\psi, \cos\theta)$. We shall assume a planar structure, i.e., that the director is throughout in the xy plane ($\theta=\pi/2$). In practice this assumption is approximately valid in schlieren textures except in the immediate vicinity of a singularity where a decrease in birefringence is observed indicating a change of θ. The theory is therefore not valid too close to a disclination.

In a cylindrical coordinate system $r\alpha z$ the components of the director are $n_r=\cos(\psi-\alpha)$, $n_\alpha=\sin(\psi-\alpha)$, $n_z=0$. From the equation of equilibrium (3.3.17) we get

$$\begin{aligned}&\tfrac{1}{2}(k_{11}+k_{33})(\psi_{,rr}+r^{-1}\psi_{,r}+r^{-2}\psi_{,\alpha\alpha})\\&\quad-\tfrac{1}{2}(k_{11}-k_{33})\{[\psi_{,rr}-r^{-1}\psi_{,r}(1-2\psi_{,\alpha})-r^{-2}\psi_{,\alpha\alpha}]\cos 2(\psi-\alpha)\\&\quad+[2r^{-1}\psi_{,r\alpha}-\psi_{,r}^2-r^{-2}\psi_{,\alpha}(2-\psi_{,\alpha})]\sin 2(\psi-\alpha)\}+k_{33}\psi_{,zz}=0,\end{aligned} \tag{3.5.1}$$

where the subscripts denote partial derivatives, e.g.,

$$\psi_{,r}=\partial\psi/\partial r,\quad \psi_{,rr}=\partial^2\psi/\partial r^2,\quad \psi_{,r\alpha}=\partial^2\psi/\partial r\partial\alpha,\quad \text{etc.}$$

We seek solutions independent of r and therefore equate to zero all derivatives with respect to r as also the sum of the coefficients belonging to the same power of r:

$$\begin{aligned}&\tfrac{1}{2}(k_{11}+k_{33})\psi_{,\alpha\alpha}+\tfrac{1}{2}(k_{11}-k_{33})[\psi_{,\alpha\alpha}\cos 2(\psi-\alpha)\\&\quad+\psi_{,\alpha}(2-\psi_{,\alpha})\sin 2(\psi-\alpha)]=0,\end{aligned} \tag{3.5.2}$$

$$k_{33}\psi_{,zz}=0. \tag{3.5.3}$$

The last equation gives

$$\psi=qz+b,$$

where q and b are functions of α; $q=0, b=0$ corresponds to the uniformly oriented nematic, and $q\neq0$ (constant), $b=0$ to the uniformly twisted cholesteric. We shall now discuss the case $q=0, b\neq0$, which describes the configuration around the disclinations. Let us make the approximation that $k_{11}=k_{22}=k_{33}$; then

$$\psi_{,\alpha\alpha}=\partial^2\psi/\partial\alpha^2=0. \tag{3.5.4}$$

The solutions of (3.5.4) are $\psi = \text{constant}$, which is of no interest, and

$$\psi = s\alpha + c, \tag{3.5.5}$$

where c is a constant. If the orientational order is apolar, it is clear that a rotation of $m\pi$ (where m is an integer) in the director orientation ψ should correspond to a rotation of 2π in the polar angle α (fig. 3.5.1). On the other hand, for a polar medium a change of $2m\pi$ in ψ should correspond to a change of 2π in α. More generally

$$s = \pm\tfrac{1}{2}, \pm 1, \pm\tfrac{3}{2} \ldots, \quad \text{with } 0 < c < \pi \text{ (apolar)},$$

$$s = \pm 1, \pm 2, \pm 3 \ldots, \quad \text{with } 0 < c < 2\pi \text{ (polar)}.$$

s is called the *strength* of the disclination.

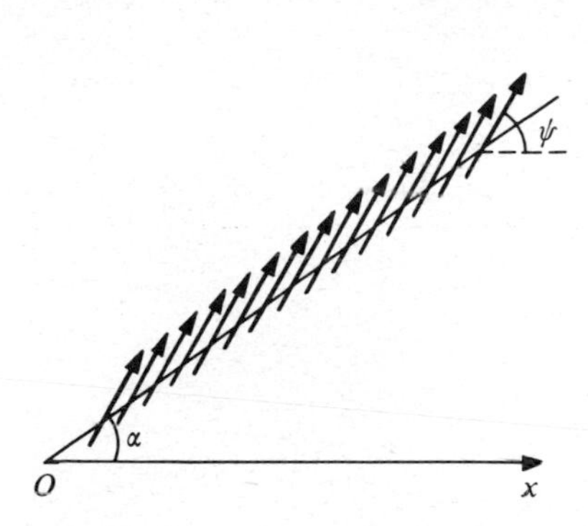

Fig. 3.5.1. Director orientation (indicated by arrows) along a polar line making an angle α. Incident light that is linearly polarized at angle ψ or $\psi \pm \pi/2$ will be extinguished by a crossed analyser and will give rise to a dark brush.

We shall now show that $s = \frac{1}{4}$(number of brushes). If the incident light is linearly polarized at an angle ψ with respect to the x axis, it is seen from fig. 3.5.1 that the polarization will be unchanged at all points on the polar line α and hence will not be transmitted by the analyzer. This will result in a black brush at an angle α. A similar situation will arise when ψ changes by $\pi/2$. The angle between two successive dark brushes is therefore $\Delta\alpha = \Delta\psi/s = \pi/2s$. Thus the number of dark brushes per singularity is $2\pi/|\Delta\alpha| = 4|s|$. Also if the polarizers are turned through angle ω the brushes rotate by an angle ω/s. The rate of rotation of the brushes of the two-brush disclination ($s = \pm\frac{1}{2}$) is therefore twice as fast as that in the four-brush variety ($s = \pm 1$). If the polarizers are kept fixed and the microscope stage is rotated by ω the brushes turn by $\omega(s-1)/s$. Observations in polarized light therefore enable one to determine s both in sign and magnitude. The existence of $|s| = \frac{1}{2}$ in nematic liquid crystals establishes the absence of polarity in this phase.

If $\theta \neq \pi/2$, half-integral values of s are not possible even if the molecular order is apolar, and

$$s = \pm 1, \pm 2, \pm 3 \ldots, \quad \text{with} \quad 0 < c < 2\pi (\theta \neq \pi/2).$$

The molecular orientation in the neighbourhood of a disclination is shown in fig. 3.5.2 for a few values of s. The curves represent the projection of the director field in the xy plane. For $s \neq 1$, a change in c merely causes a rotation of the figure by $\Delta c/(1-s)$, while for $s = 1$, the pattern itself is changed.

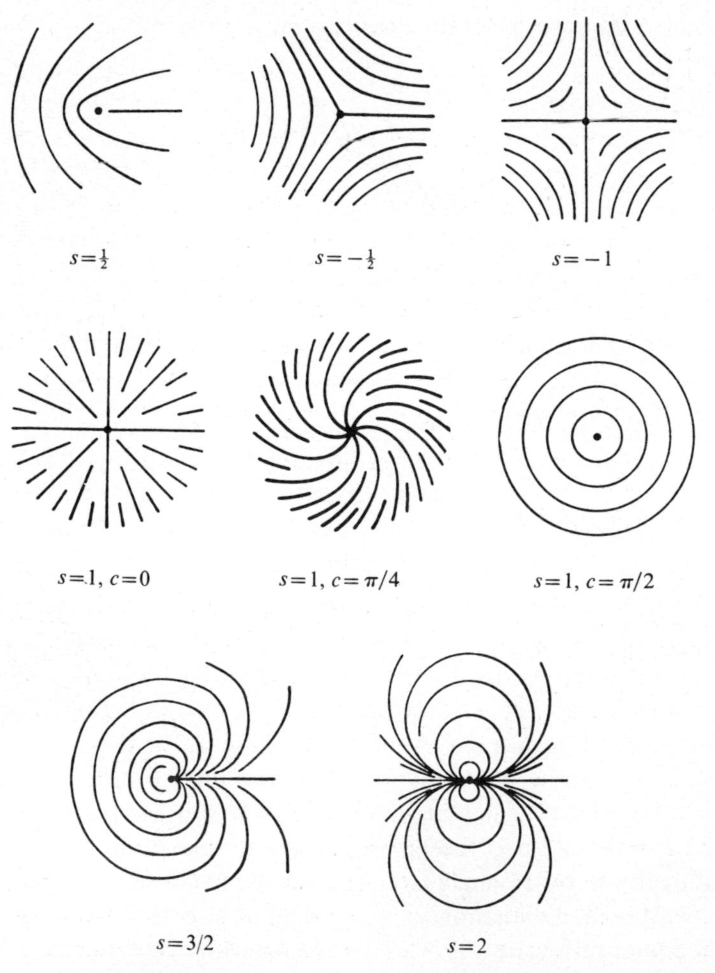

Fig. 3.5.2. Molecular orientation in the neighbourhood of a disclination. (After Frank.[3])

When $k_{11} \neq k_{22} \neq k_{33}$, (3.5.5) is no longer a solution of (3.5.2) except when

$$\psi = \alpha \pm (n\pi/2) \quad (n \text{ an integer})$$

and

$$\psi = 2\alpha + c.$$

For $s \neq 1$, a solution in the first order of approximation is [33]

$$\psi = (s\alpha + c) + \frac{k_{11} - k_{33}}{k_{11} + k_{33}} \times \left[\frac{s(2-s)}{4(s-1)^2} \sin\{2\alpha(s-1)+2\} + c' \right],$$

where c' is a constant. This results in a non-uniform rotation of the brushes when the polarizers are rotated uniformly. In principle this can be used to estimate the elastic anisotropy[3] but experimentally it is somewhat difficult to do so because the brushes are often distorted near the centre of the disclination, as the molecules in this region tend to align themselves normal to the surface, and also because of the disturbing influence of neighbouring disclinations.

When the strengths of two neighbouring disclinations are equal and opposite, the brushes connecting them are circular. By superposition of solutions of the type (3.5.5)

$$\psi = \psi_1 + \psi_2 = s_1\alpha_1 + s_2\alpha_2 + \text{const.}$$

Putting $s_1 = -s_2 = s$,

$$\psi = s\beta + \text{const.},$$

where $\beta = \alpha_1 - \alpha_2$. Thus curves of constant ψ will be arcs of circles passing through the two disclinations (fig. 3.5.3). For $s = \frac{1}{2}$, ψ changes by $\pi/2$ on going from one side of the chord to the other (since $\beta' = \pi - \beta$ or $\psi' = \pi s - \psi$) while for $s = 1$ it is unchanged. An example of such circular brushes is shown in plate 6.

3.5.2. Inversion walls

Another type of schlieren texture is sometimes seen when the surfaces impose a uniform orientation on the sample, as is the case when the glass slides have been rubbed. The black brushes appear as nearly parallel pairs, one brush of each pair generally being slightly more diffuse than the other. For special orientations of the crossed polarizers, all point

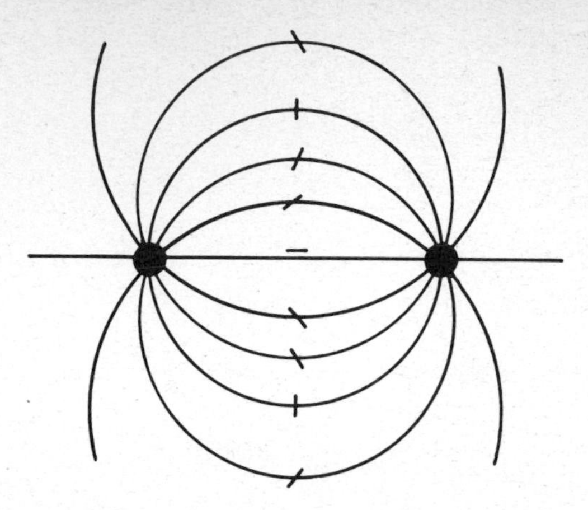

Fig. 3.5.3. Curves of equal alignment around a pair of singularities of equal and opposite strengths. The orientations marked on the circles refer to the case $s = 1$, $c = 0$.

singularities are connected by sharp black lines and at the same time large areas of the sample appear black (plates 7 and 8).

Such textures arise from inversion walls normal to the layer. The molecular orientation near an inversion wall is shown in fig. 3.5.4. The alignment is parallel to the surfaces except near the wall. The solid lines represent the alignment curves of the director, while the broken lines indicate the borders of the wall. The total change in orientation on crossing the wall is π; a change of $\pi/2$ is reached at the centre of the wall which is shown by another broken line. Between crossed polarizers two black brushes generally appear, one on either side of the central $\pi/2$ line. This explains why the brushes are usually seen in parallel pairs. From the shape of the contours in fig. 3.5.4 it is also clear why the brushes on opposite sides of the central broken line are not equally sharp. At certain points, the sharp and diffuse brushes change sides with respect to the central line. When the polarizers are parallel or perpendicular to the orientation outside the wall, a single black brush appears at the centre of the wall and large areas of the sample are black.

Since the molecular orientation changes by π on crossing the wall, the singularities located within an inversion wall have integral values of s. The inversion walls may form closed loops or start from disclinations of the type $s = \pm\frac{1}{2}$. The curves in fig. 3.5.4 are drawn for different types of singularities which from a comparison with fig. 3.5.2 can be seen to be $s = +\frac{1}{2}, -1, +1$ and $-\frac{1}{2}$ in that order from left to right. The cross-over points where the sharp and diffuse brushes change sides with respect to the central line are singularities of $s = -1$ and $+1$.

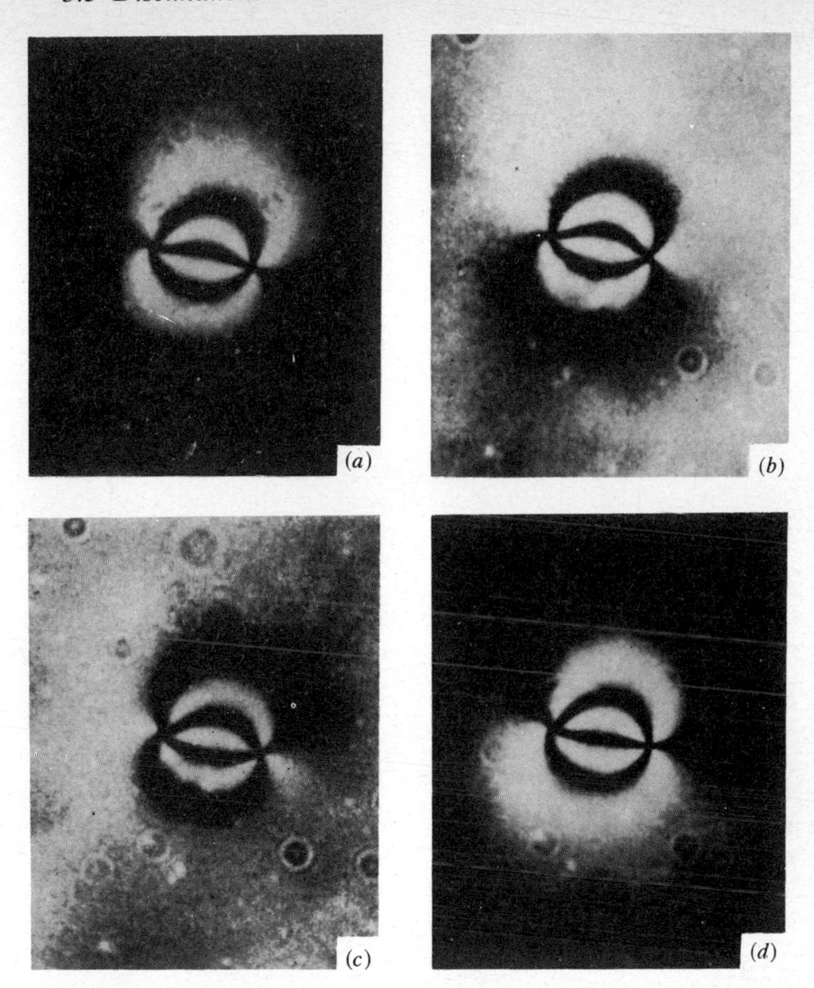

Plate 6. Brushes connecting a pair of disclinations of equal and opposite strengths, $s = 1$ and -1, in nematic MBBA. Crossed polarizers rotated clockwise by 22.5° in each successive photograph. In (*d*) the directions of extinction are parallel to the edges of the picture. (Nehring and Saupe.[33])

A second kind of inversion wall is shown in plate 9(*a*). In addition to the usual black brushes originating from points three sharp lines can be seen. These lines are inversion walls arising from a tilt of the molecules out of the plane of the layer in a sample in which the orientation is normally parallel to the surfaces. Molecules on opposite sides of the walls have opposite tilts. At the centre of the wall, they are aligned parallel to the surface and the birefringence has a maximum value.

More general types of disclinations have been discussed by Friedel and Kléman.[34]

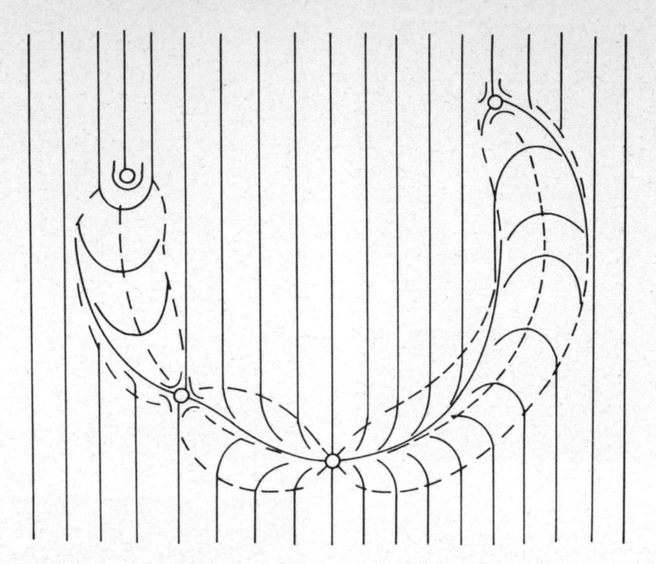

Fig. 3.5.4. Molecular alignment in the vicinity of an inversion wall of the first kind. Solid lines give the vector field of the projection of **n** in the plane of the layer, broken lines are selected curves of equal alignment. (After Nehring and Saupe.[33])

3.5.3. Interaction between disclinations

The deformation energy of an isolated disclination in a circular layer of radius R and of unit thickness is [33,35]

$$W = \int_0^R \int_0^{2\pi} F(r) r \, \mathrm{d}r \, \mathrm{d}\alpha \tag{3.5.6}$$

where, assuming a single constant approximation ($k = k_{11} = k_{22} = k_{33}$) and $\theta = \pi/2$,

$$F = \tfrac{1}{2}(k/r^2)\psi_{,\alpha}^2 = \tfrac{1}{2}(k/r^2)s^2.$$

Of course, this expression is not valid close to the centre of the disclination where $\theta \neq (\pi/2)$. Moreover, there is supposed to be a singularity at the core whose energy is not known. To allow for this, we postulate a radius r_c around the disclination and integrate (3.5.6) for distances greater than r_c to obtain

$$W = W_c + \pi k s^2 \log(R/r_c), \tag{3.5.7}$$

where W_c is the energy of the central region. As $R \to \infty$, $W \to \infty$ logarithmically, i.e., an isolated disclination in an infinitely extended layer has infinite energy. However, such a situation does not arise in practice owing

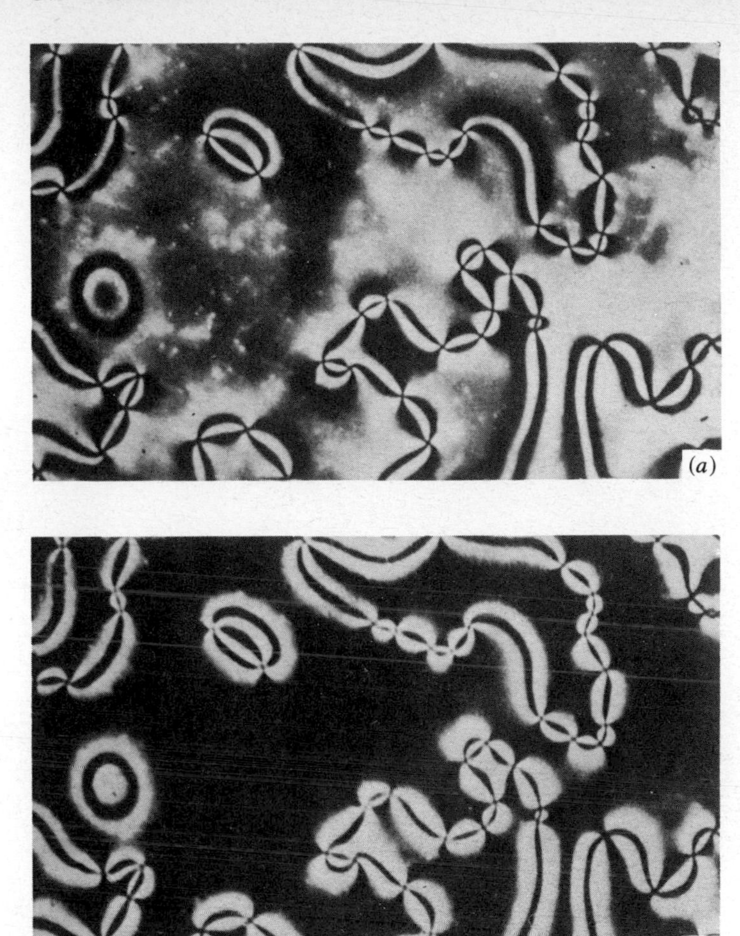

Plate 7. Inversion lines of the first kind in nematic MBBA. Crossed polarizers rotated clockwise by 67.5° on going from (a) to (b). (Nehring and Saupe.[33])

to the occurrence of pairs of disclinations of opposite signs, as we shall see below.

Consider two disclinations of strength s_1 and s_2 situated at $\mathbf{a}_1$ and $\mathbf{a}_2$ respectively. Using the superposition principle, the energy density at any point $\mathbf{r}$ is

$$F(r)=\tfrac{1}{2}k\left[\left(\sum_{i=1,2} s_i\right)\sum_{i=1,2}\frac{s_i}{(\mathbf{r}-\mathbf{a}_i)^2}-s_1s_2\frac{\mathbf{r}_{12}^2}{(\mathbf{r}-\mathbf{a}_1)^2(\mathbf{r}-\mathbf{a}_2)^2}\right], \tag{3.5.8}$$

where $\mathbf{r}_{12}=\mathbf{a}_2-\mathbf{a}_1$. In general, if there are a finite number of disclinations

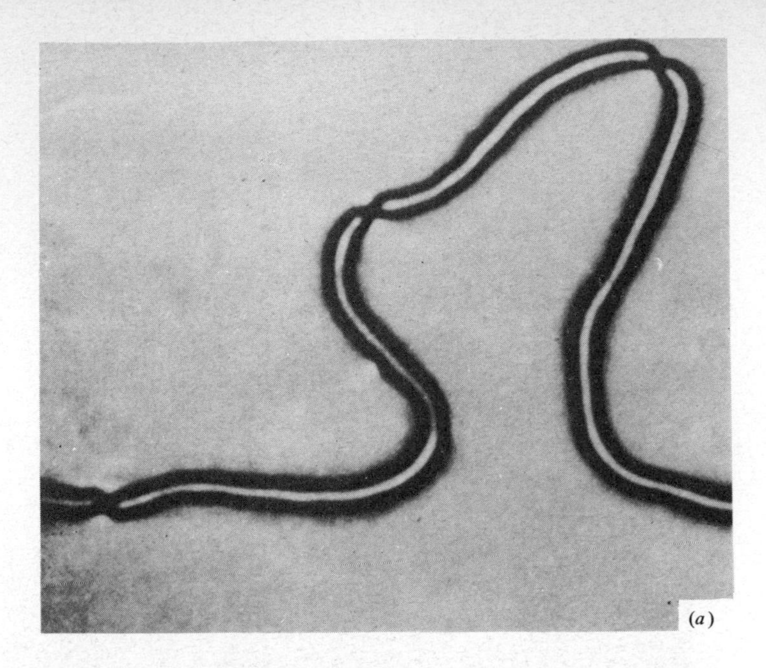

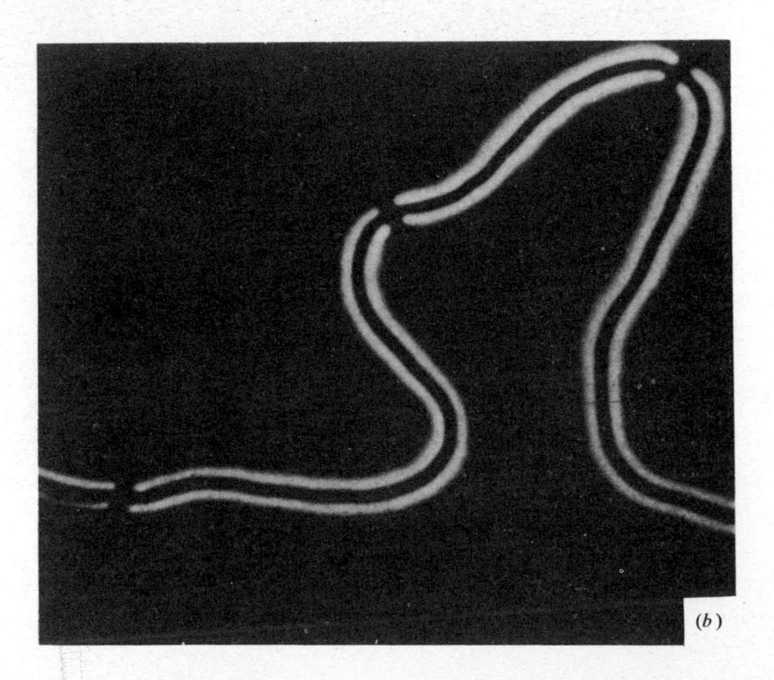

Plate 8. Narrow inversion line of the first kind in an uncovered sample of nematic MBBA. Crossed polarizers rotated by 45° on going from (*a*) to (*b*). (Nehring and Saupe.[33])

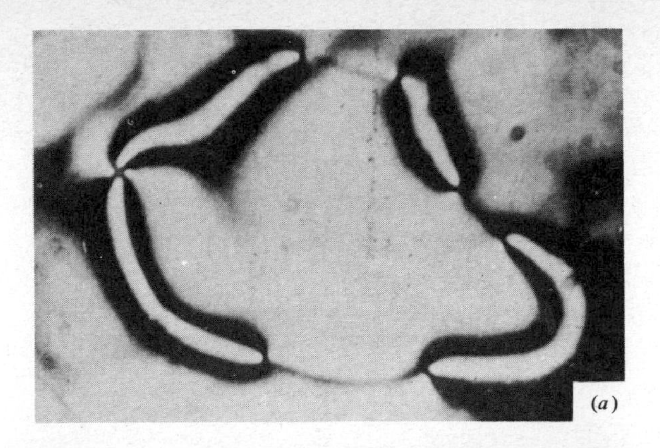

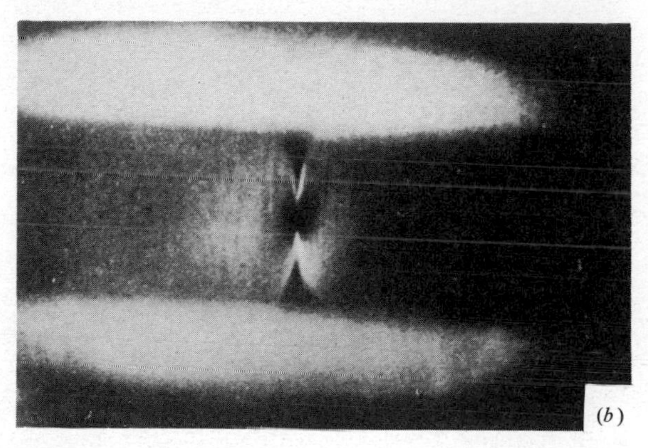

Plate 9. (*a*) Inversion walls of the second kind in MBBA. Crossed polarizers. (Nehring and Saupe.[33]) (*b*) Point singularity in a capillary filled with MBBA. (Saupe.[35])

in a layer and the sum of their strengths $\sum s_i = 0$, the first term in the square brackets of (3.5.8) vanishes and the total deformation energy is finite. This result was first pointed out by Ericksen.[36]

The second term in the square brackets of (3.5.8) represents the interaction between the two disclinations. Integrating as before, the total interaction energy per unit thickness of film

$$W_{12} = -2\pi k s_1 s_2 \log(r_{12}/r_c).$$

The assumption here is that $r_c \ll r_{12} \ll R$. The force between the two singularities is therefore $-2\pi k s_1 s_2/r_{12}$. Accordingly, singularities of opposite signs attract and those of like signs repel, the force being inversely proportional to the distance.[33] The behaviour is analogous to the force between two current-carrying conductors.

3.5.4. The core structure

The structure very close to the centre of the disclination has been studied in some detail for the particular case $s = 1$. It has been shown by Meyer[37] and by Williams, Pieranski and Cladis[38] that the so-called line singularity can almost certainly be replaced by a non-singular continuous structure of lower energy. As mentioned earlier, the birefringence decreases in this region showing that the molecules tend to align themselves normal to the surface. We seek solutions which fit with a planar structure ($\theta = \pi/2$) at $r \geqslant r_{\pi/2}$. Let us write in cylindrical coordinates $n_r = \sin\theta\cos(\psi - \alpha)$, $n_\alpha = \sin\theta\sin(\psi - \alpha)$, $n_z = \cos\theta$. The equation of equilibrium for $s = 1$ becomes

$$\frac{\mathrm{d}^2\theta}{\mathrm{d}r^2} + \frac{1}{r}\frac{\mathrm{d}\theta}{\mathrm{d}r} - \frac{\cos\theta\sin\theta}{r^2} = 0.$$

The simplest assumption is that $\theta = 0$ at $r = 0$ and $\theta = \pi/2$ at $r = r_{\pi/2}$. The solution is then

$$\tan(\theta/2) = r/r_{\pi/2}. \qquad (3.5.9)$$

The deformation now involves both splay and bend. Calculations show that W_c of (3.5.7) is finite – with the one-constant approximation it is $3\pi k$[38] – whereas a line singularity ($s = 1$) leads to a logarithmic divergence of the splay energy as $r \to 0$ (see fig. 3.5.2). Structures of this type have been realized experimentally in thin capillaries (fig. 3.5.5(a)). Williams *et al.*[38] have also established the existence of singular points along the capillary axis (fig. 3.5.5(b)) and plate 9 (b)). The problem of point singularities has been treated theoretically by Saupe.[35]

3.6. Flow properties

3.6.1. Miesowicz's experiment

We shall now discuss the application of the Ericksen–Leslie theory to some practical problems in viscometry.[39] Probably the first precise determination of the anisotropic viscosity of a nematic liquid crystal was by Miesowicz.[40] He oriented the sample by applying a strong magnetic field and measured the viscosity coefficients in the following three geometries using an oscillating plate viscometer:

(i) **n** parallel to the flow;
(ii) **n** parallel to the velocity gradient;
(iii) **n** perpendicular to the flow and to the velocity gradient.

In the presence of a strong field, the magnetic coherence length is quite small and one may without sensible error neglect boundary effects and

Fig. 3.5.5. (*a*) Structure in a thin capillary: the wall alignment is homeotropic and changes by 90° from wall to axis. (*b*) Point singularities in a capillary. (After Saupe.[35])

director gradients. The observational data for the three geometries can then be readily interpreted on the basis of the equations given in § 3.3.

The apparent viscosity for any geometry

$$\eta = \frac{\text{shear stress}}{\text{velocity gradient}} = \frac{t_{ji}}{2d_{ij}}.$$

If we take the flow to be along x and the velocity gradient along y,

$$v = v_x, \qquad v_{i,j} = v_{x,y},$$

$$w_{xy} = \tfrac{1}{2}v_{x,y}, \qquad w_{yx} = -\tfrac{1}{2}v_{x,y},$$

and

$$d_{xy} = d_{yx} = \tfrac{1}{2}v_{x,y}.$$

Since director gradients are neglected, the elastic part of the stress tensor $t_{ji}^0 = 0$. Thus, for $\mathbf{n} = (1, 0, 0)$ we have from (3.3.5)

$$t_{yx} = \mu_3 N_y n_x + \mu_4 d_{yx} + \mu_6 n_x^2 d_{xy}$$

$$= \tfrac{1}{2}(\mu_3 + \mu_4 + \mu_6)v_{x,y}$$

or

$$\eta_1 = \tfrac{1}{2}(\mu_3 + \mu_4 + \mu_6).$$

Similarly

$$\eta_2 = \tfrac{1}{2}(-\mu_2 + \mu_4 + \mu_5)$$

$$\eta_3 = \tfrac{1}{2}\mu_4.$$

Miesowicz's results for two compounds are presented in table 3.6.1.

TABLE 3.6.1. *Miesowicz's viscosity coefficients (measurements in 10^{-2} poise)*

	Molecules parallel to flow direction η_1	Molecules parallel to velocity gradient η_2	Molecules perpendicular to flow direction and to velocity gradient η_3
p-Azoxyanisole 122 °C	2.4 ± 0.05	9.2 ± 0.4	3.4 ± 0.3
p-Azoxyphenetole 144.4 °C	1.3 ± 0.05	8.3 ± 0.4	2.5 ± 0.3

3.6.2. Tsvetkov's experiment

In this experiment, a tube containing a nematic liquid crystal is suspended in a uniform magnetic field acting in a horizontal plane and is spun at a constant angular velocity Ω about a vertical axis. If the axis of rotation is along z and the magnetic field along y, the components of the bulk velocity of the fluid are

$$v_x = -\Omega y, \qquad v_y = \Omega x \quad \text{and} \quad v_z = 0.$$

Since the director lies in the xy plane $\mathbf{n} = (\cos\varphi, \sin\varphi, 0)$. If the diameter of the tube is large enough, wall effects and director gradients may be neglected. Therefore (3.3.2) may be written as

$$\rho_1 \ddot{n}_i = g'_i + G_i,$$

where g'_i is given by (3.3.13) and G_i by (3.4.1). Therefore

$$\rho_1 \ddot{\varphi} = \lambda_1(\dot{\varphi} - \Omega) - \chi_a H^2 \sin\varphi \cos\varphi, \tag{3.6.1}$$

where $\lambda_1 = \mu_2 - \mu_3$. Below a critical angular velocity Ω_c, (3.6.1) has the simple solution

$$\sin 2\varphi = \Omega/\Omega_c, \tag{3.6.2}$$

where

$$\Omega_c = -(\chi_a H^2/2\lambda_1). \tag{3.6.3}$$

Thus when $\Omega < \Omega_c$ the director makes a constant angle φ with the field. This has been accurately verified by Leslie, Luckhurst and Smith[41] from a study of the electron spin resonance spectrum of a paramagnetic probe dissolved in a nematic liquid crystal which is spun in a magnetic field. The spacing between the hyperfine lines was found to be in quantitative accord with (3.6.2).

When $\Omega > \Omega_c$ (3.6.1) does not have a steady state solution. Assuming director inertia to be negligible

$$\dot{\varphi} - \Omega + \Omega_c \sin 2\varphi = 0. \tag{3.6.4}$$

For $\Omega > \Omega_c$,

$$\tan\left(\varphi - \frac{\pi}{4}\right) = \left(\frac{\Omega - \Omega_c}{\Omega + \Omega_c}\right)^{1/2} \tan[(\Omega^2 - \Omega_c^2)^{1/2} t - t_o], \tag{3.6.5}$$

where

$$t_0 = \tan^{-1}\left(\frac{\Omega + \Omega_c}{\Omega - \Omega_c}\right)^{1/2}$$

assuming the initial condition that $\varphi = 0$ when $t = 0$.[41] Hence the director rotates with a mean angular velocity

$$\omega = (\Omega^2 - \Omega_c^2)^{1/2}. \tag{3.6.6}$$

An alternative method of performing the experiment is to have a stationary sample in a rotating magnetic field. In point of fact Tsvetkov[42] (and more recently Gasparoux and Prost[43]) used this method and measured the torque exerted by the fluid on the cylinder as a function of Ω. With increasing Ω the torque at first increases linearly, reaches a maximum and then starts to decrease. From (3.3.5), the stress tensor

$$t'_{r\varphi} = [-\mu_2 + (\mu_2 + \mu_3)\sin^2 \varphi](\Omega - \dot{\varphi}).$$

The total torque on a cylinder of length L and radius R is therefore

$$\begin{aligned}\tau &= R^2L\int_0^{2\pi} t'_{r\varphi}\,\mathrm{d}\varphi \\ &= -\pi R^2 L\lambda_1(\Omega-\dot{\varphi}) \\ &= -V\lambda_1\Omega_c \sin 2\varphi\end{aligned}$$

from (3.6.4), where $V=\pi R^2L$ is the volume of the cylinder. When $\Omega<\Omega_c$,

$$\tau=-V\lambda_1\Omega. \tag{3.6.7}$$

The torque increases linearly with the angular velocity and offers a direct method of determining λ_1. When $\Omega=\Omega_c$,

$$\tau=\tfrac{1}{2}V\chi_a H^2. \tag{3.6.8}$$

When $\Omega>\Omega_c$ (3.6.5) yields the relation

$$\begin{aligned}\sin 2\varphi &= \frac{2\tan\varphi}{1+\tan^2\varphi} \\ &= \frac{2[\Omega_c/\Omega+(1-\Omega_c^2/\Omega^2)^{1/2}\tan\{(\Omega^2-\Omega_c^2)^{1/2}t-t_0\}]}{1+[\Omega_c/\Omega+(1-\Omega_c^2/\Omega^2)^{1/2}\tan\{(\Omega^2-\Omega_c^2)^{1/2}t-t_0\}]^2}.\end{aligned}$$

The mean value of the torque is therefore

$$\bar{\tau}=-\frac{\lambda_1 V\Omega_c^2}{\Omega+(\Omega^2-\Omega_c^2)^{1/2}}. \tag{3.6.9}$$

When $\Omega\gg\Omega_c$,

$$\bar{\tau}=-\frac{\lambda_1 V}{2\Omega}\Omega_c^2. \tag{3.6.10}$$

Above the critical angular velocity, the torque decreases with increasing Ω. The predictions are generally in agreement with observations[43] (fig. 3.6.1). However, the shape of the experimental curve at higher angular velocities appears to be rather sensitive to the nature of the solid surface in contact with the liquid crystal, showing that a complete theory has to take into account boundary effects and the production and migration of disclination walls at the surface.[44]

3.6.3. Poiseuille flow

We shall next consider the rigorous theory of Poiseuille flow, i.e., the steady laminar flow, caused by a pressure gradient, of an incompressible

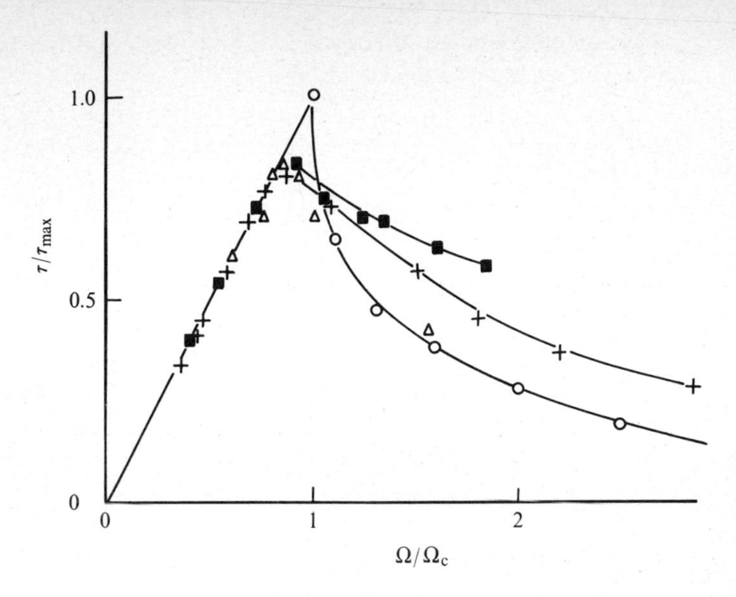

Fig. 3.6.1. Variation of the torque as a function of the angular velocity of the rotating magnetic field. ○ theoretical values. ■ experimental values for PAA, $T = 112$ °C, $H = 2900$ G (Tsvetkov[42]); + experimental values for MBBA, $T = 24$ °C, $H = 2230$ G; △ experimental values for the same compound, $T = 22$ °C, $H = 2230$ G, but using a solid teflon cylinder immersed in the fluid to measure the torque. (After Gasparoux and Prost.[43])

fluid through a tube of circular cross-section.[45] We shall suppose that the tube is of infinite length so that end effects can be ignored. Let us choose a cylindrical polar coordinate system $r\varphi z$ with z along the axis of the tube. In the steady state the only component of the velocity gradient is $v_{z,r} = \mathrm{d}v/\mathrm{d}r$. It is natural to expect that the director is everywhere in the rz plane making an angle $\theta(r)$ with z. Thus we seek for the components of the director and velocity fields the solutions

$$n_r = \sin\theta(r), \qquad n_\varphi = 0, \qquad n_z = \cos\theta(r).$$

$$v_r = 0, \qquad v_\varphi = 0, \qquad v_z = v(r).$$

If a magnetic field with components $(H_r, 0, H_z)$ is applied, the external body force

$$G_r = \chi_a(H_r \sin\theta + H_z \cos\theta)H_r,$$

$$G_z = \chi_a(H_r \sin\theta + H_z \cos\theta)H_z,$$

where χ_a is the diamagnetic anisotropy per unit volume. Substituting in (3.3.1) and (3.3.2) we have in the steady state

$$2f(\theta)\frac{d^2\theta}{dr^2}+\frac{df}{d\theta}\left(\frac{d\theta}{dr}\right)^2+\frac{2f}{r}\frac{d\theta}{dr}-\frac{k_{11}\sin 2\theta}{r^2}+(\lambda_1+\lambda_2\cos 2\theta)\frac{dv}{dr}$$

$$+2\chi_a(H_r\sin\theta+H_z\cos\theta)(H_r\cos\theta-H_z\sin\theta)=0, \quad (3.6.11)$$

$$\frac{dv}{dr}=\frac{1}{g(\theta)}\left(-\frac{ar}{2}+\frac{b}{r}\right), \quad (3.6.12)$$

where

$$f(\theta)=k_{11}\cos^2\theta+k_{33}\sin^2\theta, \quad (3.6.13)$$

$$g(\theta)=\tfrac{1}{2}[2\mu_1\sin^2\theta\cos^2\theta+(\mu_5-\mu_2)\sin^2\theta$$
$$+(\mu_6+\mu_3)\cos^2\theta+\mu_4], \quad (3.6.14)$$

$a=-\dfrac{dp}{dz}$, and b is a constant.

Equations (3.6.11) and (3.6.12) are also applicable to flow through the annular space between two coaxial cylinders. For flow through a capillary we put $b=0$ to avoid the singularity in (3.6.12) at $r=0$. The two equations can then be used to obtain the orientation and velocity profiles. At high flow rates, the contributions of the elastic terms tend to become small and

$$\lambda_1+\lambda_2\cos 2\theta\simeq 0$$

in the absence of a magnetic field. The director orientation then approaches an asymptotic value given by

$$\cos 2\theta_0=-(\lambda_1/\lambda_2), \quad (3.6.15)$$

or, assuming Parodi's relation (3.3.15),

$$\tan^2\theta_0=\mu_3/\mu_2. \quad (3.6.16)$$

The amount of fluid flowing per second

$$Q=2\pi\int_0^R v(r)r\,dr$$

and the apparent viscosity

$$\eta=\pi aR^4/8Q.$$

By scaling the radius and the time as $r'=hr$ and $t'=kt$ respectively, with $k=h^2$, it is easily shown[(46)] that *in the absence of a magnetic field Q/R is a unique function of aR^3*. Consequently η plotted versus Q/R should be a universal curve for all tube radii and flow rates. This has been confirmed experimentally[47] (fig. 3.6.2). The apparent viscosity increases slightly at lower pressure gradients.

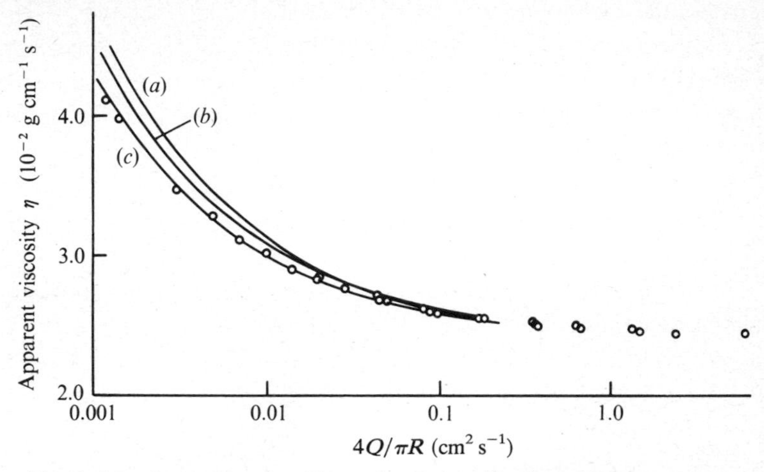

Fig. 3.6.2. Apparent viscosity η for Poiseuille flow of PAA at 122 °C (homeotropic wall orientation) plotted against the ratio of the flow rate to the radius of the tube. ○ experimental data of Fischer and Fredrickson,[47] (*a*) values obtained from table I, (*b*) values obtained from table II with $\mu_1 = 0$, (*c*) values obtained from table II with $\mu_1 = -0.038$. (After Tseng, Silver and Finlayson.[48])

Table I	*Table II*
$\mu_1 = 0.043$ (g cm^{-1} s^{-1})	$\mu_1 = 0$ or -0.038 (g cm^{-1} s^{-1})
$\mu_2 = -0.069$	$\mu_2 = -0.068$
$\mu_3 = -0.002$	$\mu_3 = 0.000$
$\mu_4 = 0.068$	$\mu_4 = 0.068$
$\mu_5 = 0.047$	$\mu_5 = 0.048$
$\mu_6 = -0.023$	$\mu_6 = -0.020$
$\mu_2 - \mu_3 = \lambda_1 = -0.067$	$\mu_2 - \mu_3 = \lambda_1 = -0.068$
$\mu_5 - \mu_6 = \lambda_2 = 0.0705$	$\mu_5 - \mu_6 = \lambda_2 = 0.068$
$\theta_0 = 9.1°$	$\theta_0 = 0$

Equations (3.6.11) and (3.6.12) with $H = 0$ have been solved numerically by Tseng, Silver and Finlayson[48] assuming the boundary conditions

$$v(R) = 0,$$

$$\theta(R) = -\frac{\pi}{2}$$

and

$$\theta(0)=0.$$

The computed apparent viscosity versus shear rate is shown in fig. 3.6.2. The agreement with the experimental data can be seen to be quite good. Calculations have also been made of the effect of an axial magnetic field.[49] The apparent viscosity decreases appreciably in the presence of the field but an experimental study of this effect has not yet been reported. Curves for η, the orientation and velocity profiles as functions of shear rate and magnetic field are presented in figs. 3.6.3, 3.6.4 and 3.6.5.

Imperfect alignment or weak anchoring can be a serious source of error in the determination of η.[47] For example, in fig. 3.6.3 η for $\theta(R)=-(\pi/4)$ is seen to be significantly lower than that for $\theta(R)=-(\pi/2)$.

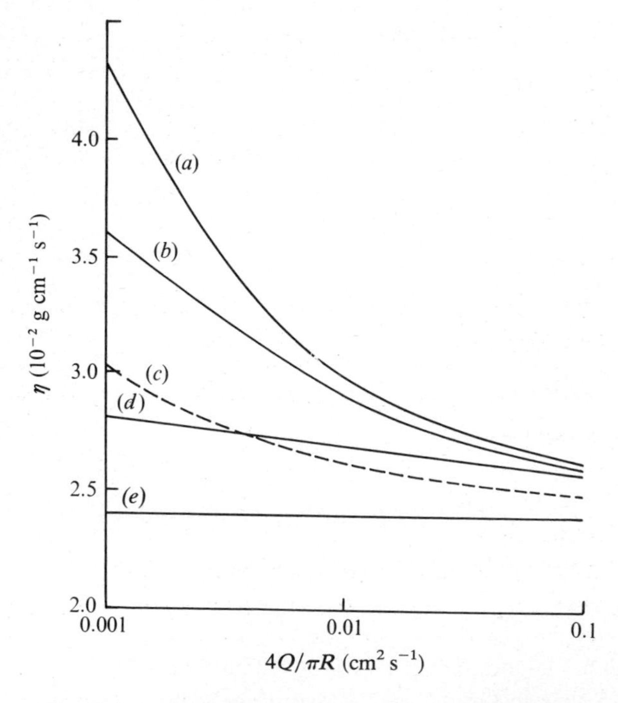

Fig. 3.6.3. Apparent viscosity η for Poiseuille flow of PAA versus $4Q/\pi R$ computed for a tube of radius $R=55.5$ μm. The values of $\chi_a H^2$ and $\theta(R)$, the orientation at the wall, are respectively (*a*) 0, $-\pi/2$, (*b*) 1.23, $-\pi/2$, (*c*) 0, $-\pi/4$, (*d*) 24.2, $-\pi/2$, (*e*) ∞, $-\pi/2$ or 0, 0. (After reference 49.)

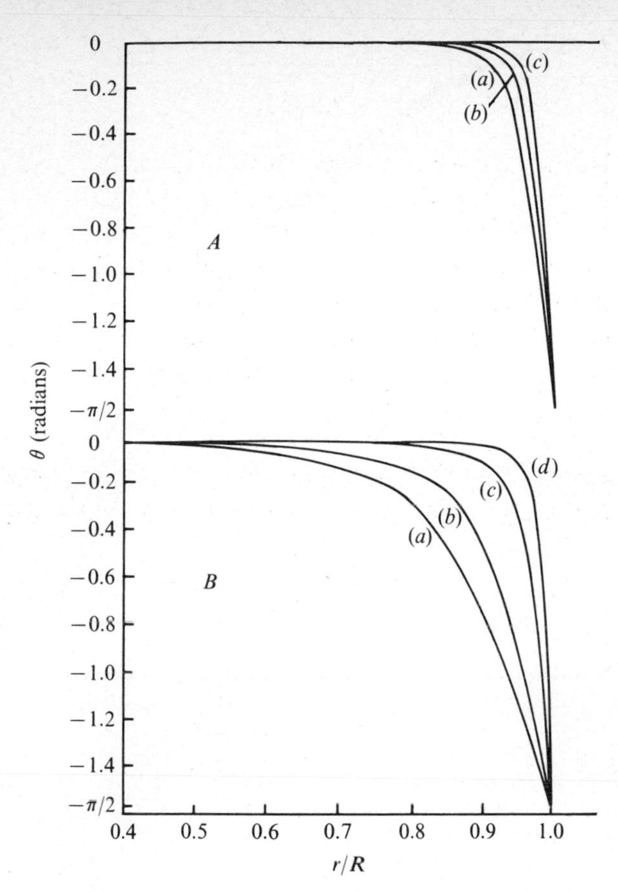

Fig. 3.6.4. Orientation profile for Poiseuille flow for different shear rates. Wall alignment homeotropic. The values of $\chi_a H^2$ are 24.2 for A and 1.23 for B. The values of $4Q/\pi R$ in A are (*a*) 0.003045, (*b*) 0.03354, (*c*) 0.1345, and in B they are (*a*) 0.001245, (*b*) 0.004167, (*c*) 0.032438, and (*d*) 0.1372. (After reference 49.)

3.6.4. Shear flow

Consider now the steady laminar flow of a nematic fluid between two parallel plates. If the flow is along x and the velocity gradient along y the components of the velocity and the director are

$$v_x = v(y), \qquad v_y = 0, \qquad v_z = 0,$$

$$n_x = \cos\theta(y), \qquad n_y = \sin\theta(y), \qquad n_z = 0,$$

where θ is the angle made by the director with x. Proceeding as before, the differential equations for the velocity and the director orientation in

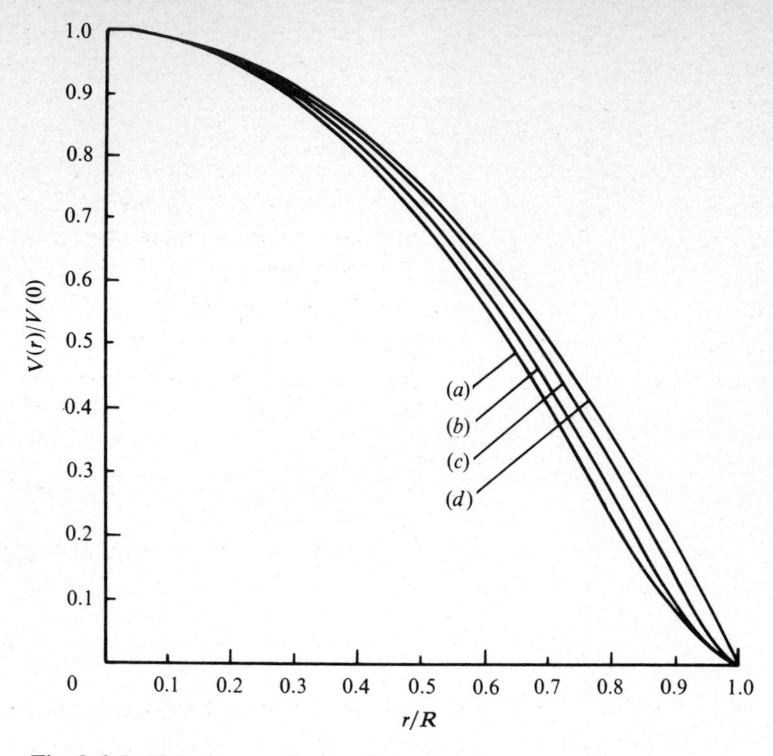

Fig. 3.6.5. Velocity profiles for Poiseuille flow; the curves (*a*) and (*b*) are for different values of $4Q/\pi R$ but for the same field, while for curves (*b*) and (*c*) $4Q/\pi R$ is nearly the same but the fields are different. The values of $\chi_a H^2$ and $4Q/\pi R$ are respectively (*a*) 1.23, 0.001245, (*b*) 1.23, 0.004167, (*c*) 24.2, 0.003045 and (*d*) $\chi_a H^2 = \infty$ or $4Q/\pi R = \infty$ (truly parabolic). (After reference 49.)

the presence of a magnetic field $(H_x, H_y, 0)$ turn out to be[7]

$$2f(\theta)\frac{\mathrm{d}^2\theta}{\mathrm{d}y^2}+\frac{\mathrm{d}f}{\mathrm{d}\theta}\left(\frac{\mathrm{d}\theta}{\mathrm{d}y}\right)^2+(ay+c)\frac{\lambda_1+\lambda_2\cos 2\theta}{g(\theta)}$$

$$+2\chi_a(H_x\cos\theta+H_y\sin\theta)(H_y\cos\theta-H_x\sin\theta)=0, \quad (3.6.17)$$

$$g(\theta)\frac{\mathrm{d}v}{\mathrm{d}y}=ay+c, \quad (3.6.18)$$

where $f(\theta)$ and $g(\theta)$ are defined by (3.6.13) and (3.6.14) respectively and $a = \mathrm{d}p/\mathrm{d}x$ is the pressure gradient and c the constant shear stress applied to the fluid. If both plates are stationary $a \neq 0$, and taking $y = 0$ half-way between the plates, $c = 0$. On the other hand, if there is no pressure gradient, and the flow is caused by one of the plates moving at a uniform velocity in its own plane, $a=0$ and $c\neq 0$. Equations (3.6.17) and (3.6.18) can be solved to yield the apparent viscosity, velocity and orientation profiles under different boundary conditions.[50]

If the plate separation is large enough, boundary effects and elastic terms can justifiably be neglected in (3.6.17). For large shear rates and zero magnetic field the director orientation approaches the value θ_0, defined by (3.6.15). But if a magnetic field of moderate strength is applied, the orientation profile is modified slightly. Gähwiller[51] has studied this behaviour by measuring the change in birefringence. He used capillaries (5 cm long) of rectangular cross section (4 mm × 0.3 mm) and measured the rate of flow due to a pressure gradient. If **H** is along the flow direction and shear rates are large, we obtain from (3.6.17) and (3.6.18)

$$(\lambda_1+\lambda_2\cos 2\theta)\frac{\mathrm{d}v}{\mathrm{d}y}=\chi_a H^2\sin 2\theta. \tag{3.6.19}$$

Gähwiller assumed that the velocity profile may be approximated by the usual parabolic dependence

$$v(y)=v_0[1-(4y^2/d^2)], \tag{3.6.20}$$

where v_0 is the velocity half-way between the plates at $y=0$ and d the plate separation. From (3.6.19) and (3.6.20)

$$\tan\theta\simeq 8\mu_3 v_0 y/\chi_a H^2 d^2. \tag{3.6.21}$$

The phase difference between the two perpendicularly polarized components when light is incident normal to the plates is then

$$\delta_\parallel=(2\pi/\lambda)\int_{-d/2}^{d/2}(n(\theta)-n_o)\,\mathrm{d}y,$$

where

$$1/n^2=(\sin^2\theta/n_o^2)+(\cos^2\theta/n_e^2),$$

where n_o and n_e are the ordinary and extraordinary refractive indices. If the magnetic field is applied along the velocity gradient,

$$\cot\theta\simeq 8\mu_2 v_0 y/\chi_a H^2 d^2 \tag{3.6.22}$$

and $\delta_\perp$ can similarly be calculated. In the absence of a magnetic field,

$$\delta_0=(2\pi/\lambda)(n(\theta_0)-n_o)d.$$

Thus the three measurements δ_0, $\delta_\parallel$ and $\delta_\perp$ yield μ_3/μ_2, μ_3/χ_a and μ_2/χ_a.

At high magnetic fields the experiment reduces in effect to Miesowicz's method except that Gähwiller extended it to arbitrary orientations of the magnetic field. If θ is the angle between the director and the flow direction and φ that between the projection of the director on the yz plane and the velocity gradient,

$$\begin{aligned}\eta(\theta, \varphi) &= \mu_1 \sin^2\theta \cos^2\theta \cos^2\varphi - \tfrac{1}{2}\mu_2 \sin^2\theta \cos^2\varphi \\ &\quad + \tfrac{1}{2}\mu_3 \cos^2\theta + \tfrac{1}{2}\mu_4 + \tfrac{1}{2}\mu_5 \sin^2\theta \cos^2\varphi + \tfrac{1}{2}\mu_6 \cos^2\theta \\ &= \eta_1 \cos^2\theta + (\eta_2 + \mu_1 \cos^2\theta) \sin^2\theta \cos^2\varphi \\ &\quad + \eta_3 \sin^2\theta \sin^2\varphi.\end{aligned}$$

By choosing θ and φ appropriately, one can determine η_1, η_2, η_3 and μ_1.

Using these two sets of data, Gähwiller was able to determine all 5 independent viscosity coefficients as well as χ_a. Some of his results are presented in figs. 3.6.6 and 3.6.7.

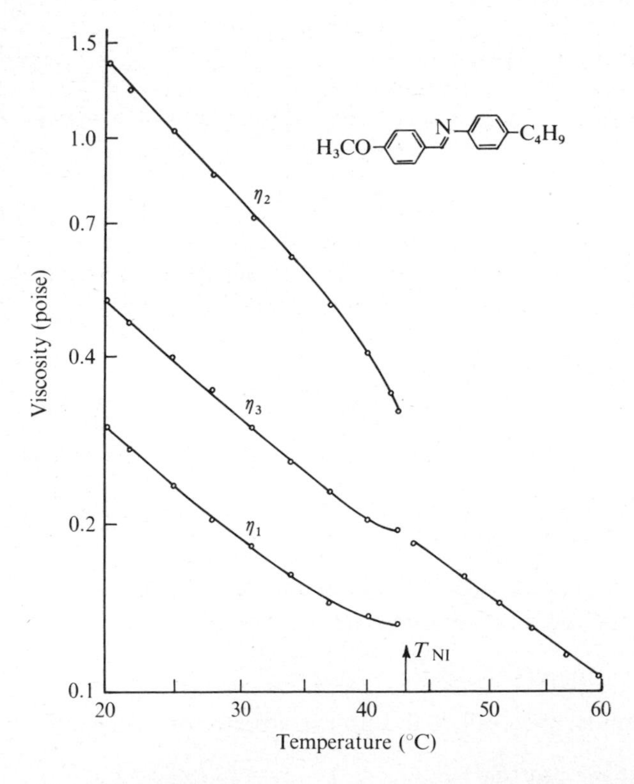

Fig. 3.6.6. The viscosity coefficients η_1, η_2 and η_3 of MBBA as functions of temperature. The temperature scale is linear in T^{-1}. (After Gähwiller.[51])

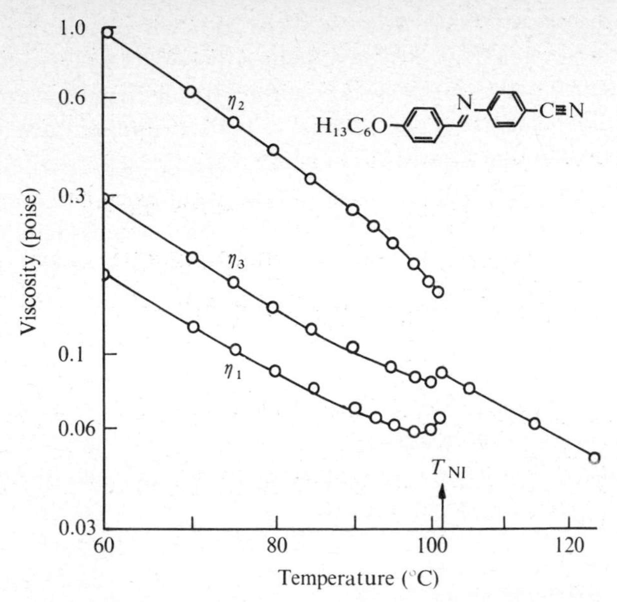

Fig. 3.6.7. The viscosity coefficients η_1, η_2 and η_3 of *p*-n-hexyloxybenzylidene-*p*′-aminobenzonitrile (HBAB) as functions of temperature. The temperature scale is linear in T^{-1}. (After Gähwiller.[51])

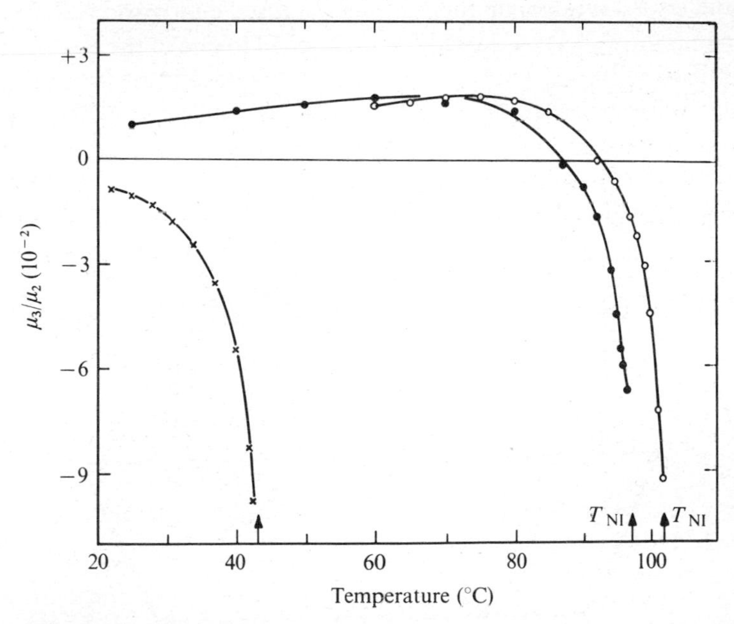

Fig. 3.6.8. The ratio μ_3/μ_2 versus temperature for MBBA (×), HBAB (○) and 1:1:1 molar mixture of HBAB, *p*-n-butoxybenzylidene-*p*′-aminobenzonitrile and *p*-n-octanoyloxybenzylidene-*p*′-aminobenzonitrile (●). (After Gähwiller.[51])

Gähwiller discovered that materials of strong positive dielectric anisotropy exhibit an unusual type of instability at lower temperatures. The limiting value of the director orientation at relatively high flow rates, θ_0, decreases rapidly and becomes zero at a certain temperature, below which the steady laminar flow breaks up into many irregular domains (fig. 3.6.8). He interpreted this phenomenologically as a reversal in the sign of μ_3 below the critical temperature. The effect is probably associated with the very strong permanent dipolar interaction in such materials (see § 2.5.3). On the other hand, in MBBA in which the permanent dipolar interaction is negligible the flow alignment is stable throughout the nematic range.

When $|\lambda_1/\lambda_2| > 1$, (3.6.15) is evidently not valid. Steady state solutions now exist in which the director is rotated by many turns on going from one plate to the other.[80] Such a cycloidal configuration induced by flow has not yet been observed experimentally.

3.7. Reflexion of shear waves

The viscosity coefficients may also be determined by studying the reflexion of ultrasonic shear waves at a solid–nematic interface. The technique was developed by Martinoty and Candau.[52] A thin film of a nematic liquid crystal is taken on the surface of a fused quartz rod with obliquely cut ends (fig. 3.7.1). A quartz crystal bonded to one of the ends generates a transverse wave. At the solid–nematic interface there is a transmitted wave, which is rapidly attenuated, and a reflected wave which is received at the other end by a second quartz crystal. The reflexion coefficient, obtained by measuring the amplitudes of reflexion with and without the nematic sample, directly yields the effective coefficient of viscosity.

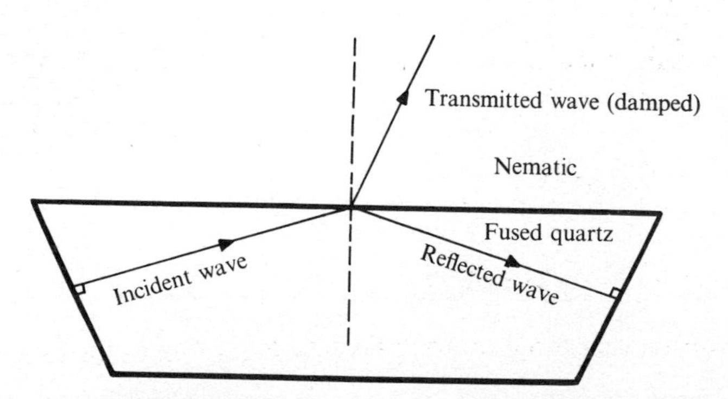

Fig. 3.7.1. Experimental arrangement for studying the reflexion of ultrasonic shear waves at a solid–nematic interface.

To explain the principle of the method we shall consider the simpler case of normal incidence. Let a shear wave be incident along z with its vibration direction along x, and let the nematic director be anchored firmly at the interface ($z=0$) along y. The only non-vanishing component of the velocity of the fluid is $v_x = v$, and the velocity gradient is along z. From (3.3.5) the stress across the interface is (neglecting director gradients)

$$t_{zx} = \tfrac{1}{2}\mu_4(\partial v/\partial z), \tag{3.7.1}$$

and from (3.3.1)

$$t_{zx,z} = \rho\dot{v}. \tag{3.7.2}$$

Now the velocities associated with the incident, reflected and transmitted waves may be written as

$$v_{\mathrm{i}} = A_{\mathrm{i}} \exp(-\Gamma_{\mathrm{s}} z) \exp(\mathrm{i}\omega t),$$

$$v_{\mathrm{r}} = A_{\mathrm{r}} \exp(\Gamma_{\mathrm{s}} z) \exp(\mathrm{i}\omega t),$$

$$v_{\mathrm{t}} = A_{\mathrm{t}} \exp(-\Gamma_{\mathrm{n}} z) \exp(\mathrm{i}\omega t),$$

respectively, where Γ may be complex and the subscripts s and n stand for solid and nematic. Therefore, in the nematic medium

$$t_{zx} = -Z_{\mathrm{n}} A_{\mathrm{t}} \exp(-\Gamma_{\mathrm{n}} z) \exp(\mathrm{i}\omega t),$$

where $Z_{\mathrm{n}} = \tfrac{1}{2}\mu_4\Gamma_{\mathrm{n}}$ is the mechanical impedance, and

$$\rho\dot{v}_{\mathrm{t}} = \tfrac{1}{2}\mu_4(\partial^2 v/\partial z^2).$$

We thus have

$$\tfrac{1}{2}\mu_4\Gamma_{\mathrm{n}}^2 = \mathrm{i}\rho\omega,$$

or

$$\Gamma_{\mathrm{n}} = (1+\mathrm{i})(\rho\omega/\mu_4)^{1/2} \tag{3.7.3}$$

and

$$Z_{\mathrm{n}} = \tfrac{1}{2}(1+i)(\rho\omega\mu_4)^{1/2}. \tag{3.7.4}$$

The complex reflexion coefficient is given by

$$r = \frac{A_{\mathrm{r}}}{A_{\mathrm{i}}} = \frac{Z_{\mathrm{s}} - Z_{\mathrm{n}}}{Z_{\mathrm{s}} + Z_{\mathrm{n}}},$$

where Z_s is the mechanical impedance of the solid which may be assumed to be real quantity. If

$$r = |r| \exp(-i\theta),$$

$$Z_n = Z_s \frac{1 - |r| \exp(i\theta)}{1 + |r| \exp(i\theta)}$$

$$= Z_s \frac{1 - |r|^2 + 2i|r| \sin\theta}{1 + |r|^2 + 2|r| \cos\theta}. \tag{3.7.5}$$

Since according to (3.7.4) the real and imaginary parts of Z_n are of equal magnitude, it follows that

$$\sin\theta = \frac{1 - |r|^2}{2|r|}. \tag{3.7.6}$$

Thus a measurement of $|r|$ at once gives Z_n, which in turn yields μ_4. Similarly, for the director oriented along x at the interface

$$Z_n = \tfrac{1}{2}(1 + i)\{\rho\omega[(\mu_3 + \mu_4 + \mu_6) - \mu_3(1 + (\lambda_2/\lambda_1))]\}^{1/2},$$

and for the director along z at the interface

$$Z_n = \tfrac{1}{2}(1 + i)\{\rho\omega[(\mu_4 + \mu_5 - \mu_2) + \mu_2(1 - (\lambda_2/\lambda_1))]\}^{1/2}.$$

The theory may easily be generalized to the case of oblique incidence. Martinoty and Candau found that the viscosity coefficients determined by the ultrasonic technique compare fairly well with those derived from capillary flow.

3.8. Dynamics of the Freedericksz effect

3.8.1. Twist deformation

We shall now extend the theory of the Freedericksz effect to study the dynamical behaviour when the magnetic field is switched on or off suddenly.[53] The analysis is particularly simple for a twist deformation (fig. 3.4.1(*b*)) because the torsion exerted on the director does not result in a translational motion of the centres of gravity of the molecules. Neglecting director inertia in (3.3.2) we obtain the following equation of motion for this geometry:

$$k_{22} \frac{\partial^2 \theta}{\partial z^2} + \chi_a H^2 \sin\theta \cos\theta = \lambda_1 \frac{\partial \theta}{\partial t},$$

where λ_1 is the twist viscosity defined by (3.3.14). If θ is small,

$$\xi^2 \frac{\partial^2 \theta}{\partial z^2} + \theta - \tfrac{2}{3}\theta^3 = \lambda \frac{\partial \theta}{\partial t}, \tag{3.8.1}$$

where

$$\xi^2 = k_{22}/\chi_a H^2, \qquad \lambda = \lambda_1/\chi_a H^2.$$

The most general solution satisfying the boundary conditions $\theta = 0$ at $z = \pm d/2$ is

$$\theta = \sum_n C_n(t) \cos\{(2n+1)(\pi z/d)\}.$$

Neglecting higher harmonics and remembering that θ has a maximum value θ_m at $z = 0$, we take

$$\theta = \theta_m(t) \cos(\pi z/d).$$

Equation (3.8.1) then gives

$$[1-(H_c^2/H^2)]\theta_m - (\theta_m^3/2) = \lambda \frac{\partial \theta_m}{\partial t},$$

or

$$\theta_m^2(t) = \frac{\theta^2(\infty)}{1+\{\theta^2(\infty)/\theta^2(0)-1\} \exp\{-(2\chi_a H_c^2/\lambda_1)((H^2/H_c^2)-1)t\}}.$$

Thus $\theta_m(t)$ attains the value $\theta(\infty)$ with a time constant $\tau(H)$ given by

$$\tau^{-1}(H) = (\chi_a/\lambda_1)(H^2 - H_c^2). \tag{3.8.2}$$

If the field is now reduced to a value less than H_c, the decay rate is still given by the same expression only with a negative sign.

If the field is switched off from $H > H_c$ to zero,

$$k_{22}\frac{\partial^2 \theta}{\partial z^2} = \lambda_1 \frac{\partial \theta}{\partial t},$$

and the decay rate

$$\tau^{-1}(0) = (k_{22}/\lambda_1)(\pi^2/d^2).$$

The twist viscosity can be determined from a measurement of τ.[53]

Typically, $\tau(0)$ for a film of 25 μm is about 10^{-1} s. This gives an idea of the order of magnitude of the relaxation time for most nematic liquid crystal devices. An exception to this is the so-called 'fast turn off' electrohydrodynamic mode which gives rise to the oscillating domains (see § 3.10); in this case the spatial periodicity of the pattern is only about 1 μm so that the relaxation time is just a few milliseconds.

3.8.2. Homeotropic to planar transition: backflow effects

The other two geometries used in the Freedericksz experiment are more interesting as they result in a new effect, namely, hydrodynamic flow induced by orientational deformation. This is the inverse of the more familiar property of flow alignment that has been discussed at length in previous sections.

Let us consider the homeotropic to planar transition (fig. 3.4.1(c)). For this geometry, $\mathbf{n}=(\sin\theta, 0, \cos\theta)$, $\theta=\theta(z)$, $\mathbf{v}=v_x(z)$, and $v_z(z)=0$. Setting $k_{11}=k_{33}=k$, $\cos\theta\approx 1$ and $\sin\theta\approx\theta$, we get from (3.3.2)

$$\xi^2\frac{\partial^2\theta}{\partial z^2}+\theta=\lambda\frac{\partial\theta}{\partial t}+\lambda\lambda'\frac{\partial v_x}{\partial z}, \tag{3.8.3}$$

where

$$\xi^2=k/\chi_a H^2, \qquad \lambda=\lambda_1/\chi_a H^2 \quad \text{and} \quad \lambda'=(\lambda_2-\lambda_1)/2\lambda_1.$$

Neglecting inertial effects, (3.3.1) reduces in the present case to

$$t_{zx,z}=0,$$

where, using (3.3.3),

$$t_{zx}=\tfrac{1}{2}(\mu_4+\mu_5-\mu_2)\frac{\partial v_x}{\partial z}+\mu_2\frac{\partial\theta}{\partial t},$$

neglecting squares and higher powers of θ. Therefore

$$\frac{\partial}{\partial z}\left(a\frac{\partial v_x}{\partial z}+b\frac{\partial\theta}{\partial t}\right)=0, \tag{3.8.4}$$

where $a=\frac{1}{2}(\mu_4+\mu_5-\mu_2)$ and $b=\mu_2$.

The boundary conditions for $\mathbf{n}$ and $\mathbf{v}$ are $\theta=0$ and $v_x=0$ at $z=\pm d/2$. The solutions are of the form

$$\theta=\theta_0[\cos qz-\cos(qd/2)]\exp(t/\tau), \tag{3.8.5}$$

$$v_x=v_0[\sin qz-(2z/d)\sin(qd/2)]\exp(t/\tau). \tag{3.8.6}$$

Substituting in (3.8.3) and (3.8.4) we obtain the following relations,

$$\frac{\lambda}{\tau}(1-A)=1-\frac{4\psi^2}{\pi^2(H/H_c)^2} \tag{3.8.7}$$

and

$$\left(\frac{H}{H_c}\right)^2=\frac{4\psi^2}{\pi^2}\frac{(\psi/A)-\tan\psi}{\psi-\tan\psi}, \tag{3.8.8}$$

where $A=\lambda'b/a$ and $\psi=qd/2$. From numerical calculations, Pieranski, Brochard and Guyon[53] have shown that the relaxation rate can be expressed in the form

$$\tau^{-1}(H)=(\chi_a/\lambda_1^*)(H^2-H_c^2),$$

where the apparent viscosity λ_1^* is now strongly dependent on H.

The translational velocity (3.8.6) has two components in the simplest case: one linear in z and the other oscillatory, the wavelength of the latter diminishing with increasing value of the final field H. The transient velocity profile is illustrated schematically in fig. 3.8.1. The effect of this *backflow* is to relax the constraints, i.e., to reduce the apparent viscosity.

In the planar to homeotropic transition (fig. 3.4.1(a)) backflow effects are not usually so pronounced near the threshold. In this geometry, the torque exerted by the director on an elementary volume of the fluid is

$$\Gamma=\tfrac{1}{2}(\lambda_1+\lambda_2)\Omega,$$

and

$$\lambda'=(\lambda_1+\lambda_2)/2\lambda_1,$$

where Ω is the angular velocity (fig. 3.8.2). In many nematic liquids, λ_1 and λ_2 are of opposite signs and of comparable magnitude (see, for

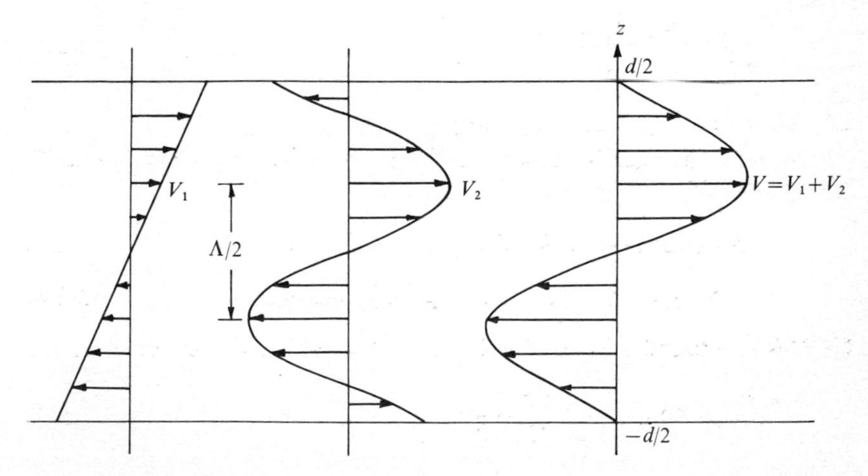

Fig. 3.8.1. Velocity profile in the homeotropic to planar transition (see fig. 3.4.1(c)). The velocity has two components, one linear in z and the other oscillatory, the wavelength Λ of the latter diminishing with increasing value of H.

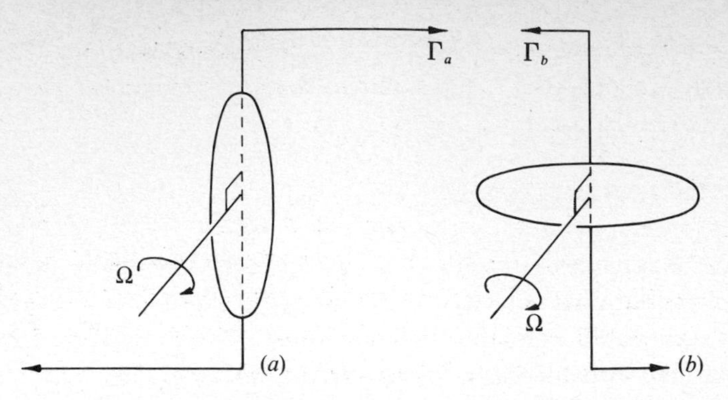

Fig. 3.8.2. Torques acting on an elementary volume of the fluid when the molecules are rotating with angular velocity Ω: (*a*) homeotropic to planar transition, $\Gamma_a = \frac{1}{2}(\lambda_1 - \lambda_2)\Omega$; (*b*) planar to homeotropic transition, $\Gamma_b = \frac{1}{2}(\lambda_1 + \lambda_2)\Omega$. As a rule, $|\Gamma_a| \gg |\Gamma_b|$.

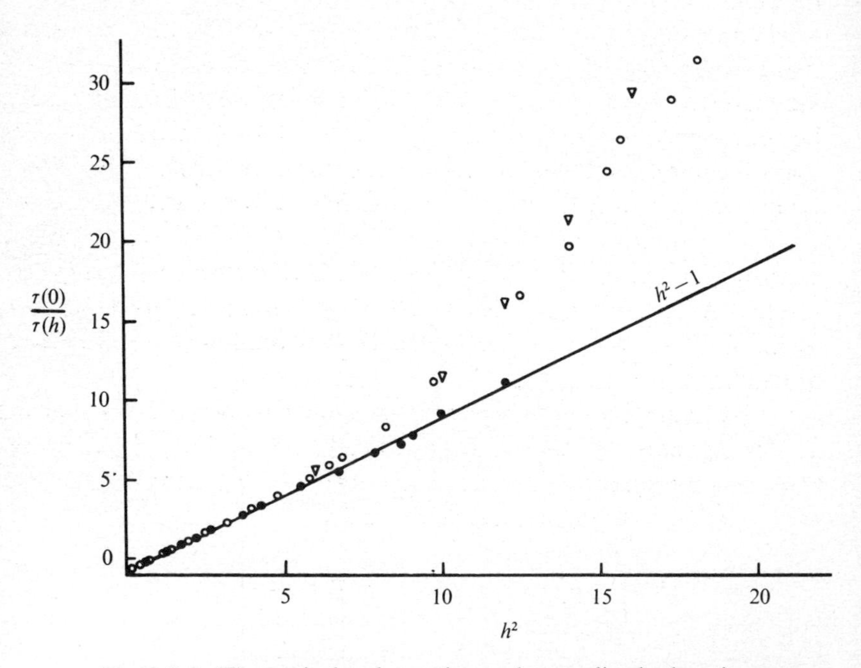

Fig. 3.8.3. Theoretical and experimental normalized relaxation rates as functions of $h^2 = H^2/H_c^2$. Open circles and triangles are respectively the experimental and theoretical values for the homeotropic to planar transition. Closed circles are the experimental values for the planar to homeotropic transition. The line represents the variation expected from (3.8.9). The departure from this line for the homeotropic to planar transition is a consequence of backflow. Material: MBBA. (After Pieranski, Brochard and Guyon.[53])

example, legend of fig. 3.6.2) so that Γ is small. In other words, A is small and the solutions of (3.8.7) and (3.8.8) reduce to $\psi = \pi/2$ and

$$\tau^{-1}(H) = (1/\lambda)(1 - H_c^2/H^2)$$

as for the twist geometry.

The marked difference in the relaxation rates for the two geometries at higher fields confirms the existence of backflow as predicted by the Leslie equations (fig. 3.8.3). Backflow has also been studied by direct observation of the motion of disclination walls separating two regions of opposite tilt in a film which is subjected to a magnetic field.[54]

3.9. Light scattering

3.9.1. Orientational fluctuations

One of the most striking features of the nematic liquid crystal is its turbidity. From systematic observations of the Rayleigh scattering from oriented samples, Chatelain[55] showed that the scattered intensity is strongly depolarized and exhibits a marked angular variation. An early model put forward to explain this phenomenon assumed the medium to be composed of swarms, about 1 μm in diameter, of aligned molecules, the orientations of the different swarms being uncorrelated. However, it is now well established[22,56–58] that the light scattering can be interpreted rigorously in terms of the small amplitude orientational fluctuations as described by the continuum theory. The intensity of this scattering turns out to be very much larger than that arising from the density fluctuations in the fluid, so much so that the latter contribution can be neglected altogether.

Let ω_0, $\mathbf{k}_0$ and $\mathbf{i}$ be respectively the angular frequency, wave-vector and unit polarization vector of the incident beam and ω_1, $\mathbf{k}_1$ and $\mathbf{f}$ the corresponding quantities for the scattered beam. The scattering process is associated with an angular frequency change

$$\omega = \omega_0 - \omega_1$$

and a wave-vector change

$$\mathbf{q} = \mathbf{k}_0 - \mathbf{k}_1.$$

We define the differential scattering cross-section per unit volume of the scatterer, per unit solid angle (Ω), per unit angular frequency change as

$$\frac{\mathrm{d}^2\sigma}{\mathrm{d}\Omega\,\mathrm{d}\omega} = \pi^2\lambda^{-4}\iint_{-\infty}^{+\infty} \langle\delta\varepsilon^2\rangle \exp[\mathrm{i}(\mathbf{q}\cdot\mathbf{r} - \omega t)]\,\mathrm{d}\mathbf{r}\,\mathrm{d}t \tag{3.9.1}$$

where $\lambda = 2\pi c/\omega_0$ is the vacuum wavelength and

$$\langle\delta\varepsilon^2\rangle = \langle\delta\varepsilon(0, 0)\delta\varepsilon(\mathbf{r}, t)\rangle$$

is the mean square fluctuation of the dielectric constant at a given point **r** and time *t*.

For the uniaxial nematic medium, the dielectric tensor at any point **r** can be written as (see § 2.3.1)

$$\varepsilon_{if} = \bar{\varepsilon} + \varepsilon_a(n_i n_f - \tfrac{1}{3}) \tag{3.9.2}$$

where $\bar{\varepsilon} = (\varepsilon_\parallel + 2\varepsilon_\perp)/3$ is the mean dielectric constant (at *optical* frequencies), $\varepsilon_a = \varepsilon_\parallel - \varepsilon_\perp$ is the dielectric anisotropy assumed to be $\ll \bar{\varepsilon}$, $n_i = \mathbf{n}\cdot\mathbf{i}$ and $n_f = \mathbf{n}\cdot\mathbf{f}$. An electric vector polarized along **i** induces a displacement $D_f = \varepsilon_{if}E_i$ along **f**. The director at **r**

$$\mathbf{n}(\mathbf{r}) = \mathbf{n}_0 + \delta\mathbf{n}$$

where, to a first approximation, $\delta\mathbf{n}\cdot\mathbf{n}_0 = 0$, since we assume the fluctuations to be of small amplitude. Thus, neglecting density fluctuations (i.e., assuming $\bar{\varepsilon}$ and ε_a to be constant) we have from (3.9.2)

$$\delta\varepsilon_{if} = \varepsilon_a[f_0(\delta\mathbf{n}\cdot\mathbf{i}) + i_0(\delta\mathbf{n}\cdot\mathbf{f})]$$

where $i_0 = \mathbf{n}_0\cdot\mathbf{i}$ and $f_0 = \mathbf{n}_0\cdot\mathbf{f}$. Therefore

$$\langle\delta\varepsilon^2\rangle = \langle\delta\varepsilon_{fi}(0, 0)\delta\varepsilon_{if}(\mathbf{r}, t)\rangle. \tag{3.9.3}$$

The fluctuations can be analysed into Fourier components. For a given Fourier component of wave-vector **q**, we may conveniently resolve $\delta\mathbf{n}$ into two components δn_1 and δn_2, the former in the **q**, $\mathbf{n}_0$ plane and the latter perpendicular to it (fig. 3.9.1). Let us therefore introduce two unit vectors

$$\mathbf{e}_2 = (\mathbf{n}_0 \times \mathbf{q})q_\perp^{-1},$$

$$\mathbf{e}_1 = \mathbf{e}_2 \times \mathbf{n}_0,$$

where $q_\perp$ is the component of **q** perpendicular to $\mathbf{n}_0$. Defining $i_\alpha = \mathbf{e}_\alpha\cdot\mathbf{i}$ and $f_\alpha = \mathbf{e}_\alpha\cdot\mathbf{f}$ ($\alpha = 1, 2$),

$$\delta\varepsilon_{if} = \varepsilon_a \sum_{\alpha=1,2} \delta n_\alpha(i_\alpha f_0 + f_\alpha i_0).$$

Hence

$$\langle\delta\varepsilon^2\rangle = \varepsilon_a^2\left[\sum_\alpha (i_\alpha f_0 + f_\alpha i_0)^2\langle\delta n_\alpha(0, 0)\delta n_\alpha(\mathbf{r}, t)\rangle\right]. \tag{3.9.4}$$

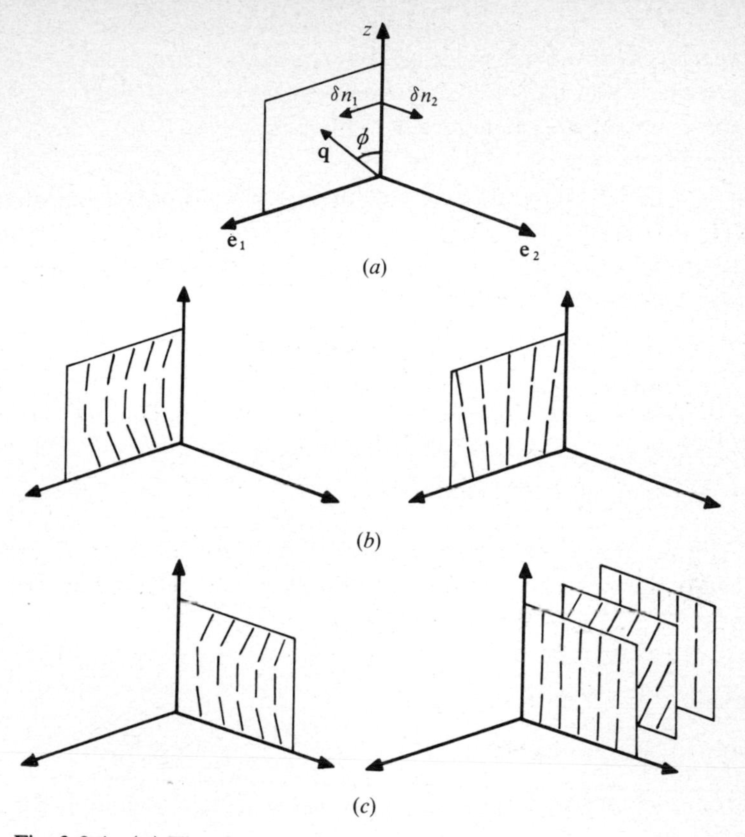

Fig. 3.9.1. (*a*) The two uncoupled modes δn_1 and δn_2; (*b*) components of the deformation in the δn_1 mode, bend and splay; (*c*) components of the deformation in the δn_2 mode, bend and twist.

The scattering cross-section is then

$$\mathrm{d}^2\sigma/\mathrm{d}\Omega\,\mathrm{d}\omega = \pi^2\lambda^{-4}\varepsilon_\mathrm{a}^2\sum_\alpha (i_\alpha f_0 + i_0 f_\alpha)^2 I_\alpha(\mathbf{q}, \omega), \tag{3.9.5}$$

where

$$I_\alpha(\mathbf{q}, \omega) = \int_{-\infty}^{\infty} \langle \delta n_\alpha(-\mathbf{q}, 0)\delta n_\alpha(\mathbf{q}, t)\rangle \exp(-\mathrm{i}\omega t)\,\mathrm{d}t \tag{3.9.6}$$

and

$$\delta n_\alpha(\mathbf{q}, t) = \int \delta n_\alpha(\mathbf{r}, t) \exp(\mathrm{i}\mathbf{q}\cdot\mathbf{r})\,\mathrm{d}\mathbf{r}. \tag{3.9.7}$$

Here $(i_\alpha f_0 + i_0 f_\alpha)^2$ is a purely geometric factor while I_α is a correlation function which describes the power spectrum of the fluctuations.

For $\alpha=1$, the director vibrates in the $\mathbf{n}_0$, $\mathbf{q}$ plane and the mode is a superposition of bend and splay. For $\alpha=2$, the vibration is normal to the $\mathbf{n}_0$, $\mathbf{q}$ plane and the mode is a superposition of bend and twist. The two modes are shown schematically in fig. 3.9.1.

3.9.2. Intensity and angular dependence of the scattering

It is of interest to consider first the intensity of the scattered light integrated over time (or frequency). The differential scattering cross-section is then

$$\frac{\mathrm{d}\sigma}{\mathrm{d}\Omega}=\pi^2\lambda^{-4}\varepsilon_\mathrm{a}^2\sum_\alpha(i_\alpha f_0+i_0 f_\alpha)^2\langle\delta n_\alpha^2(\mathbf{q})\rangle \tag{3.9.8}$$

where

$$\langle\delta n_\alpha^2(\mathbf{q})\rangle=\langle\delta n(-\mathbf{q})\delta n(\mathbf{q})\rangle.$$

Writing $\delta\mathbf{n}(\mathbf{r},t)=\delta\mathbf{n}\exp[\mathrm{i}(\mathbf{q}\cdot\mathbf{r}-\omega t)]$ and substituting in (3.3.6) we obtain the free energy of elastic deformation of the system

$$F=\tfrac{1}{2}\sum_\alpha k_\alpha(\mathbf{q})|\delta n_\alpha|^2,$$

where

$$k_\alpha(\mathbf{q})=k_{33}q_z^2+k_{\alpha\alpha}q_\perp^2.$$

From the equipartition theorem (which is certainly valid in the present problem)

$$|\delta n_\alpha|^2=\frac{k_\mathrm{B}T}{k_{33}q_z^2+k_{\alpha\alpha}q_\perp^2}, \tag{3.9.9}$$

where k_B is the Boltzmann constant. To get an idea of the order of magnitude of the scattering cross-section let us suppose that $k_{11}\simeq k_{22}\simeq k_{33}=k$; then

$$\frac{\mathrm{d}\sigma}{\mathrm{d}\Omega}\approx\left(\frac{\pi\varepsilon_\mathrm{a}}{\lambda^2}\right)^2\frac{k_\mathrm{B}T}{kq^2}. \tag{3.9.10}$$

On the other hand, the cross-section due to density fluctuation is given by the well known formula

$$\frac{\mathrm{d}\sigma'}{\mathrm{d}\Omega}=\frac{\pi^2}{\lambda^4}(\mathbf{i}\cdot\mathbf{f})^2\langle\delta\rho^2\rangle\left(\frac{\partial\bar\varepsilon}{\partial\rho}\right)^2$$

$$=\left(\frac{\pi}{\lambda^2}\rho\frac{\partial\bar\varepsilon}{\partial\rho}\right)^2\beta k_\mathrm{B}T(\mathbf{i}\cdot\mathbf{f})^2, \tag{3.9.11}$$

where β is the isothermal compressibility. Taking $\rho(\partial\bar{\varepsilon}/\partial\rho)\sim\varepsilon_a$,

$$\frac{d\sigma'}{d\sigma}\sim\beta kq^2.$$

Typically $\beta\sim10^{-11}$ cm^2 dyne^{-1}, $k\sim10^{-6}$ dyne, $q\sim10^4$ cm^{-1}, so that

$$\frac{d\sigma'}{d\sigma}\sim10^{-8}.$$

Thus the director fluctuations make the predominant contribution to the light scattering, as was first pointed out by de Gennes.[56]

A comparison of the polarization factors in (3.9.5) and (3.9.11) at once explains why the light scattering from a nematic liquid crystal is strongly depolarized. The angular dependence of the light scattering is also accounted for in a straightforward manner. Let us, for example, consider one of the geometries used by Chatelain (fig. 3.9.2). The incident and scattered beams are both normal to z; the incident beam is linearly polarized in the plane of scattering while the scattered beam is polarized along z, the optic axis of the medium. If the angle of scattering is φ,

$$\tfrac{1}{2}q_\perp\simeq k_0\sin(\varphi/2),\qquad q_z=0,\qquad i_1=\cos(\varphi/2),$$
$$i_2=\sin(\varphi/2),\qquad i_z=f_1=f_2=0,\qquad f_z=1.$$

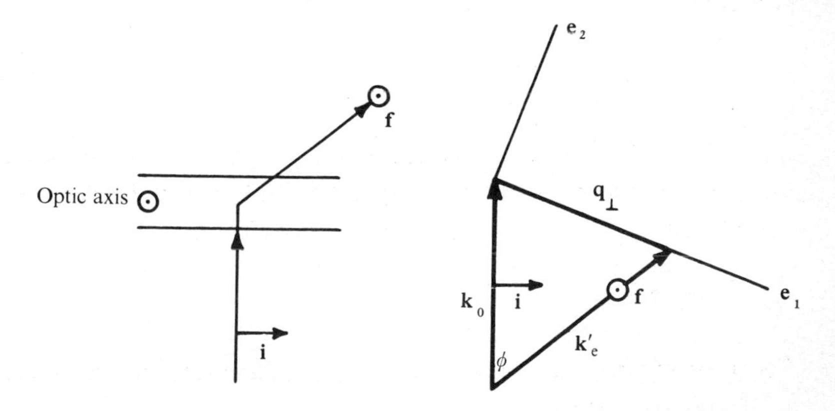

Fig. 3.9.2. A typical experimental configuration used by Chatelain in his light-scattering studies: **k** and **i** are the wave-vector and unit polarization vector of the incident light, **k**$'$ and **f** the corresponding quantities for the scattered light, the suffixes e and o denote the extraordinary and ordinary polarizations with respect to the optic axis of the nematic medium and $\mathbf{q}_\perp$ is the wave-vector change on scattering.

Therefore

$$\frac{d\sigma}{d\Omega}=\left(\frac{\pi\varepsilon_a}{\lambda^2}\right)^2\left(\frac{k_BT\cos^2(\varphi/2)}{k_{11}q^2}+\frac{k_BT\sin^2(\varphi/2)}{k_{22}q^2}\right)$$

$$=\left(\frac{\varepsilon_a}{4\lambda}\right)^2\frac{k_BT}{k_{11}}\left(\cot^2\frac{\varphi}{2}+\frac{k_{11}}{k_{22}}\right). \tag{3.9.12}$$

Fig. 3.9.3 compares this relation with the experimental data of Chatelain and as can be seen the agreement is good. In principle a measurement of the intensity of the scattering and its angular variation offers a method of determining the elastic constants.

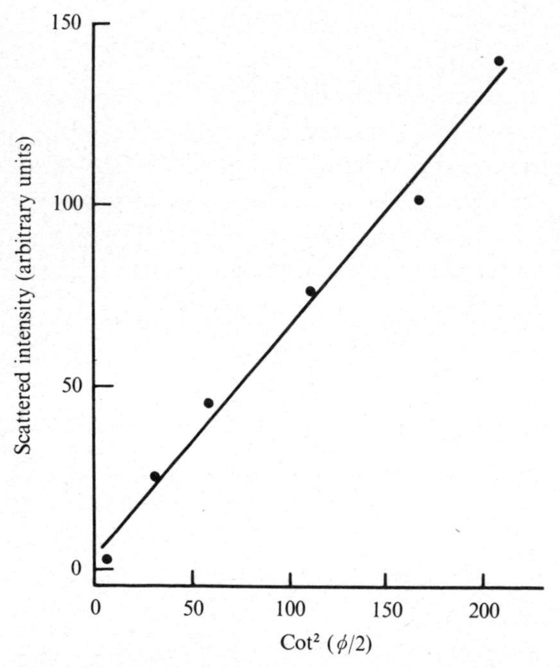

Fig. 3.9.3. Angular dependence of the intensity of scattering for PAA. Circles give the experimental values of Chatelain and the line represents the theoretical variation. (After de Gennes.[56])

3.9.3. Eigenmodes and the frequency spectrum of the scattered light

For a given mode of vibration of angular frequency ω and wave-vector $\mathbf{q}$, we can write

$$\delta\mathbf{n}(\mathbf{r},t)=\delta\mathbf{n}\exp[\mathrm{i}(\mathbf{q}\cdot\mathbf{r}-\omega t)],$$

$$\mathbf{v}(\mathbf{r},t)=\mathbf{v}\exp[\mathrm{i}(\mathbf{q}\cdot\mathbf{r}-\omega t)].$$

Substituting in the basic equations of motion (3.3.1) and (3.3.2) we obtain

$$\mathrm{i}\omega\rho v_k = \mathscr{F}_k - q_k(q_j\mathscr{F}_j/q^2), \tag{3.9.13}$$

$$\lambda_1 N_k + \lambda_2 n_j d_{kj} - \frac{\partial F}{\partial n_k} + \left(\frac{\partial F}{\partial n_{k,j}}\right)_{,j} = 0. \tag{3.9.14}$$

We ignore inertial effects as well as terms quadratic in $\delta\mathbf{n}$. $\mathscr{F}_k = \mathrm{i}q_j t'_{jk}$, where t'_{jk} is the viscous part of the stress tensor defined by (3.3.5), and F is the elastic energy density. As the fluid is supposed to be incompressible $v_{k,k}=0$ or $q_k v_k = 0$. If v_1, v_2, v_z are the components of $\mathbf{v}$ in the frame e_1, e_2, z, we have

$$d_{zz} = \mathrm{i}q_z v_z, \qquad d_{11} = \mathrm{i}q_\perp v_1 = -\mathrm{i}q_z v_z,$$

$$d_{z1} = \mathrm{i}(q_\perp^2 - q_z^2)v_z/2q_\perp = \tfrac{1}{2}\mathrm{i}(q_\perp v_z + q_z v_1),$$

$$d_{z2} = \tfrac{1}{2}\mathrm{i}q_z v_2, \qquad d_{12} = \tfrac{1}{2}\mathrm{i}q_\perp v_2,$$

$$\omega_{12} = -\tfrac{1}{2}\mathrm{i}q_\perp v_2, \qquad \omega_{1z} = -\mathrm{i}q^2 v_z/2q_\perp,$$

$$\omega_{2z} = \tfrac{1}{2}\mathrm{i}q_z v_2,$$

where $q^2 = q_\perp^2 + q_z^2$. Substituting in (3.9.13) and (3.9.14)

$$\left.\begin{aligned} &\mathrm{i}C_1(\mathbf{q})v_z + (\mathrm{i}\lambda_1\omega + k_1(\mathbf{q}))\delta n_1 = 0,\\ &\mathrm{i}C_2(\mathbf{q})v_2 + (\mathrm{i}\lambda_1\omega + k_2(\mathbf{q}))\delta n_2 = 0,\end{aligned}\right\} \tag{3.9.15}$$

$$\left.\begin{aligned} &v_z(\rho\omega - \mathrm{i}P_1(\mathbf{q})) - \mathrm{i}\omega Q_1(\mathbf{q})\delta n_1 = 0,\\ &v_2(\rho\omega - \mathrm{i}P_2(\mathbf{q})) - \mathrm{i}\omega Q_2(\mathbf{q})\delta n_2 = 0,\end{aligned}\right\} \tag{3.9.16}$$

where

$$C_1(\mathbf{q}) = \tfrac{1}{2}[(\lambda_1+\lambda_2)q_\perp^2 + (\lambda_1-\lambda_2)q_z^2]/q_\perp,$$

$$C_2(\mathbf{q}) = \tfrac{1}{2}qz(\lambda_2-\lambda_1) = \mu_2 qz,$$

$$k_\alpha(\mathbf{q}) = k_{33}q_z^2 + k_{\alpha\alpha}q_\perp^2,$$

$$P_1(\mathbf{q}) = \tfrac{1}{2}(\eta_s q_\perp^4 + \eta_v q_z^4 + \eta_m q_\perp^2 q_z^2)/q^2,$$

$$P_2(\mathbf{q}) = \tfrac{1}{2}(\eta_4 q_\perp^2 + \eta_v q_z^2),$$

$$Q_1(\mathbf{q}) = (q_\perp/q^2)(\mu_3 q_\perp^2 - \mu_2 q_z^2),$$

$$Q_2(\mathbf{q}) = \mu_2 q_z,$$

$$\eta_s = \mu_3 + \mu_4 + \mu_6,$$

$$\eta_v = -\mu_2 + \mu_4 + \mu_5,$$

$$\eta_m = 2(\mu_1+\mu_4) + \mu_5 + \mu_6 + \mu_3 - \mu_2.$$

For compatibility of (3.9.15) and (3.9.16) we have the vanishing of the determinant, and therefore

$$(\rho\omega - iP_\alpha(\mathbf{q}))(i\omega\lambda_1 + k_\alpha(\mathbf{q})) - C_\alpha(\mathbf{q})Q_\alpha(\mathbf{q}) = 0$$

or

$$i\lambda_1\rho\omega^2 + \omega(\rho k_\alpha(\mathbf{q}) + \lambda_1 P_\alpha(\mathbf{q}) - C_\alpha(\mathbf{q})Q_\alpha(\mathbf{q}))$$
$$-iP_\alpha(\mathbf{q})k_\alpha(\mathbf{q}) = 0, \qquad (3.9.17)$$

which has two roots. Typically, $k \sim 10^{-6}$ dyne, $\rho \sim 1$ gm cm^{-3}, $\eta \sim \mu \sim$ 0.1 poise,

$$P(\mathbf{q}) \sim \eta q^2, \qquad C(\mathbf{q}) \sim Q(\mathbf{q}) \sim \eta q, \qquad k_\alpha(\mathbf{q}) \sim kq^2.$$

Consequently $\rho k_\alpha(\mathbf{q})$ is negligible compared to $\lambda_1 P_1(\mathbf{q})$ and $C_\alpha(\mathbf{q})Q_\alpha(\mathbf{q})$, and therefore the two roots of (3.9.17) are

$$\omega_{s\alpha} \simeq i\frac{k_\alpha(\mathbf{q})P_\alpha(\mathbf{q})}{\lambda_1 P_\alpha(\mathbf{q}) - C_\alpha(\mathbf{q})Q_\alpha(\mathbf{q})}, \qquad (3.9.18)$$

$$\omega_{f\alpha} \simeq i\frac{\lambda_1 P_\alpha(\mathbf{q}) - C_\alpha(\mathbf{q})Q_\alpha(\mathbf{q})}{\rho\lambda_1}. \qquad (3.9.19)$$

The subscripts s and f denote 'slow' and 'fast', for

$$\omega_s \sim ikq^2/\eta,$$

$$\omega_f \sim i\eta q^2/\rho,$$

and

$$\omega_s/\omega_f \sim \rho k/\eta^2 \ll 1.$$

We observe that both modes are purely dissipative (non-propagating); ω_s involves the elastic constants while ω_f does not. The slow mode therefore represents the relaxation of the orientational motion of the director, while the fast mode may be looked upon as the diffusion of a vorticity but one in which there is no torque on the molecules. The light scattering, being dependent primarily on the orientational fluctuations, is controlled entirely by the slow mode, as we shall proceed to show.

We have resolved the director fluctuations into δn_1 and δn_2 which from the symmetry of the problem can be seen to be uncoupled. If G_1 and G_2 are the forces responsible for these tilts in the director, then in a first order theory $\delta n_1 = \chi_1 G_1$, $\delta n_2 = \chi_2 G_2$, where χ_1, χ_2 are susceptibilities; more generally

$$\delta n_\alpha(\mathbf{q}, \omega) = \chi_\alpha(\mathbf{q}, \omega)G_\alpha.$$

We have seen that due to the thermal agitation, there are spontaneous fluctuations in **n** whose mean square value is defined by

$$I_\alpha(\mathbf{q}, \omega) = \langle \delta n_\alpha(-\mathbf{q}, 0)\, \delta n_\alpha(\mathbf{q}, \omega)\rangle.$$

According to the fluctuation-dissipation theorem,[59] the relation between I_α and χ_α is

$$I_\alpha(\mathbf{q}, \omega) = \frac{k_B T}{\pi\omega}\mathrm{Im}(\chi_\alpha(\mathbf{q}, \omega)), \tag{3.9.20}$$

where Im stands for the imaginary part.

We can derive an expression for $I_\alpha(\mathbf{q}, \omega)$ from the equation of motion for the director in the presence of an external field:

$$\lambda_1 N_k + \lambda_2 n_j d_{kj} - \frac{\partial F}{\partial n_k} + \left(\frac{\partial F}{\partial n_{k,j}}\right)_{,j} - G_k = 0.$$

Accordingly (3.9.15) becomes

$$\mathrm{i}C_1(\mathbf{q})v_z + \delta n_1[\mathrm{i}\omega\lambda_1 + k_1(\mathbf{q})] = G_1,$$

$$\mathrm{i}C_2(\mathbf{q})v_2 + \delta n_2[\mathrm{i}\omega\lambda_1 + k_2(\mathbf{q})] = G_2.$$

Along with (3.9.16), this can be simplified to obtain

$$\chi_\alpha(\mathbf{q}, \omega) = \frac{\rho\omega - \mathrm{i}P_\alpha(\mathbf{q})}{(\rho\omega - \mathrm{i}P_\alpha(\mathbf{q}))(\mathrm{i}\lambda_1\omega + k_\alpha(\mathbf{q})) - C_\alpha(\mathbf{q})Q_\alpha(\mathbf{q})\omega}$$

or

$$I_\alpha(\mathbf{q}, \omega) = \frac{k_B T}{\pi\lambda_1 u_{f\alpha}\rho}\left[\frac{P_\alpha(\mathbf{q})}{\omega^2 + u_{s\alpha}^2} - \frac{C_\alpha(\mathbf{q})Q_\alpha(\mathbf{q})}{(\omega^2 + u_{f\alpha}^2)\lambda_1}\right], \tag{3.9.21}$$

where

$$u_{s\alpha} = -\mathrm{i}\omega_{s\alpha} \quad \text{and} \quad u_{f\alpha} = -\mathrm{i}\omega_{f\alpha}. \tag{3.9.22}$$

Thus I_α is a superposition of two Lorentzians. However as $u_{f\alpha} \gg u_{s\alpha}$, we may ignore the second term in the square brackets of (3.9.21) and rewrite the power spectrum as

$$I_\alpha(\mathbf{q}, \omega) = \frac{k_B T}{\pi k_\alpha(\mathbf{q})}\left(\frac{u_{s\alpha}}{\omega^2 + u_{s\alpha}^2}\right). \tag{3.9.23}$$

The light scattering is therefore determined entirely by the slow mode. The integrated intensity

$$\int I_\alpha(\mathbf{q}, \omega)\, \mathrm{d}\omega = \frac{k_B T}{k_\alpha(\mathbf{q})},$$

in agreement with (3.9.10) Also from (3.9.18), (3.9.19) and (3.9.22)

$$\frac{k_1(\mathbf{q})}{u_{s1}} = \lambda_1 - \frac{2(\mu_3 q_\perp^2 - \mu_2 q_z^2)^2}{\eta_s q_\perp^4 + \eta_m q_\perp^2 q_z^2 + \eta_v q_z^4}, \tag{3.9.24}$$

$$\frac{k_2(\mathbf{q})}{u_{s2}} = \lambda_1 - \frac{2\mu_2^2 q_z}{\mu_4 q_\perp^2 + \eta_v q_z^2}. \tag{3.9.25}$$

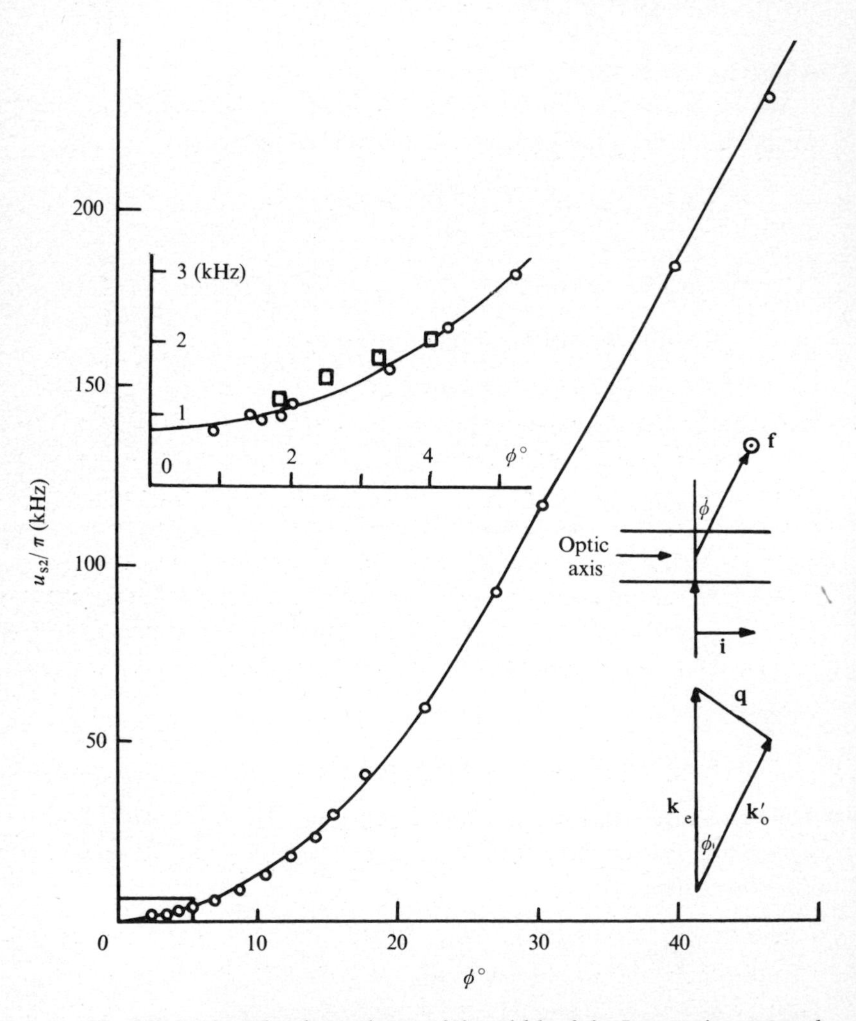

Fig. 3.9.4. Angular dependence of the width of the Lorentzian spectral density for mode 2 in PAA at 125 °C. Open circles denote experimental values in the $[\mathbf{k}_e, \mathbf{k}'_o]$ configuration and open squares the values in the $[\mathbf{k}_o, \mathbf{k}'_e]$ case. The curve is obtained by a least squares fit with the theory. (After the Orsay Liquid Crystals Group.[58])

The integrated intensity gives the elastic constants k_{ii} while the half width yields $u_{s\alpha}$. It is therefore possible to measure the viscosity coefficients from an analysis of the scattered light using appropriate geometries.

As an example, we present in fig. 3.9.4 a convenient geometry for isolating mode 2. The director is aligned parallel to the walls of the glass plates. The incident beam is polarized parallel to the director and the scattered beam perpendicular to it. If n_e and n_o are the extraordinary and ordinary indices of the liquid crystal, and φ the scattering angle,

$$k_e = 2\pi n_e/\lambda, \qquad k'_o = 2\pi n_o/\lambda,$$

$$q_z = k'_o \sin\varphi, \qquad q_\perp = k_e - k'_o \cos\varphi.$$

For small angles of scattering,

$$\varphi \to 0, \quad q_z \to 0, \quad q_\perp \to 2\pi(n_e - n_o)/\lambda,$$

and from (3.9.25)

$$u_{s2} = k_{22} q_\perp^2/\lambda_1,$$

which enables λ_1 to be determined. By going to higher angles, it is possible to obtain $\mu_4/2\mu_2^2$ and $\eta_v/2\mu_2^2$. A typical curve for the angular dependence of the width of the Lorentzian spectral density is shown in fig. 3.9.4. The experiments are rather difficult because of the large amount of stray radiation, especially in the forward direction, arising from defects in the alignment of the specimen. However, by very careful techniques using a laser light beat spectrometer, and employing various geometries Léger-Quercy[58] has been able to determine four viscosity coefficients of PAA, μ_2 to μ_5. The values of η_1, η_2 and η_3 calculated from these coefficients are in reasonably good agreement with those determined by Miesowicz (see table 3.6.1).

3.10. Electrohydrodynamics

3.10.1. The experimental situation

From dielectric studies in the radio frequency region,[60] it has long been known that PAA is negatively anisotropic, i.e., $\varepsilon_a = \varepsilon_\parallel - \varepsilon_\perp < 0$. However, in a number of early investigations on the effect of an external DC electric field it was noticed that the PAA molecules align themselves parallel to the field, rather than perpendicular to it as would be expected of a material of negative dielectric anisotropy. This observation gave rise to some controversy in the 1930s but it has since been confirmed by the systematic experiments of Carr,[61] who proved that the anomalous alignment is due to the anisotropy of the electrical conductivity of the liquid crystal. His studies showed that there is a critical frequency of the applied field below which the alignment of PAA is anomalous, and that this frequency increases with the conductivity of the material, ranging from 2 to 100 kHz in the samples examined by him.

Macroscopic motion of the fluid induced by electric fields was observed many years ago by Freedericksz and Zolina,[19] Tsvetkov and Mikhailov,[62] Bjornstahl[63] and Naggiar.[64] Tsvetkov also noted that the flow decreases with increasing frequency of the applied field, and probably recognized the fact that the phenomenon may be connected in some way with the electrical conductivity. More recently, Williams[65] discovered that a thin layer of a nematic material of negative dielectric anisotropy between conducting glass plates forms regular striations when a DC voltage of sufficient magnitude is applied. At higher voltages, the regular pattern gives way to turbulence accompanied by intense scattering of light, which has come to be known as *dynamic scattering* and has found practical applications in display devices.[66] Similar observations have been reported by other authors.[67]

The experimental arrangement for observing the *Williams domains* is shown in fig. 3.10.1. The nematic film of negative dielectric anisotropy (for example, PAA or MBBA) is aligned with the director parallel to the glass surfaces which are coated with a transparent conducting material. When a DC or low frequency AC field is applied between the transparent electrodes, there appear above a threshold voltage a regular set of parallel striations perpendicular to the initial unperturbed orientation of the director (plate 10(*a*)). Dust particles are seen to undergo periodic motion in the field of view proving that the domains are due to hydrodynamic motion. The distortion of the director orientation caused by this motion results in a focussing action for light polarized parallel to the

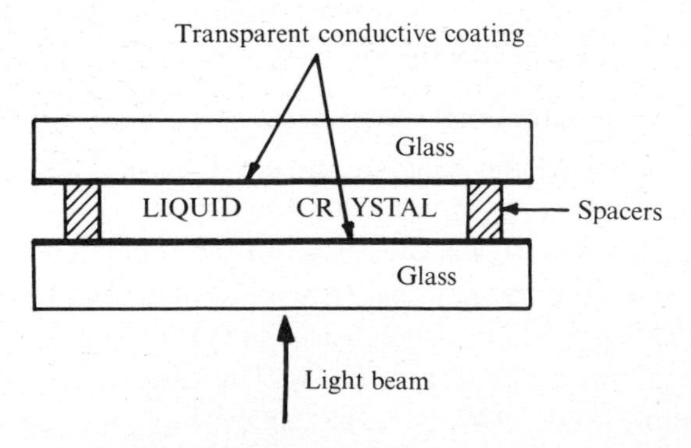

Fig. 3.10.1. Experimental arrangement for observing Williams domains.

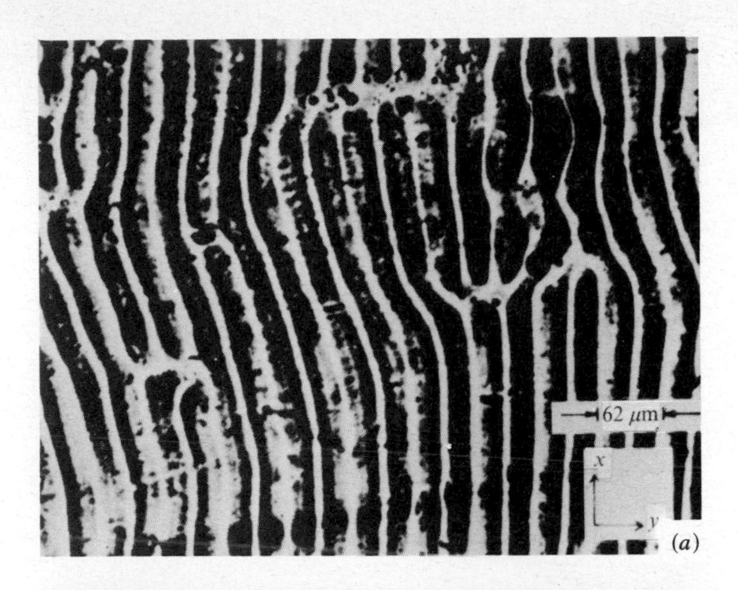

Plate 10. Electrohydrodynamic alignment patterns in nematic liquid crystals. (*a*) Williams domains in a 38 μm thick sample of *p*-azoxyanisole. 7.8 V, 100 Hz. (Penz.[81]) (*b*) Chevron pattern of oscillating domains in MBBA. Sample thickness ~100 μm. Distance between bright lines ~5 μm. 260 V, 120 Hz. (Orsay Liquid Crystals Group.[69])

director[68] (fig. 3.10.2). This is responsible for the appearance of a set of bright lines with a spacing approximately equal to the film thickness when the microscope is focussed at the top surface. The lines are shifted by about half the spacing when the focal plane is moved down to the bottom surface. The pattern disappears when the light is polarized perpendicular to the director.

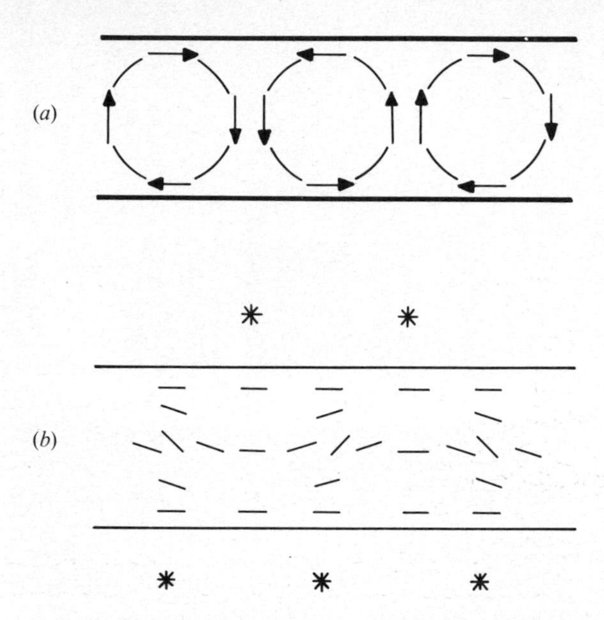

Fig. 3.10.2. (*a*) Flow and (*b*) orientation patterns of Williams domains. The periodic orientation pattern and the consequent refractive index variation has a focussing action for light polarized in the plane of the paper. This gives rise to the bright domain lines as indicated by the stars above and below the sample. (After Penz.[68])

The threshold voltage is usually a few volts and is practically independent of the sample thickness. It is however strongly dependent on the frequency[69] (fig. 3.10.3). There is a cut-off frequency ω_c above which the domains do not appear, the value of ω_c increasing with the conductivity of the sample. Below ω_c, i.e., in the so-called *conduction regime*, the regular Williams pattern becomes unstable at about twice the threshold voltage and the medium goes over to the dynamic scattering mode. Above ω_c, in the *dielectric regime*, another type of domain pattern is observed. Parallel striations, again perpendicular to the initial orientation of the director but with a much shorter spacing (a few microns), are formed in the midplane of the sample. When the field is increased very slightly above the

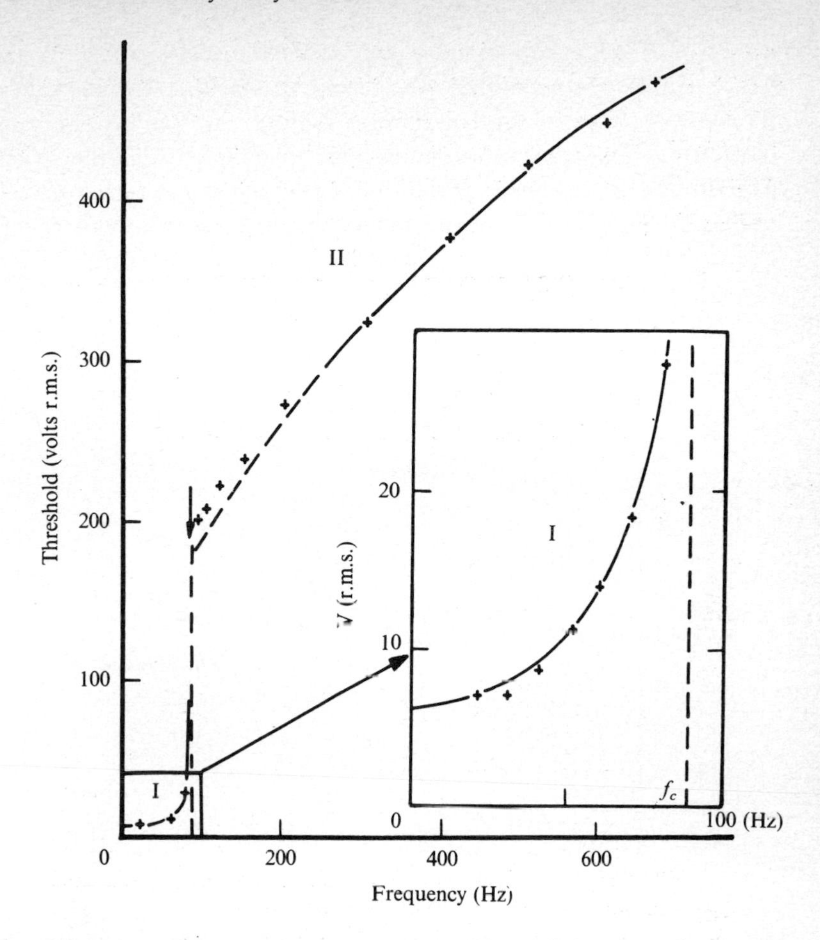

Fig. 3.10.3. Threshold voltage of the AC instabilities versus frequency for MBBA. Sample thickness 100 μm. Region I: conducting regime (stationary Williams domains); region II: dielectric regime ('chevrons'). Full line is the theoretical curve. The cut-off frequency $f_c = 89$ Hz. (After the Orsay Liquid Crystals Group.[69])

threshold, the striations bend and move to form a *chevron pattern* (plate 10(*b*)). In this regime, the threshold is determined by a *critical field strength* rather than a critical voltage. Both the threshold field strength and the spatial periodicity of the pattern are frequency dependent – the former increasing with frequency (as $\omega^{1/2}$) and the latter diminishing with it. The relaxation time of the oscillating chevron pattern is a few milliseconds while that of the stationary Williams pattern is typically about 0.1 s for a thickness of 25 μm. The oscillating domain regime is therefore sometimes called the *fast turn-off mode*. The chevron pattern also gives way to turbulence at about twice the threshold field. An applied magnetic field parallel to the initial orientation of the director increases the threshold voltage in the conduction regime, but has no effect in the dielectric regime except to increase the spacing between the striations. The threshold curve has a pronounced sigmoid shape with square wave excitation (fig. 3.10.4) indicating that at high electric fields there is a quenching of the conductive instability even when $\omega < \omega_c$.

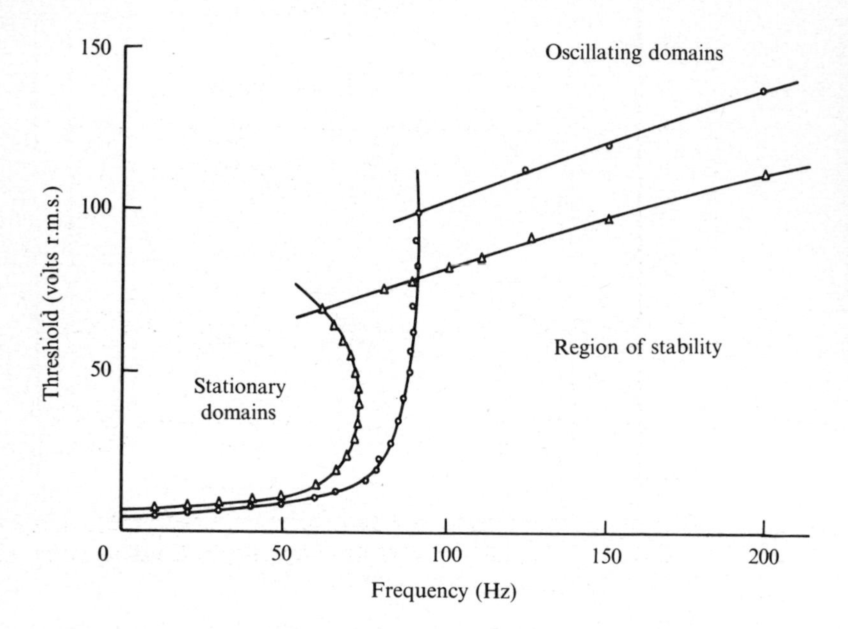

Fig. 3.10.4. Threshold voltage versus frequency for MBBA. Sample thickness 50 μm. ○ sinusoidal excitation; △ square wave excitation. (After the Orsay Liquid Crystals Group.[69])

DC and very low frequency AC voltages produce electrohydrodynamic instabilities in the isotropic phase also ($T > T_{NI}$), the threshold being comparable to that in the nematic phase. It has been suggested that this is due to charge injection at the electrodes. A frequency of about 10 Hz is usually enough to suppress this effect showing that charge injection is not the primary mechanism for the AC field instabilities in the nematic phase.

If the initial alignment of the director is homeotropic, domains as well as turbulence can be produced. The threshold voltage for the domains is somewhat higher than in the case of parallel alignment, but the patterns persist even when the voltage is reduced to a lower value.

We have so far discussed only materials of negative dielectric anisotropy. Electrohydrodynamic distortions are observed even in weakly positive materials,[70] but only when the initial orientation of the director is perpendicular to the applied field. Striations appear above a threshold voltage but vanish at still higher voltages and there is no dynamic scattering. The frequency dependence of the threshold voltage is shown in fig. 3.10.5.

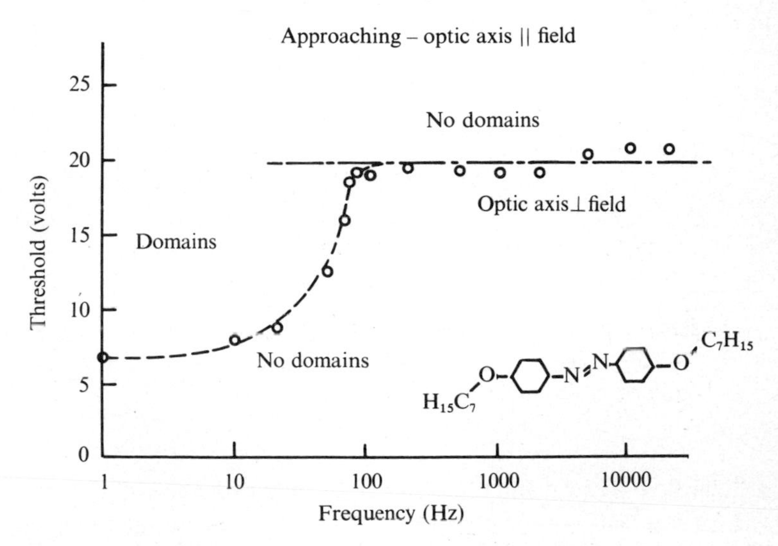

Fig. 3.10.5. Frequency dependence of the threshold voltage in *p,p′*-di-n-heptoxyazobenzene, a nematic of positive dielectric anisotropy. (After Gruler and Meier.[70])

3.10.2. Helfrich's theory

The basic mechanism for the electric-field-induced instabilities is now quite well understood. The current carriers in the nematic phase are ions whose mobility is greater along the preferred axis of the molecules than perpendicular to it. The ratio of the conductivities $\sigma_{\|}/\sigma_{\perp}$ is usually about 4/3. Because of this anisotropy, space charge can be formed by ion segregation in the liquid crystal itself, as was first pointed by Carr.[61] The manner in which the space charge can build up due to a bend fluctuation is shown schematically in fig. 3.10.6. The applied field acts on the charges to give rise to material flow in alternating directions which in turn exerts a torque on the molecules. This is reinforced by the dielectric torque due to the transverse field created by the space charge distribution. Under appropriate conditions, these torques may offset the normal elastic and dielectric torques and the system may become unstable. The resulting

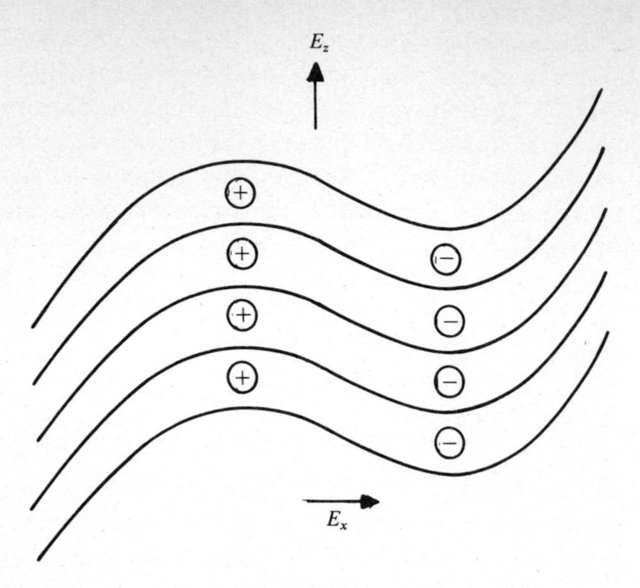

Fig. 3.10.6. Charge segregation in an applied field E_z caused by a bend fluctuation in a nematic of positive conductivity anisotropy. The resulting transverse field is E_x.

cellular flow pattern and director orientations are sketched in fig. 3.10.2. Even a conductivity of the order $10^{-9}\ \Omega^{-1}\ \mathrm{cm}^{-1}$ is enough to produce this type of fluid motion. Indeed unless very special precautions are taken, the impurity conductivity is usually greater than this value.

With a DC field, there may be injection of charge carriers at the solid–liquid interface but its role in the electrohydrodynamics of the nematic phase is not yet fully understood. However, as remarked earlier, a frequency of about 10 Hz is enough to suppress charge injection. We shall therefore neglect it in the present discussion.

We shall now outline the theory of electrohydrodynamic instabilities proposed by Helfrich[71] and extended by Dubois-Violette, de Gennes and Parodi[72] and Smith *et al.*[73] We consider a nematic film of thickness d lying in the xy plane and subjected to an electric field E_z along z. Let the initial unperturbed orientation of the director be along x, and let there also be a stabilizing magnetic field along the same direction. We consider a bend fluctuation in which the director is in the xz plane and makes an angle φ with x. We ignore wall effects and assume that the deflection φ is a function of x only.

Due to the anisotropy of conductivity, space charges will develop as indicated in fig. 3.10.6 till the transverse electric field stops the transverse current. The local transverse field in the steady state is easily seen to be

$$E_x = -\left(\frac{\sigma_a \cos\varphi \sin\varphi}{\sigma_\parallel \cos^2\varphi + \sigma_\perp \sin^2\varphi}\right) E_z,$$

where $\sigma_a = \sigma_\parallel - \sigma_\perp > 0$, $\sigma_\parallel$ and $\sigma_\perp$ being the principal electrical conductivities along and perpendicular to the local director axis. Also, as soon as the electric field is applied there will be a transverse field $E_{\varepsilon x}$ due to the dielectric anisotropy. Since the transverse displacement is zero,

$$E_{\varepsilon x} = -[\varepsilon_a \cos\varphi \sin\varphi/(\varepsilon_\parallel \cos^2\varphi + \varepsilon_\perp \sin^2\varphi)]E_z.$$

The space charge per unit area produced on any plane normal to the x axis is

$$\pm(\varepsilon_\parallel \cos^2\varphi + \varepsilon_\perp \sin^2\varphi)(E_x - E_{\varepsilon x})/4\pi,$$

the positive and negative signs standing for the two sides of the plane under consideration. The applied field E_z acting on these charges causes a flow along the z axis; the resulting shear stress is evidently

$$t'_{xz} = (\varepsilon_\parallel \cos^2\varphi + \varepsilon_\perp \sin^2\varphi)(E_x - E_{\varepsilon x})E_z/4\pi. \qquad (3.10.1)$$

From the Ericksen–Leslie theory, we know that the viscous torque is

$$\mathbf{\Gamma}_{\text{visc}} = \mathbf{n} \times \mathbf{g}', \qquad (3.10.2)$$

where $\mathbf{g}'$ is given by (3.3.13). This is the frictional torque exerted by the molecules on the hydrodynamic motion. Clearly in the present geometry (3.10.2) reduces to

$$\Gamma_{\text{visc},y} = -\lambda_1\left(\dot{\varphi} - \frac{\lambda_1 - \lambda_2}{2\lambda_1}\frac{\partial v_z}{\partial x}\right), \qquad (3.10.3)$$

where $\lambda_1 = \mu_2 - \mu_3$ and $\lambda_2 = \mu_5 - \mu_6$. Also, in the present geometry, the viscous stress tensor t'_{ji} given by (3.3.5) can be simplified to

$$t'_{xz} = \left(\frac{\lambda_1 - \lambda_2}{2\lambda_1}\right)\Gamma_{\text{visc},y} + \eta'\frac{\partial v_z}{\partial x}, \qquad (3.10.4)$$

where, making use of Parodi's relation (3.3.15),

$$\eta' = [\lambda_1(2\mu_4 + \mu_5 + \mu_6) - \lambda_2^2]/4\lambda_1.$$

Setting $\Gamma_y = \Gamma_{\text{elast},y} + \Gamma_{\text{diel},y} + \Gamma_{\text{mag},y} - \Gamma_{\text{visc},y} = -A\varphi$, where A is a force constant, the condition for instability may be written as[74]

$$A = -\Gamma_y/\varphi = 0. \qquad (3.10.5)$$

The elastic, dielectric and magnetic torques can be evaluated from the functional derivative

$$\Gamma_y(\mathbf{r}) = \frac{\delta \mathscr{F}}{\delta \varphi(\mathbf{r})},$$

where

$$\mathscr{F} = \int_V F(\mathbf{r})\, \mathrm{d}V,$$

and F is given by (3.3.6). We then have

$$\begin{aligned}\Gamma_{\text{elast},y} &= -(k_{33}\cos^2\varphi + k_{11}\sin^2\varphi)(\mathrm{d}^2\varphi/\mathrm{d}x^2)\\ &\quad -(k_{11}-k_{33})\sin\varphi\cos\varphi(\mathrm{d}\varphi/\mathrm{d}x)^2,\\ \Gamma_{\text{diel},y} &= -(4\pi)^{-1}[\varepsilon_\text{a}\cos\varphi\sin\varphi(E_z^2 - E_x^2)\\ &\quad -\varepsilon_\text{a}(\sin^2\varphi - \cos^2\varphi)E_xE_z],\\ \Gamma_{\text{mag},y} &= (\chi_\text{a}\cos\varphi\sin\varphi)H^2.\end{aligned}$$

Since we are interested in the threshold conditions, we retain only first order terms in φ, i.e.,

$$E_x = -\frac{\sigma_\text{a}}{\sigma_\parallel}\varphi E_z,$$

$$E_{\varepsilon x} = -\frac{\varepsilon_\text{a}}{\varepsilon_\parallel}\varphi E_z, \text{ etc.}$$

Also assuming a spatially periodic fluctuation of the form

$$\varphi = \varphi_0 \cos k_x x,$$

we have

$$\left.\begin{aligned}\Gamma_{\text{elast},y} &= -k_{33}k_x^2\varphi,\\ \Gamma_{\text{diel},y} &= -(4\pi)^{-1}\varepsilon_\text{a}E_z^2[\varphi + (E_x/E_z)]\\ &= -(4\pi)^{-1}\varepsilon_\text{a}E_z^2(\sigma_\perp/\sigma_\parallel)\varphi,\\ \Gamma_{\text{mag},y} &= \chi_\text{a}H^2\varphi.\end{aligned}\right\} \quad (3.10.6)$$

3.10.3. DC excitation

In the DC case, we may set $\dot{\varphi} = 0$ and the analysis becomes simple. Using (3.10.1), (3.10.3) and (3.10.4),

$$\Gamma_{\text{visc},y} = \frac{\varepsilon_\parallel}{4\pi}\left(\frac{\sigma_\perp}{\sigma_\parallel} - \frac{\varepsilon_\perp}{\varepsilon_\parallel}\right)\frac{2\lambda_1^2}{(\lambda_1-\lambda_2)(\lambda_1+\eta_0)}E_z^2\varphi,$$

where $\eta_0 = \eta'[2\lambda_1/(\lambda_1 - \lambda_2)]^2$. Applying (3.10.5), we obtain the threshold field

$$E_{th}(DC) = \frac{E_0}{(\zeta^2 - 1)^{1/2}}, \tag{3.10.7}$$

where

$$E_0^2 = -\frac{4\pi\varepsilon_\parallel}{\varepsilon_a \varepsilon_\perp}(\chi_a H^2 + k_{33} k_x^2) \tag{3.10.8}$$

and ζ is a dimensionless quantity, called the *Helfrich parameter*, given by

$$\zeta^2 = \left(1 - \frac{\varepsilon_\parallel \sigma_\perp}{\varepsilon_\perp \sigma_\parallel}\right)\left(1 - \frac{\varepsilon_\parallel}{\varepsilon_a} \frac{2\lambda_1^2}{(\lambda_1 - \lambda_2)(\lambda_1 + \eta_0)}\right). \tag{3.10.9}$$

Typically ζ^2 is a small number; for example, for MBBA it is about 3. For DC (and low frequency AC), $k_x \simeq \pi/d$. (A numerical solution[75] of the two-dimensional problem with appropriate boundary conditions has confirmed that *at the threshold* k_x is indeed π/d.) Thus when $H = 0$, we have a *voltage* threshold independent of film thickness given by

$$V_{th}^2(DC) = \frac{V_0^2}{(\zeta^2 - 1)}, \tag{3.10.10}$$

where

$$V_0^2 = -\frac{4\pi^3}{\varepsilon_a} \frac{\varepsilon_\parallel}{\varepsilon_\perp} k_{33}.$$

We have treated the distortion as a pure bend but this is not exactly true. Since the thickness of the sample is of the same order as the periodicity of the distortion, the orientation of the director will vary in the z direction also. There will therefore be a non-negligible splay component. To allow for this, it has been suggested[73] that the bend elastic constant k_{33} should be replaced by

$$k_{33} + k_{11}(k_z/k_x)^2$$

where the wave-vector in the z direction $k_z \sim \pi/d$. With this correction, the DC threshold given by (3.10.10) is in good agreement with the experimental value. For MBBA it is about 8 volts.

3.10.4. Square wave excitation

In the AC case, the time dependence of φ cannot be neglected. Using (3.10.3), (3.10.5) and (3.10.6)

$$\Gamma_{visc,y} = -\frac{\varepsilon_a}{4\pi}\left[\left(\frac{\varepsilon_\perp}{\varepsilon_\parallel}E_0^2 + E_z^2\right)\varphi + E_z E_x\right]. \tag{3.10.11}$$

In addition we have to take into consideration the conservation of charge. The charge balance equation reads

$$\dot{q}+\partial J_x/\partial x=0, \tag{3.10.12}$$

where q is the excess charge per unit volume and J_x the electric current given by

$$J_x=\sigma_{\|}E_x+\sigma_{\mathrm{a}}E_z\varphi$$

retaining only the first order terms in φ and E_x. We suppose that diffusion currents make a negligible contribution. From the relation

$$\frac{\partial D_x}{\partial x}=\frac{\partial}{\partial x}[\varepsilon_{\|}E_x+\varepsilon_{\mathrm{a}}E_z\varphi]=4\pi q,$$

where D_x is the x component of the displacement, we obtain

$$\frac{\partial E_x}{\partial x}=\frac{4\pi q}{\varepsilon_{\|}}-\frac{\varepsilon_{\mathrm{a}}}{\varepsilon_{\|}}E_z\psi, \tag{3.10.13}$$

where $\psi=\partial\varphi/\partial x$ is the local curvature. Therefore

$$\dot{q}+(q/\tau)+\sigma_{\mathrm{H}}E_z\psi=0, \tag{3.10.14}$$

where τ is the dielectric relaxation time given by

$$1/\tau=4\pi\sigma_{\|}/\varepsilon_{\|}$$

and

$$\sigma_{\mathrm{H}}=\sigma_{\|}\left(\frac{\varepsilon_{\perp}}{\varepsilon_{\|}}-\frac{\sigma_{\perp}}{\sigma_{\|}}\right).$$

Neglecting the inertial term (3.3.1) may be written as

$$t_{ji,j}+f_i=0 \tag{3.10.15}$$

where f_i is the body force per unit volume, which in this case is equal to qE_z. We are interested only in the z component of this equation. Since the director orientation φ is assumed to be a function of x only, t^0_{xz} vanishes, and (3.10.15) reduces to

$$\frac{\partial}{\partial x}t'_{xz}+qE_z=0$$

or, from (3.10.4),

$$\left(\frac{\lambda_1-\lambda_2}{2\lambda_1}\right)\frac{\partial}{\partial x}\Gamma_{\text{visc},y}+\eta'\frac{\partial^2 v_z}{\partial x^2}+qE_z=0. \qquad (3.10.16)$$

At the threshold, we have the condition

$$\partial\Gamma_y/\partial x=0. \qquad (3.10.17)$$

From (3.10.3) and (3.10.16)

$$-\frac{\partial\Gamma_{\text{visc},y}}{\partial x}=\frac{\lambda_1\eta_0}{\lambda_1+\eta_0}\left(\dot{\psi}+\frac{1}{\eta_0}\frac{2\lambda_1}{\lambda_1-\lambda_2}qE_z\right).$$

From (3.10.6) and (3.10.13),

$$\frac{\partial\Gamma_{\text{diel},y}}{\partial x}=-\frac{\varepsilon_{\text{a}}E_z}{4\pi\varepsilon_\parallel}(4\pi q+\varepsilon_\perp E_z\psi)-\frac{\varepsilon_{\text{a}}\varepsilon_\perp}{4\pi\varepsilon_\parallel}E_z^2-\frac{\varepsilon_{\text{a}}}{\varepsilon_\parallel}E_z q,$$

$$\frac{\partial}{\partial x}(\Gamma_{\text{elast},y}+\Gamma_{\text{mag},y})=\frac{\varepsilon_{\text{a}}}{4\pi}\frac{\varepsilon_\perp}{\varepsilon_\parallel}E_0^2\psi.$$

Therefore, from (3.10.17) we obtain the following equation for the curvature

$$\dot{\psi}+\gamma\psi+\frac{q}{\eta}E_z=0 \qquad (3.10.18)$$

where

$$\gamma=1/T=\left(\frac{1}{\lambda_1}+\frac{1}{\eta_0}\right)\left(-\frac{\varepsilon_{\text{a}}\varepsilon_\perp}{4\pi\varepsilon_\parallel}E_z^2+\chi_{\text{a}}H^2+k_{33}k_x^2\right)$$
$$=(\Lambda E_z^2+\Lambda_0)\quad\text{say,} \qquad (3.10.19)$$

T is the decay time for curvature, and η is an effective viscosity coefficient given by

$$\frac{1}{\eta}=\frac{1}{\eta_0}\left(\frac{2\lambda_1}{\lambda_1-\lambda_2}\right)-\frac{\varepsilon_{\text{a}}}{\varepsilon_\parallel}\left(\frac{1}{\lambda_1}+\frac{1}{\eta_0}\right).$$

(3.10.14) and (3.10.18) are two coupled equations which cannot be solved analytically for any general value of $E_z(t)$ since γ itself depends on E_z and t. However for a square wave

$$E_z(t)=\begin{cases}+E, & 0<t<\pi/\omega\\ -E, & \pi/\omega<t<2\pi/\omega.\end{cases}$$

In this case γ remains constant in any half-period and the solutions are simpler and enable a physical interpretation of many of the observed phenomena. If the solutions are taken in the form[73]

$$\psi = C_\psi \exp(At/\tau), \qquad q = C_q \exp(At/\tau),$$

the general solutions become

$$\psi(t) = a \exp(A_1 t/\tau) + b \exp(A_2 t/\tau), \tag{3.10.20}$$

$$q(t) = -\frac{\sigma_H E\tau}{1+A_1} a \exp(A_1 t/\tau) - \frac{\sigma_H E\tau}{1+A_2} b \exp(A_2 t/\tau), \tag{3.10.21}$$

where

$$A_{1,2} = -\tfrac{1}{2}(1+\gamma\tau) \pm [\tfrac{1}{4}(1-\gamma\tau)^2 + \zeta^2(\gamma\tau - \Lambda_0\tau)]^{1/2}.$$

These solutions are valid as long as E is constant. We observe that as

$$q = -\sigma_H \psi E\tau/(1+A_1),$$

a change of sign of E implies that q or ψ changes sign but not both, i.e., if (q, ψ) is a solution for a half-period, it will be $(q, -\psi)$ or $(-q, \psi)$ for the other half-period. Subject to these conditions we obtain two sets of solutions given by (3.10.22) and (3.10.23) below:

$$\begin{aligned}\psi &= \left[1 - \exp\left(\frac{s\tau - A_1}{2\nu\tau}\right)\right] \exp\left(\frac{A_2 t}{\tau}\right) + \left[\exp\left(\frac{s\tau - A_1}{2\nu\tau}\right) - \exp\left(\frac{A_2 - A_1}{2\nu\tau}\right)\right] \exp\left(\frac{A_1 t}{\tau}\right), \\ \frac{q}{E\sigma_H\tau} &= \frac{1 - \exp\left(\dfrac{s\tau - A_1}{2\nu\tau}\right)}{-1 - A_2} \exp\left(\frac{A_2 t}{\tau}\right) + \frac{\exp\left(\dfrac{s\tau - A_1}{2\nu\tau}\right) - \exp\left(\dfrac{A_2 - A_1}{2\nu\tau}\right)}{-1 - A_1} \exp\left(\frac{A_1 t}{\tau}\right),\end{aligned} \tag{3.10.22}$$

where $\nu = \omega/2\pi$ is the frequency of the voltage. Here q changes sign with E, but ψ does not, which is the situation in the conduction regime.

$$\psi = \left[1+\exp\left(\frac{s\tau - A_1}{2\nu\tau}\right)\right]\exp\left(\frac{A_2 t}{\tau}\right) - \left[\exp\left(\frac{s\tau - A_1}{2\nu\tau}\right)\right.$$

$$\left.+\exp\left(\frac{A_2 - A_1}{2\nu\tau}\right)\right]\exp\left(\frac{A_1 t}{\tau}\right),$$

$$\frac{q}{E\sigma_{\mathrm{H}}\tau} = \frac{1+\exp\left(\dfrac{s\tau - A_1}{2\nu\tau}\right)}{-1-A_2}\exp\left(\frac{A_2 t}{\tau}\right) + \left[\exp\left(\frac{s\tau - A_1}{2\nu\tau}\right)\right.$$

$$\left.+\exp\left(\frac{A_2 - A_1}{2\nu\tau}\right)\right]\left(\frac{1}{1+A_1}\right)\exp\left(\frac{A_1 t}{\tau}\right). \qquad (3.10.23)$$

In this case ψ changes sign with E, but q does not. This corresponds to the dielectric regime in which the director oscillates.

In the above equations, s is a real number which determines whether the system is stable or not, and which takes the value 0 at the threshold. Setting $s = 0$ and defining

$$\Delta = [(1-\gamma\tau)^2 + 4\zeta^2 E^2 \tau\Lambda]^{1/2} > 0$$

it can be easily shown that in the conduction regime

$$\Delta \sinh\left[\frac{\pi}{2\omega}\left(\frac{1}{T}+\frac{1}{\tau}\right)\right] = \left(1-\frac{\tau}{T}\right)\sinh\frac{\pi\Delta}{2\omega\tau}, \qquad (3.10.24)$$

and in the dielectric regime

$$\Delta \sinh\left[\frac{\pi}{2\omega}\left(\frac{1}{T}+\frac{1}{\tau}\right)\right] = \left(\frac{\tau}{T}-1\right)\sinh\frac{\pi\Delta}{2\omega\tau}. \qquad (3.10.25)$$

These equations show clearly that the problem falls naturally into two distinct parts. For $T > \tau$ (3.10.25) has no solution and consequently there is no dielectric regime, while for $T < \tau$ (3.10.24) has no solution and there is no conduction regime. For $T = \tau$, neither equation has a solution (except when $\omega = 0$).

The theoretical variation of q and ψ over a full period of the exciting wave is illustrated in fig. 3.10.7 for two values of ω, one at low frequency in the conduction regime ($\omega\tau \ll 1$, $T \gg \tau$) and the other at high frequency in the dielectric regime ($\omega\tau \gg 1$, $T \ll \tau$). In the former case one obtains essentially stationary domains and oscillating charges and in the latter case vice versa. The transition from the conduction to the dielectric regime occurs at a critical frequency ω_c such that $\omega_c\tau \sim 1$.

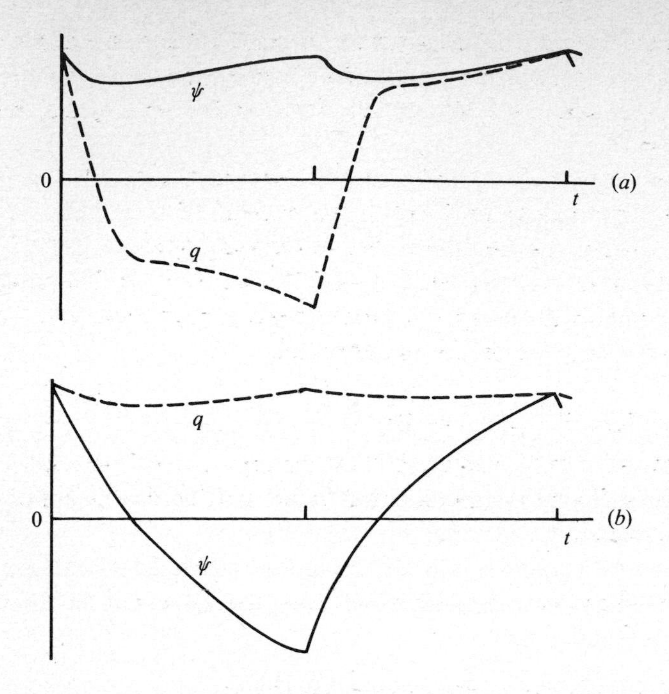

Fig. 3.10.7. Time dependence of the charge q and the curvature ψ over one period of the square wave excitation. (a) Conduction regime ($\omega\tau \ll 1$, $T \gg \tau$). The charges oscillate but the domains are stationary. (b) Dielectric regime ($\omega\tau \gg 1$, $T \ll \tau$). The charges are stationary and the domains oscillate. (After Smith *et al.*[73])

Certain other interesting conclusions can be drawn from the above equations. For example, even at frequencies less than ω_c, there can be quenching of the conductive instability at high fields. This is because $T = (\Lambda E_z^2 + \Lambda_0)^{-1}$ decreases with increasing E_z, and when it becomes equal to τ there can be restabilization. At higher fields, $T < \tau$ and the system goes over to the dielectric regime. This gives a physical insight into the origin of the sigmoid shape of the experimental threshold curve. Indeed, theoretically the threshold field as a function of frequency for conductive instability may be expected to form a closed loop.

Another result that follows from the theory is that the conduction regime can be suppressed altogether by using very thin samples. We have seen that at low frequencies, $k_x \sim \pi/d$. Now a decrease in sample thickness increases Λ_0, where from (3.10.19)

$$\Lambda_0 = \left(\frac{1}{\lambda_0} + \frac{1}{\eta_0}\right)(\chi_a H^2 + k_{33} k_x^2).$$

This in turn reduces the curvature relaxation T and when $T<\tau$ the conductivity instability is eliminated and only a dielectric instability is possible. A similar quenching can be achieved by applying a stabilizing magnetic field.

Greubel and Wolff[76] and Vistin[77] observed that for thin and pure samples the spatial periodicity of the pattern in DC excitation decreases with increasing voltage above the threshold. Smith *et al.*[73] have suggested that this *variable grating mode*[76] is due to non-linear (saturation) effects at higher voltages in samples that are so thin and pure that the conductivity instability has been quenched.

3.10.5. Sinusoidal excitation

The behaviour is qualitatively similar when the exciting field is sinusoidal. Putting $E_z = E_M \cos \omega t$, the coupled equations for the charge and the curvature become[72]

$$\dot{q} + \frac{q}{\tau} + \sigma_H \psi E_M \cos \omega t = 0, \tag{3.10.26}$$

$$\dot{\psi} + \Lambda(E_0^2 + E_M^2 \cos^2 \omega t)\psi + \frac{E_M q}{\eta} \cos \omega t = 0, \tag{3.10.27}$$

where

$$\Lambda = -\frac{\varepsilon_a \varepsilon_\perp}{4\pi\varepsilon_\parallel}\left(\frac{1}{\lambda_1} + \frac{1}{\eta_0}\right).$$

We have seen in the previous section that in the conduction regime, q changes sign with the field but ψ does not. Let us make the simplifying assumption that ψ is sensibly constant over a period. Since $\omega \sim 1/\tau$ and $T \gg \tau$, we may replace $\cos^2 \omega t$ by its average value $\frac{1}{2}$. Also expanding $q(t)$ as

$$q(t) = \sum_{n=0}^{\infty} (q'_n \cos n\omega t + q''_n \sin n\omega t),$$

we have

$$\langle q(t) \cos \omega t \rangle_{\text{period}} = \tfrac{1}{2} q'_1.$$

Therefore from (3.10.27),

$$\psi = -\frac{E_M q'}{\Lambda \eta (E_M^2 + 2E_0^2)}. \tag{3.10.28}$$

Integration of (3.10.26) yields

$$q = -\frac{\sigma_H E_M \psi \tau}{1+\omega^2\tau^2}(\cos \omega t + \omega\tau \sin \omega t),$$

so that

$$q_1' = -\frac{\sigma_H E_M \psi \tau}{1+\omega^2\tau^2}.$$

Using (3.10.17) the threshold field is given by

$$E_{th}^2(AC) = \langle E^2(t)\rangle = \tfrac{1}{2}E_M^2 = \frac{E_0^2(1+\omega^2\tau^2)}{\zeta^2 - 1 - \omega^2\tau^2}. \quad (3.10.29)$$

Thus the threshold field increases with ω. Though the analysis is not strictly valid when $\omega T \sim 1$, calculations show that (3.10.29) holds good quite well over the range 0 to ω_c where

$$\omega_c = (\zeta^2 - 1)^{1/2}/\tau, \quad (3.10.30)$$

which represents a critical 'cut-off' frequency beyond which the system goes over to the dielectric regime (fig. 3.10.3). To discuss the effect of the magnetic field, we write (3.10.8) as

$$E_0^2 = -\frac{4\pi\varepsilon_\parallel}{\varepsilon_a \varepsilon_\perp}\frac{k_{33}\pi^2}{d^2}\left(1 + \frac{d^2}{\pi^2\xi^2}\right)$$

where ξ is the magnetic coherence length (see § 3.4.1). For low magnetic fields, $\xi \gg d$, we get a *voltage threshold* independent of the thickness of the sample:

$$V_{th}^2(\omega) = \frac{V_0^2(1+\omega^2\tau^2)}{\zeta^2 - 1 - \omega^2\tau^2}, \quad (3.10.31)$$

where

$$V_0 = \left(-\frac{4\pi\varepsilon_\parallel}{\varepsilon_a\varepsilon_\perp}k_{33}\pi^2\right)^{1/2}.$$

For higher magnetic fields,

$$V_{th}(\omega, H) = V_{th}(\omega, 0)\left(1 + \frac{d^2}{\pi^2\xi^2}\right)^{1/2}$$

$$= V_{th}(\omega, 0)\left(1 + \frac{H^2}{H_c^2}\right)^{1/2} \quad (3.10.32)$$

where $H_c = (\pi/d)(k_{33}/\chi_a)^{1/2}$. The threshold voltage therefore increases with H in agreement with observations. For very high magnetic fields, $H \gg H_c$, theory predicts a field threshold independent of thickness but proportional to H.

When $\omega\tau \gg 1$, q may be taken as constant over a period. Expanding $\psi(t)$ as a Fourier series

$$\psi(t) = \sum_{n=0}^{\infty} (\psi_n' \cos n\omega t + \psi_n'' \sin n\omega t)$$

and replacing $\psi(t) \cos \omega t$ by its average value $\frac{1}{2}\psi_1'$,

$$q = -\tfrac{1}{2}\sigma_H E_M \tau \psi_1'.$$

Equation (3.10.27) can now be integrated but solutions can only be obtained by numerical techniques.[72] Calculations show that in the dielectric regime (a) the threshold field E_{th} is independent of the wave-vector k_x, which results in a *field* threshold (as distinct from a voltage threshold as in the conduction regime), (b) E_{th}^2 varies linearly as ω, (c) for a given H, k_x^2 varies linearly as ω, and (d) for a given ω, $\chi_a H^2 + k_{33}k_x^2$ is a constant. These predictions have been confirmed experimentally.

Other aspects of the problem, for example, the dependence of the instabilities on the magnitude and sign of the dielectric anisotropy and on the conductivity, the behaviour in the vicinity of the threshold on either side, the effect of external stabilizing fields, etc., have been discussed extensively by Dubois-Violette *et al.*[72,73]

3.11. Curvature electricity

If the molecule possesses shape polarity in addition to a permanent dipole moment, then the possibility exists that a splay or bend deformation will polarize the material, and conversely that an electric field will induce a deformation (fig. 3.11.1). In a first order theory, the polarization **P** should be proportional to the distortion:

$$\mathbf{P} = e_1[\mathbf{n}(\nabla \cdot \mathbf{n})] + e_3[\mathbf{n} \cdot \nabla \mathbf{n}]$$

where e_1 and e_3 are the *curvature electric* coefficients corresponding to splay and bend respectively. This effect, which is the analogue of piezoelectricity in solids, was first proposed by Meyer.[78]

Experimental evidence for this phenomenon has been reported by Schmidt, Schadt and Helfrich.[79] A thin film of nematic MBBA was homeotropically aligned between two glass plates. Aluminium foils were used as spacers as well as electrodes and an electric field was applied along x normal to the undisturbed orientation of the director $(0, 0, n_z)$. The

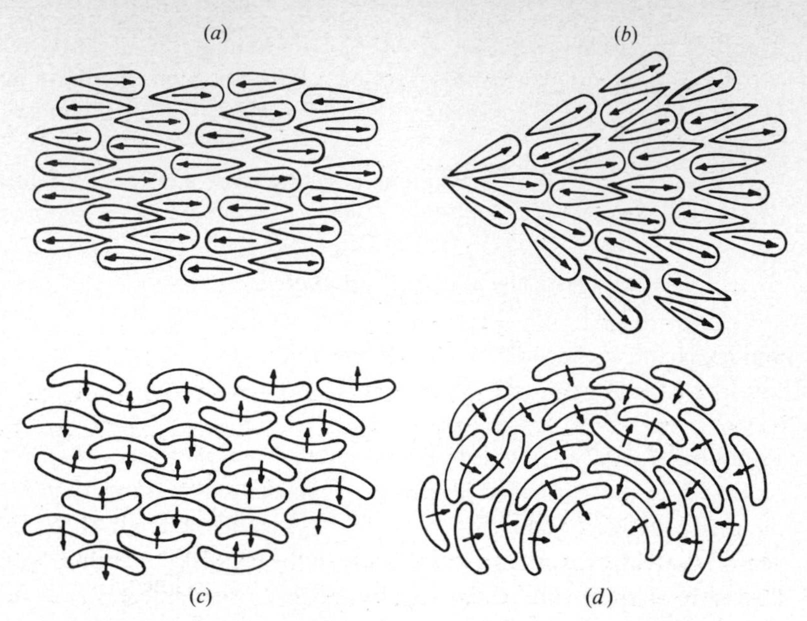

Fig. 3.11.1. Meyer's model of curvature electricity. The nematic medium composed of polar molecules is non-polar in the undeformed state (*a* and *c*) but polar under splay (*b*) or bend (*d*). (After Meyer.[78])

deformation was detected by optical observations normal to the plates. Since MBBA is dielectrically negative the field only stabilizes the initial orientation. Also, the fact that the deformation is observable even at quite low voltages and that it does not possess a threshold rules out the possibility of electrohydrodynamic alignment. The measured distortion may therefore be attributed to curvature electricity.

In the present geometry, the electric field produces a bend so that

$$P_x \simeq e_3 \frac{\partial \varphi}{\partial z}.$$

Equating the electric and elastic energies

$$P_x E = k_{33}\left(\frac{\partial \varphi}{\partial z}\right)^2,$$

or

$$\frac{\partial \varphi}{\partial z} = \left(\frac{e_3}{k_{33}}\right) E.$$

For light polarized parallel to the field, the refractive index at any point is

$$n = [\{(\sin^2 \varphi)/n_e^2\} + \{(\cos^2 \varphi)/n_o^2\}]^{-1/2},$$

or

$$n-n_o=\tfrac{1}{2}n_o[1-(n_o^2/n_e^2)]\varphi^2$$

since φ is small. The total phase retardation between polarizations parallel and perpendicular to the field is

$$\delta=\frac{2\pi}{\lambda}\int_{-d/2}^{d/2}(n-n_o)\,dz$$

$$=\frac{2\pi}{\lambda}n_o\left[1-\left(\frac{n_o^2}{n_e^2}\right)\right]\left(\frac{\partial\varphi}{\partial z}\right)^2\frac{d^3}{24},$$

where d is the film thickness. Schmidt *et al.* assumed the wall anchoring to be weak and ignored its influence on $\partial\varphi/\partial z$ near the boundaries. Under such circumstances

$$\delta\propto E^2d^3, \qquad (3.11.1)$$

which is in agreement with their observations (figs. 3.11.2 and 3.11.3). The experimental value of e_3 for MBBA was found to be 3.7×10^{-5} dyne$^{1/2}$ at 22 °C.

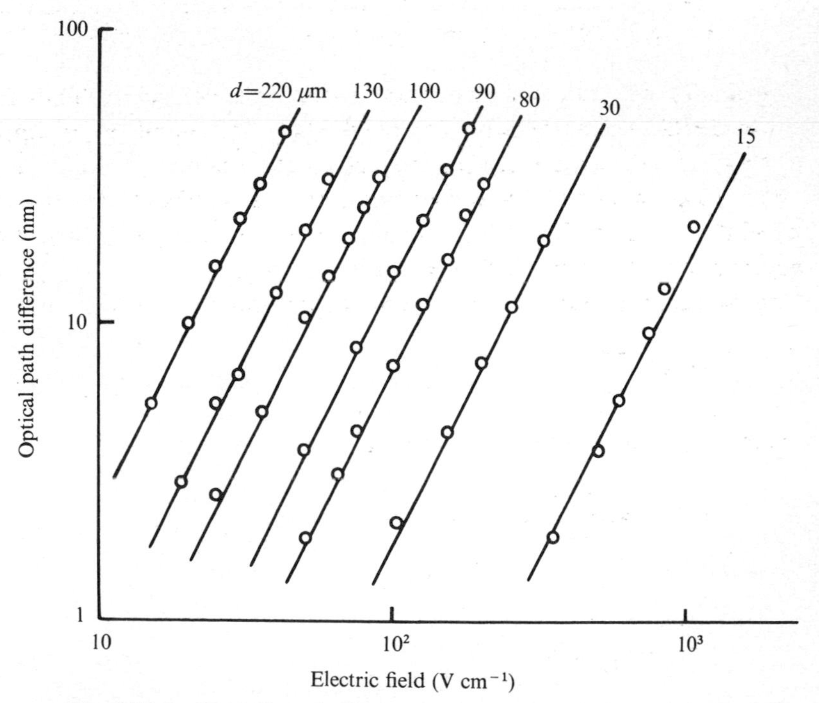

Fig. 3.11.2. Optical path difference versus electric field in MBBA for various values of the sample thickness. The lines satisfy the square law predicted by (3.11.1). (After Schmidt, Schadt and Helfrich.[79])

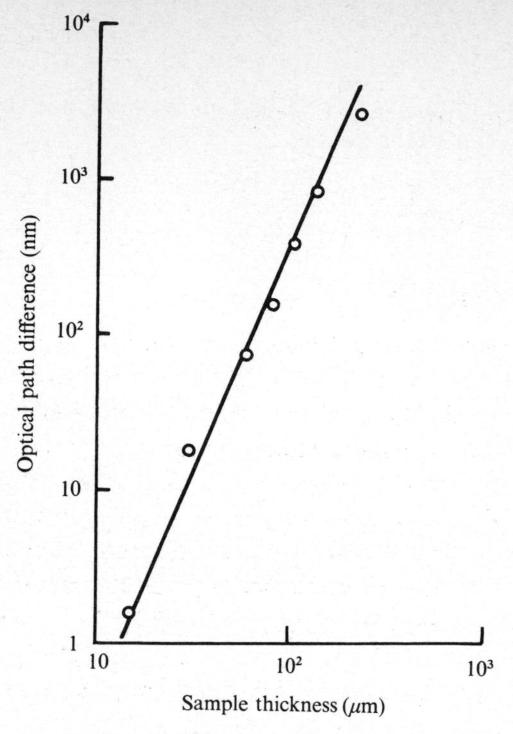

Fig. 3.11.3. Optical path difference versus sample thickness for a constant field $E = 320\ \text{V cm}^{-1}$. The best fit fulfils the cubic law predicted by (3.11.1). (After Schmidt, Schadt and Helfrich.[79])

4

Cholesteric liquid crystals

4.1. Optical properties

The unique optical properties of the cholesteric phase were recognized by both Reinitzer and Lehmann at the time of their early investigations which culminated in the discovery of the liquid crystalline state. When white light is incident on a 'planar' sample (whose optic axis is perpendicular to the glass surfaces) selective reflexion takes place, the wavelengths of the reflected maxima varying with angle of incidence in accordance with Bragg's law. At normal incidence, the reflected light is strongly circularly polarized; one circular component is almost totally reflected over a spectral range of some 100 Å, while the other passes through practically unchanged. Moreover, contrary to usual experience, the reflected wave has the same sense of circular polarization as that of the incident wave.

Along its optic axis, the medium possesses a very high rotatory power, usually of the order of several thousands of degrees per millimetre. In the neighbourhood of the region of reflexion, the rotatory dispersion is anomalous and the sign of the rotation opposite on opposite sides of the reflected band. The behaviour is rather similar to that of an optically active molecule in the vicinity of an absorption. Following the theoretical work of Mauguin,[1] Oseen[2] and de Vries[3] these remarkable properties can now be explained quite rigorously in terms of the spiral structure represented schematically in fig. 1.1.2.

4.1.1. Propagation along the optic axis for wavelengths $\ll$ pitch

Basic theory We shall first consider the propagation of light along the optic axis for wavelengths much smaller than the pitch so that reflexion and interference effects may be neglected. The problem was investigated by Mauguin[1] with a view to explaining the optical rotation produced by twisting a nematic about an axis perpendicular to the preferred direction of the molecules. He used the Poincaré sphere[4] and 'rolling cone' method, but we shall adopt an identically equivalent formalism, viz, the Jones calculus.[5,6]

The basic principle underlying the Jones method is that any elliptic vibration can be represented by the column vector

$$\mathbf{D}=\begin{bmatrix}A_1\\A_2\end{bmatrix}$$

where A_1 and A_2 are the resolved components of the electric displacement vector $\mathbf{D}$ along x and y, and are in general complex quantities. The intensity is $|A_1|^2+|A_2|^2$, while the complex ratio A_1/A_2 describes its polarization state. The azimuth Λ (i.e., the angle which the major axis of the ellipse makes with x) and the ellipticity ω are related as follows: if $\tan\alpha=|A_1|/|A_2|$ and Δ is the phase difference between A_1 and A_2,

$$\tan 2\Lambda=\cos\Delta\tan 2\alpha,$$

$$\sin 2\omega=\sin\Delta\sin 2\alpha.$$

The effect of an optical system – comprised, say, of birefringent, absorbing and dichroic plates – is to change A_1 and A_2, so that

$$\mathbf{D}'=\mathbf{J}\mathbf{D},$$

where $\mathbf{J}$ is a 2×2 matrix with complex elements. In what follows, we shall adopt the convention of describing the optical system as viewed by an observer looking at the source of light.

Our aim is to evaluate the matrix $\mathbf{J}$ for the cholesteric liquid crystal when light is incident along the helical axis. As demonstrated by Jones,[7] a problem of this type can be solved by regarding the medium as being composed of a large number of infinitesimally thin sections, each section representing an optical element, in this case a linearly birefringent or retardation plate. For the purposes of this calculation, therefore, we treat the liquid crystal (fig. 1.1.2) as a pile of very thin birefringent (quasi-nematic) layers with the principal axes of the successive layers turned through a small angle β.

Let the principal axes of the first layer be inclined at an angle β with respect to x, y. If light is incident normal to the layers, i.e., along z, the Jones retardation matrix for the first layer referred to its own principal axes is given by

$$\mathbf{G}=\begin{bmatrix}\exp(-\mathrm{i}\gamma) & 0\\ 0 & \exp(\mathrm{i}\gamma)\end{bmatrix},$$

where γ is half the phase difference between the waves linearly polarized along the principal axes after passing through a single layer of thickness p, i.e., $\gamma = \pi\delta np/\lambda$, $\delta n = n_a - n_b$ being the layer birefringence and λ the vacuum wavelength. The retardation matrix with respect to x, y is then

$$\mathbf{J}_1 = \mathbf{SGS}^{-1},$$

where

$$\mathbf{S} \equiv \mathbf{S}(\beta) = \begin{bmatrix} \cos\beta & -\sin\beta \\ \sin\beta & \cos\beta \end{bmatrix}$$

and $\mathbf{S}^{-1}$ is the inverse of $\mathbf{S}$ so that $\mathbf{SS}^{-1} = \mathbf{S}^{-1}\mathbf{S} = \mathbf{E}$, the unit matrix.

If $\mathbf{D}_0$ is the complex column vector with respect to x, y representing the incident light, the emergent light after passing through the first layer is

$$\mathbf{D}_1 = \mathbf{J}_1\mathbf{D}_0,$$

where, as we are interested at present only in the state of polarization of the emergent beam, we neglect the phase factor $\exp(-\mathrm{i}\eta)$, where $\eta = \pi(n_a + n_b)p/\lambda$. (Throughout, we follow the convention of representing the phase factor at any point $+z$ by $\exp(-\mathrm{i}2\pi nz/\lambda)$.) Let $\mathbf{D}_1$ be now incident on a second birefringent layer whose principal axes are inclined at 2β with respect to x, y. The Jones matrix for this layer is

$$\mathbf{S}(2\beta)\mathbf{GS}^{-1}(2\beta) = \mathbf{S}^2\mathbf{GS}^{-2},$$

and the emergent vector is

$$\mathbf{D}_2 = \mathbf{S}^2\mathbf{GS}^{-2}\mathbf{D}_1 = \mathbf{S}^2\mathbf{GS}^{-2}\mathbf{SGS}^{-1}\mathbf{D}_0$$
$$= \mathbf{S}^2(\mathbf{GS}^{-1})^2\mathbf{D}_0 = \mathbf{J}_2\mathbf{D}_0,$$

where $\mathbf{J}_2 = \mathbf{S}^2(\mathbf{GS}^{-1})^2$ is the appropriate Jones matrix for this system of two layers. In general, if we have a pile of m layers, where the principal axis of the sth layer is inclined at $s\beta$ with respect to x, y ($s = 1, 2, \ldots m$), the Jones matrix for the pile is evidently

$$\mathbf{J}_m = \mathbf{S}^m(\mathbf{GS}^{-1})^m = \begin{bmatrix} a & b \\ c & d \end{bmatrix} \text{ say.} \tag{4.1.1}$$

It can be shown from the theory of matrices[8,9] that

$$(\mathbf{GS}^{-1})^m = \frac{\sin m\theta}{\sin\theta}(\mathbf{GS}^{-1}) - \frac{\sin(m-1)\theta}{\sin\theta}\mathbf{E} \tag{4.1.2}$$

where

$$\cos\theta = \cos\beta\cos\gamma.$$

Now, as stated earlier, the layer thickness is assumed to be very small, say a few Å, while the pitch P is taken to be at least a few wavelengths of light, so that both $\beta(=2\pi p/P)$ and γ are small quantities. Therefore

$$\theta^2 \simeq \beta^2 + \gamma^2. \tag{4.1.3}$$

From (4.1.1) and (4.1.2)

$$a = \cos m\beta \cos m\theta + \frac{\tan\beta}{\tan\theta}\sin m\beta \sin m\theta - \mathrm{i}\frac{\sin m\theta}{\sin\theta}\sin\gamma\cos(m+1)\beta, \tag{4.1.4}$$

$$b = \frac{\tan\beta}{\tan\theta}\cos m\beta \sin m\theta - \sin m\beta\cos m\theta - \mathrm{i}\frac{\sin m\theta}{\sin\theta}\sin\gamma\sin(m+1)\beta, \tag{4.1.5}$$

$$c = -b^* \quad \text{and} \quad d = a^*,$$

where a^* and b^* are respectively the complex conjugates of a and b.

It is a standard result in optics[6] that such a system can, in general, be replaced by a rotator and a retarder. If ρ is the rotation produced by the system, 2φ the phase retardation and ψ the azimuth of the principal axes of the retarder,

$$\mathbf{J}_m = \begin{bmatrix}\cos\psi & -\sin\psi\\ \sin\psi & \cos\psi\end{bmatrix}\begin{bmatrix}\cos\rho & -\sin\rho\\ \sin\rho & \cos\rho\end{bmatrix} \times \begin{bmatrix}\exp(-\mathrm{i}\varphi) & 0\\ 0 & \exp(\mathrm{i}\varphi)\end{bmatrix}\begin{bmatrix}\cos\psi & \sin\psi\\ -\sin\psi & \cos\psi\end{bmatrix}. \tag{4.1.6}$$

From (4.1.1) and (4.1.6)

$$a = \cos\varphi\cos\rho - \mathrm{i}\sin\varphi\cos(2\psi+\rho), \tag{4.1.7}$$

$$b = -\cos\varphi\sin\rho - \mathrm{i}\sin\varphi\sin(2\psi+\rho), \tag{4.1.8}$$

$$c = -b^* \quad \text{and} \quad d = a^*. \tag{4.1.9}$$

Equating the real and imaginary parts of (4.1.4) and (4.1.7) and of (4.1.5) and (4.1.8), we obtain after simplification[10]

$$\rho = m(\beta - \theta') \text{ radians}, \tag{4.1.10}$$

$$\varphi = \cos^{-1}(\sec^2 m\theta'/\sec^2 m\theta)^{1/2}, \tag{4.1.11}$$

$$\psi = \tfrac{1}{2}[(m+1)\beta - \rho], \tag{4.1.12}$$

where

$$\theta' = \frac{1}{m}\tan^{-1}(\tan\beta\tan m\theta/\tan\theta). \tag{4.1.13}$$

In these equations m represents the total number of layers in the system. Since the layer thickness is taken to be a few Å, it turns out in actual practice that even with the thinnest specimens employed, m is usually a very large number. We shall therefore assume m to be large throughout this discussion.

Optical rotatory power When $\beta \gg \gamma$, $\theta' \simeq \theta$. This condition is satisfied when $\frac{1}{2}P\delta n \ll \lambda$, i.e., when the pitch is not too large. The optical rotation produced by m layers is then

$$\rho = m(\beta-\theta) = m[\beta-(\beta^2+\gamma^2)^{1/2}] = -m\gamma^2/2\beta,$$

and the phase retardation $2\varphi \simeq 0$. Thus the system behaves in effect as a pure rotator. If m is the number of layers per turn of the helix, $m\beta = 2\pi$ and $mp = P$, so that the rotatory power in radians per unit length

$$\rho = -\pi(\delta n)^2 P/4\lambda^2, \qquad (4.1.14)$$

the negative sign indicating that the sense of the rotation is opposite to that of the helical twist of the structure.[11] Typically, $\delta n \sim 0.05$ for a cholesteric; taking $P = 5\ \mu$m and $\lambda = 0.5\ \mu$m, $\rho \sim 2000°\ \mathrm{mm}^{-1}$. This equation has been verified experimentally in considerable detail by Robinson[12] in lyotropic systems and by Cano and Chatelain[13] in thermotropic systems.

Robinson discovered that solutions of some polypeptides in organic solvents, for example, poly-γ-benzyl-L-glutamate (PBLG) in dioxan, methylene chloride, chloroform etc., spontaneously adopted the cholesteric mesophase above a certain concentration. Under suitable conditions the solutions exhibited equi-spaced alternate bright and dark lines when observed through a microscope (see plate 13). The lines may be interpreted as a view of the structure at right angles to the screw axis, so that the periodicity of the lines is equal to half the pitch. Robinson confirmed this interpretation by observations between crossed polaroids and also by the use of a quartz wedge; the retardation plotted against distance in a direction perpendicular to the lines had an oscillating value, as is indeed to be expected from the structure. The pitch for any given polypeptide depended on factors such as concentration, solvent, temperature etc. When viewed along the screw axis no lines were seen, but a very high optical rotatory power was present. The rotation in every solution, with a very wide range of values of P was found to be proportional to $1/\lambda^2$ (fig. 4.1.1). Robinson substituted the observed values of ρ and P in (4.1.14), and calculated the layer birefringence δn per volume fraction of the polypeptide in solution. The birefringence was remarkably constant

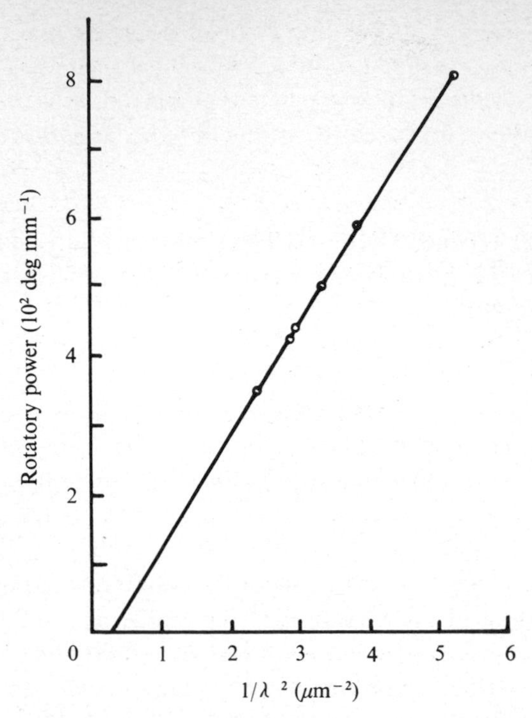

Fig. 4.1.1. Rotatory dispersion of a cholesteric liquid crystal for wavelengths ≪ pitch. Solution of poly-γ-benzyl-L-glutamate (PBLG) in chloroform (18 g/100 g). (After Robinson.[12])

despite the widely varying values of ρ and P. He then prepared a solution with equal quantities of the D and L forms (PBDG and PBLG) which too under certain conditions adopted the spontaneously birefringent phase, only, in this case, it was not the twisted cholesteric, but the untwisted nematic structure. He was therefore able to measure the birefringence directly and the value agreed well with that derived from (4.1.14).

Similar studies have been carried out by Cano and Chatelain[13] on mixtures of nematic and cholesteric liquid crystals. The birefringence of the nematic being very large, δn of the mixture could be assumed without sensible error to be equal that of the nematic itself. The pitch P of the mixture was measured directly from the Grandjean–Cano steps formed in a wedge (see § 4.2). The values of δn and P when inserted in (4.1.14) gave a rotatory power in quantitative accord with observations.

When β is comparable to or less than γ, the system is no longer a pure rotator. The rotatory power is then given by (4.1.10). The optical behaviour in this regime has been investigated for a mixture of right-handed cholesteryl chloride and left-handed cholesteryl myristate.[10,14]

Such a mixture adopts the helical structure of a cholesteric liquid crystal, but the pitch is sensitive to composition and temperature. For a given composition, there occurs an inversion of the rotatory power as the temperature is varied, indicating a change of handedness of the structure. Sackmann *et al.*[15] have made a direct determination of the pitch as a function of temperature by allowing a laser beam to be incident at right angles to the screw axis and observing the diffraction maxima in transmission (see § 4.1.6). The inverse pitch varies almost linearly with temperature, passing through zero at $T_N = 42.5$ °C (fig. 4.1.20). Using these data, some illustrative curves, derived from (4.1.10), of the variation of ρ with temperature for this mixture are given in fig. 4.1.2. Here p was assumed to be 10 Å, and δn at 20° and 55 °C estimated by inserting the observed

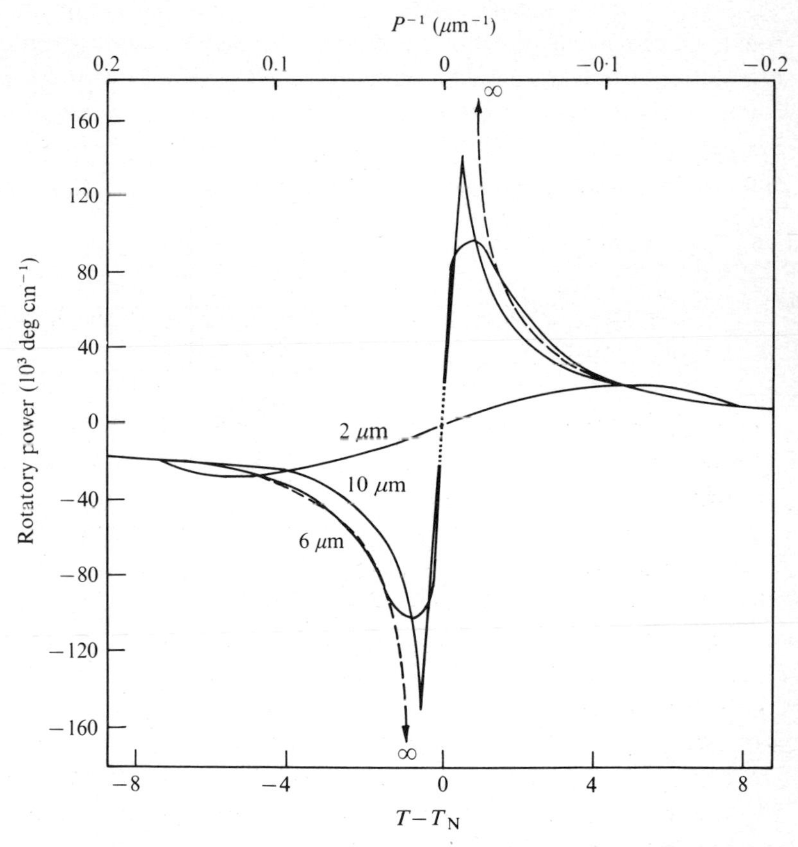

Fig. 4.1.2. Theoretical variation of the rotatory power with inverse pitch (or temperature) calculated from (4.1.10) for samples of thicknesses 2, 6 and 10 μm (full lines). The dashed curve gives the rotatory power predicted by (4.1.14). Throughout, a clockwise rotation as seen by an observer looking at the source of light is taken to be positive.

value of ρ at these temperatures in (4.1.14). This is a valid procedure since (4.1.14) holds good at temperatures much above or below T_N. The layer birefringence δn at intermediate temperatures was obtained by interpolating linearly between 20 and 55 °C. We find two interesting results. Firstly, for a given sample thickness the rotatory power shows positive and negative peaks at certain values of the pitch. Secondly, the peaks increase in height and get closer to T_N with increasing sample thickness. The predicted trends have been confirmed experimentally[10,14] (fig. 4.1.3). However the theory does not describe the behaviour exactly over a very small range on either side of T_N shown by dotted lines in fig. 4.1.2. Precise measurements in this region also appear to be rather difficult because it has been observed[14] that in the vicinity of T_N a monodomain sample tends to break up into many domains, the preferred molecular direction in different domains assuming different orientations in a plane parallel to the supporting surfaces. Moreover, the rotatory power changes from a negative to a positive value in a very small temperature interval around T_N so that unless the temperature is kept truly constant measurements become practically impossible. The optical behaviour in this regime ($\beta \ll \gamma$) is discussed below.

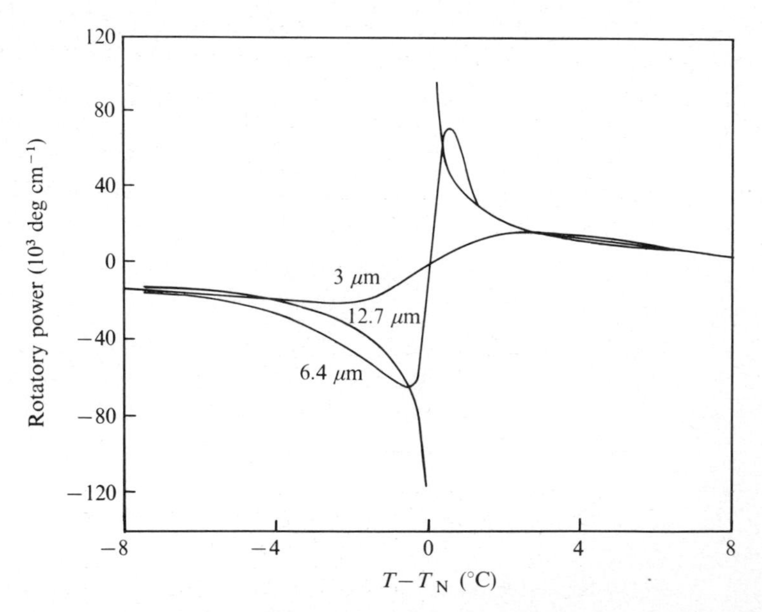

Fig. 4.1.3. Experimental rotatory power as a function of temperature in a 1.6:1 by weight mixture of cholesteryl chloride and cholesteryl myristate. Sample thicknesses 3, 6.4 and 12.7 μm. $\lambda = 0.5893$ μm. (After reference 23.)

The twisted nematic device When $\beta \ll \gamma$, i.e., when P is extremely large, (4.1.1), (4.1.4) and (4.1.5) yield

$$\mathbf{J}_m \simeq \begin{bmatrix} \cos m\beta & -\sin m\beta \\ \sin m\beta & \cos m\beta \end{bmatrix} \begin{bmatrix} \exp(-im\gamma) & 0 \\ 0 & \exp(im\gamma) \end{bmatrix}. \qquad (4.1.15)$$

This implies that at any point in the medium there are two linear vibrations polarized along the local principal axes. The polarization directions of these two vibrations rotate with the principal axes as they travel along the axis of twist and the phase difference between them is the same as that in the untwisted nematic medium. This result was first derived by Mauguin[1] and is sometimes referred to as the adiabatic approximation.

It is this optical property that is made use of in the *twisted nematic device.*[16] A homogeneously aligned film of a nematic of positive dielectric anisotropy is sandwiched between two transparent conducting plates. A 90° twist is then imposed on the liquid crystal by turning one of the plates in its own plane about an axis perpendicular to the film. An incident linear vibration polarized parallel to the director at the entrance side of the film emerges with its vibration direction turned through 90°. With a pair of polarizers, such a film can be used as an electrically controllable light shutter. For example, between parallel polarizers light will be extinguished when the electric field is off, but when the field is switched on the molecules in the bulk of the sample orient themselves perpendicular to the glass plates (fig. 4.1.4) and light will be transmitted. In analogy with

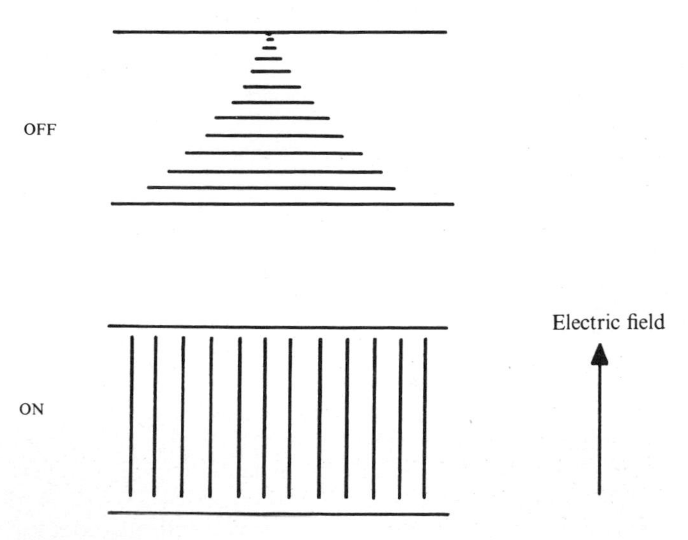

Fig. 4.1.4. The twisted nematic cell.

(3.4.14), the threshold voltage (for $2\varphi_0 = \pi/2$) is given by

$$V_c^2 = E_c^2 d^2 = [\pi^2/(\varepsilon_\parallel - \varepsilon_\perp)][k_{11} + \tfrac{1}{4}(k_{33} - 2k_{22})],$$

where $\varepsilon_\parallel - \varepsilon_\perp$ is the dielectric anisotropy. Typically V_c is just a few volts.

By working out the distortion for fields above the threshold value and applying the Jones matrix method, the intensity of transmission may be calculated as a function of the field. Fig. 4.1.5 shows the results of such a calculation by Van Doorn.[17] It is seen that the threshold for optical transmission is higher than that for elastic distortion as determined by the relative capacitance change, the difference being dependent on the sample thickness. The reason for this will be clear from the theory discussed in the preceding few pages. As long as $\gamma \gg \beta$, the adiabatic approximation (4.1.15) holds good and the optical behaviour of the sample remains unchanged; it is only when the distortion becomes large

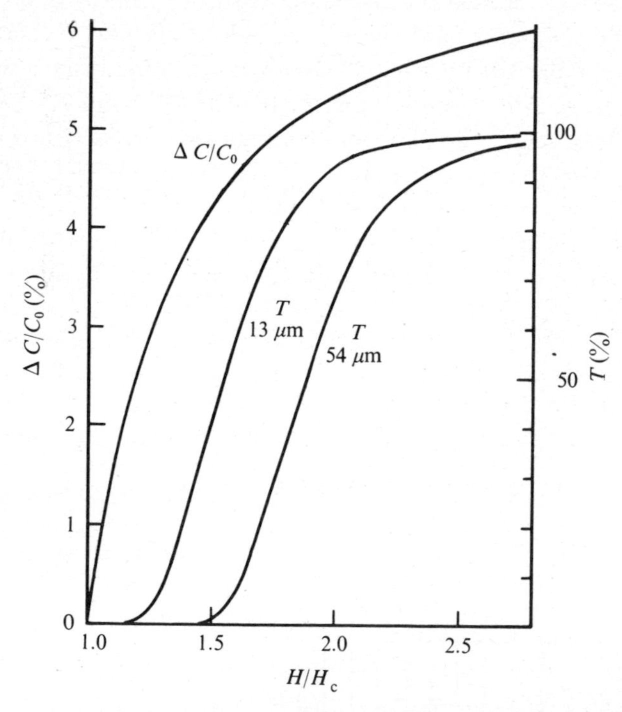

Fig. 4.1.5. Computed relative capacitance change $\Delta C/C_0$ and optical transmission T between parallel polarizers (both parallel to the director at one of the boundaries) of a twisted nematic film as functions of H/H_c. The total twist angle is $\pi/2$. Film thicknesses 13 and 54 μm. The threshold for optical transmission increases with the thickness of the film. (After Van Doorn.[17])

enough and the effective γ reduced to a sufficiently low value over an appreciable thickness of the sample that the state of polarization of the emergent beam is affected and the distortion reveals itself optically. It must be emphasized however that both β and γ vary with the z coordinate as well as with the field.

The twisted nematic has important applications in display technology.[16,18] Unlike the dynamic scattering mode in which conductivity plays a crucial role (see § 3.10), the twisted nematic is a pure *field effect* device. It is therefore advantageous to use high purity (low conductivity) materials; the power consumption is accordingly very much less than in the dynamic scattering mode and electrolytic decomposition is also reduced to a minimum. If the twist angle is exactly 90° there is equal probability of the medium acquiring a right-handed or left-handed twist and disclination walls are formed. To avoid this, the twist angle is made very slightly different from 90° or alternatively a small quantity of a cholesteric impurity is added in order to favour one sense of twist throughout the film.

By the use of pleochroic filters, multi-colour displays have also been developed.[19] The dynamics of the twisted nematic cell has been discussed recently by Baur, Steib and Meier[20] and by Berreman.[21]

Absorbing systems: circular dichroism When linearly dichroic dye molecules are dissolved in a cholesteric liquid crystal the medium exhibits circular dichroism because of the helical arrangement of the solute molecules in the structure.[22] The theory developed above can be extended to take into account the effect of absorption by treating the layers as both linearly birefringent and linearly dichroic.[23,24] Assuming that principal axes of linear birefringence and linear dichroism are the same, the Jones matrix of any layer with reference to its principal axes is

$$\mathbf{G}=\begin{bmatrix}\exp(-\mathrm{i}\gamma) & 0\\ 0 & \exp(\mathrm{i}\gamma)\end{bmatrix}\begin{bmatrix}\exp(-k_a p) & 0\\ 0 & \exp(-k_b p)\end{bmatrix}$$

$$=\exp(-\xi)\begin{bmatrix}\exp(-\mathrm{i}\hat{\gamma}) & 0\\ 0 & \exp(\mathrm{i}\hat{\gamma})\end{bmatrix},$$

where k_a, k_b are the principal absorption coefficients of the layer, and

$$\xi=\tfrac{1}{2}(k_a+k_b)p,$$

$$\hat{\gamma}=\gamma-\tfrac{1}{2}\mathrm{i}(k_a-k_b)p=\gamma-\mathrm{i}\mu,\ \text{say}.$$

Proceeding as before, the Jones matrix for m layers is

$$\mathbf{J}_m=\mathbf{S}^m(\mathbf{G}\mathbf{S}^{-1})^m. \tag{4.1.16}$$

If λ_1 and λ_2 are the eigenvalues of $(\mathbf{GS})^{-1}$, it can be shown that

$$(\mathbf{GS}^{-1})^m = \frac{\lambda_1^m - \lambda_2^m}{\lambda_1 - \lambda_2}\mathbf{GS}^{-1} - \lambda_1\lambda_2\frac{\lambda_1^{m-1} - \lambda_2^{m-1}}{\lambda_1 - \lambda_2}\mathbf{E}, \qquad (4.1.17)$$

where

$$\lambda_1 = \exp(-\xi)\exp(\mathrm{i}\hat{\theta}),$$

$$\lambda_2 = \exp(-\xi)\exp(-\mathrm{i}\hat{\theta}),$$

and

$$\cos\hat{\theta} = \cos\hat{\gamma}\cos\beta.$$

Using (4.1.16) and (4.1.17)

$$\mathbf{J}_m = \exp(-m\xi)\mathbf{S}^m\left[\frac{\sin m\hat{\theta}}{\sin\hat{\theta}}\mathbf{GS}^{-1} - \frac{\sin(m-1)\hat{\theta}}{\sin\hat{\theta}}\mathbf{E}\right]. \qquad (4.1.18)$$

Such a system can be resolved uniquely into a rotator, a retarder, a circularly dichroic plate and a linearly dichroic plate. The unique matrix resolution is given by

$$\mathbf{J}_m = \exp(-\chi)\,\mathbf{\Psi R\Sigma\Phi K\Sigma}\,\mathbf{\Psi}^{-1}, \qquad (4.1.19)$$

where

$$\mathbf{\Psi} = \begin{bmatrix} \cos\psi & -\sin\psi \\ \sin\psi & \cos\psi \end{bmatrix},$$

$$\mathbf{R} = \begin{bmatrix} \cos\rho & -\sin\rho \\ \sin\rho & \cos\rho \end{bmatrix},$$

$$\mathbf{\Sigma} = \begin{bmatrix} \cosh\sigma/2 & \mathrm{i}\sinh\sigma/2 \\ -\mathrm{i}\sinh\sigma/2 & \cosh\sigma/2 \end{bmatrix},$$

$$\mathbf{\Phi} = \begin{bmatrix} \exp(-\mathrm{i}\varphi) & 0 \\ 0 & \exp(\mathrm{i}\varphi) \end{bmatrix},$$

$$\mathbf{K} = \begin{bmatrix} \exp(-\kappa) & 0 \\ 0 & \exp(\kappa) \end{bmatrix}.$$

Here ρ is the rotation, σ the circular dichroism, 2φ the linear phase retardation, 2κ the linear dichroism, χ the attenuation coefficient and ψ the azimuth of the retardation plate (or, equivalently, of the linearly dichroic plate). From (4.1.18) and (4.1.19),

$$\left.\begin{aligned} &\rho - \mathrm{i}\sigma = m(\beta - \hat{\theta}'),\\ &\varphi - \mathrm{i}\kappa = \cos^{-1}\left(\frac{\sec^2 m\hat{\theta}'}{\sec^2 m\hat{\theta}}\right)^{1/2},\\ &\psi = \tfrac{1}{2}[(m+1)\beta - \rho],\\ &\chi = m\xi, \end{aligned}\right\} \tag{4.1.20}$$

with

$$\hat{\theta}' = \frac{1}{m}\tan^{-1}\left(\frac{\tan\beta\tan m\hat{\theta}}{\tan\hat{\theta}}\right).$$

When β is much larger than γ

$$\varphi = \kappa \simeq 0,$$

and

$$\rho - \mathrm{i}\sigma = -m\hat{\gamma}^2/2\beta.$$

Hence

$$\rho = -m(\gamma^2 - \mu^2)/2\beta,$$

and

$$\sigma = -m\gamma\mu/\beta.$$

Therefore, the linear dichroism of the layers not only results in circular dichroism but also makes a small contribution to the optical rotation which is opposite in sign to that due to linear birefringence. However this contribution is usually negligibly small. A further consequence of the theory is that σ changes sign whenever β or μ changes sign. This has been confirmed experimentally by Sackmann and Voss.[22]

When β is very small the medium behaves as an absorbing twisted nematic. At intermediate values of β, the complete expressions (4.1.20) have to be used. The parameter σ exhibits a marked dependence on pitch and on sample thickness. In actual practice σ is not usually measured directly: the experimental procedure generally consists of measuring the transmitted intensities I_R and I_L for right- and left-circular light (of equal intensity) incident on the medium and then evaluating the quantity

$$D_0 = \frac{I_R - I_L}{I_R + I_L + 2(I_R I_L)^{1/2}}.$$

To evaluate D_0 theoretically we use the following relations which describe the nature of the emergent light when right- or left-circular light is incident:

$$\begin{bmatrix} A_1 \\ A_2 \end{bmatrix} = 2^{-1/2}\mathbf{J}_m \begin{bmatrix} 1 \\ \mathrm{i} \end{bmatrix} \quad \text{for right-circular light,}$$

$$\begin{bmatrix} B_1 \\ B_2 \end{bmatrix} = 2^{-1/2}\mathbf{J}_m \begin{bmatrix} 1 \\ -\mathrm{i} \end{bmatrix} \quad \text{for left-circular light.}$$

Then

$$I_{\mathrm{R}} = |A_1|^2 + |A_2|^2,$$

$$I_{\mathrm{L}} = |B_1|^2 + |B_2|^2.$$

Fig. 4.1.6 presents the theoretical variation of the dichroic power D $(=D_0/t$, where t is the sample thickness) versus the inverse pitch. The calculations are for the cholesteryl chloride–cholesteryl myristate mixture, using the same parameters as in fig. 4.1.2. Additionally it is assumed that the linear dichroism $(k_a - k_b)p = 10^{-4}$ at $\beta = 3 \times 10^{-3}$ on the low temperature side of T_{N}, and that the layer dichroism and the layer birefringence decrease at the same rate with temperature. D changes sign on crossing the nematic point and also exhibits an interesting variation

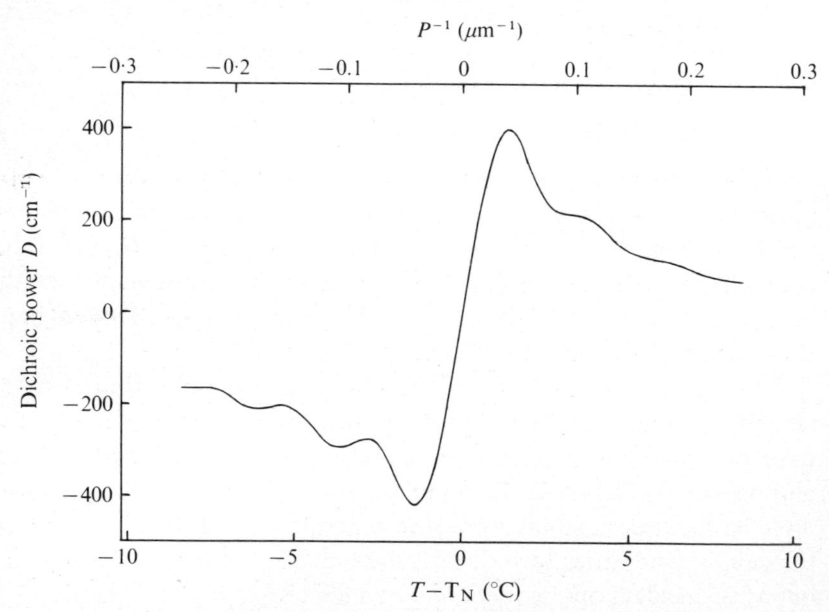

Fig. 4.1.6. Theoretical dependence of the dichroic power on inverse pitch (or temperature) for a sample of thickness 6 μm. (After reference 24.)

with sample thickness (table 4.1.1). In this respect the medium behaves quite differently from normal optically active substances in which ρ and D are independent of the thickness.

These predictions are in qualitative agreement with observations[24] (fig. 4.1.7). The weak oscillations in the theoretical curves were not detected, presumably because of slight inhomogeneities in the sample or variations in its thickness.

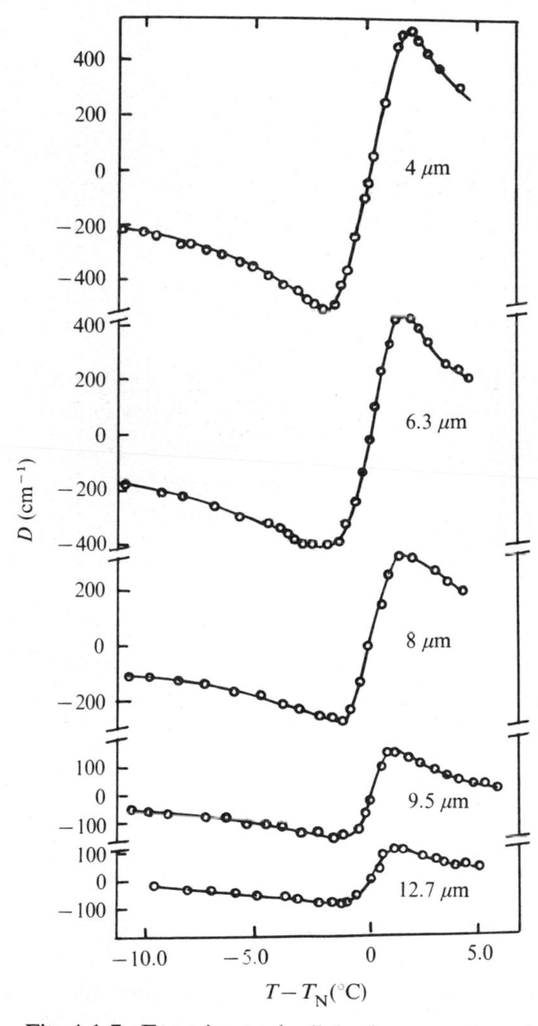

Fig. 4.1.7. Experimental dichroic power versus temperature in a 1.64 : 1 by weight mixture of cholesteryl chloride and cholesteryl myristate mixture containing 1.48 per cent by weight of β-carotene for samples of thickness 4, 6.3, 8, 9.5 and 12.7 μm. $\lambda = 0.5$ μm. (After reference 24.)

TABLE 4.1.1. *Theoretical dependence of peak dichroic power on sample thickness*

Sample thickness (μm)	Negative peak value of D (cm^{-1})
2	299.8
3	382.5
4	430.2
5	441.3
7	366.5
9	348.6
11	341.0
13	305.9
15	298.5
17	272.8
19	253.6

4.1.2. Propagation along the optic axis for wavelengths ~ pitch: analogy with Darwin's dynamical theory of X-ray diffraction

When the wavelength is comparable to the pitch, the optical properties are modified profoundly. Before discussing the rigorous electromagnetic treatment of the problem it is instructive to examine it first from the standpoint of X-ray diffraction theory.[9,25] Since the dynamical theory of X-ray diffraction from perfect crystals and its applications are now quite thoroughly understood, this approach may be useful in elucidating the optical behaviour of cholesterics and in looking for new optical analogues of certain well established X-ray effects. An example of a new phenomenon reported recently is the *Borrmann effect* in cholesterics.[26]

Kinematical theory of reflexion The theory discussed in § 4.1.1 shows that as long as the pitch is not too large compared with the wavelength, i.e., when $\frac{1}{2}P\delta n \ll \lambda$, the liquid crystal can be treated to a good approximation as a pure rotator for light propagating along the helical axis. In other words, right- and left-circular waves* travel without change of form but at slightly different velocities. The refractive indices for the two components

* Right- and left-circular polarizations are defined from the point of view of an observer looking at the source of light. If the electric vector rotates clockwise with progress of time, then it is right-circular. Thus, at any instant of time, the tip of the electric vector forms a right-handed screw in space for right-circular polarization.

are respectively

$$n_R = n - [(\delta n)^2 P/8\lambda]$$

$$n_L = n + [(\delta n)^2 P/8\lambda]$$

and the rotatory power

$$\rho = -\pi(\delta n)^2 P/4\lambda^2 \tag{4.1.21}$$

where $\delta n = n_a - n_b$ and $n = \frac{1}{2}(n_a + n_b)$.

We shall now give a simple interpretation of how under certain conditions reflexion of one of the circularly polarized components takes place. Let right-circular light given by $\mathbf{D}_0 = [\begin{smallmatrix}1\\ i\end{smallmatrix}]$ referred to x, y be incident along z. We shall suppose that the structure is right-handed, i.e., β is positive. To calculate the reflexion coefficient at the boundary between the $(s+1)$th and $(s+2)$th layers, we resolve the incident light vector along the principal axes of the $(s+1)$th layer which are inclined at angle $(s+1)\beta$ with respect to x, y. The resolved components are

$$\begin{bmatrix}\xi\\ \eta\end{bmatrix} = \begin{bmatrix}1\\ i\end{bmatrix} \exp[i\{(s+1)\beta - \varphi_{s+1}\}], \tag{4.1.22}$$

where $\varphi_{s+1} = 2\pi n_R(s+1)p/\lambda$, p being the thickness of each layer. At the boundary, the ξ vibration emerges from a medium of refractive index n_a and the η vibration from a medium of refractive index n_b. Qualitatively, it is obvious that since the principal axes of the $(s+2)$th layer are rotated slightly with respect to those of the $(s+1)$th layer, one of the components of (4.1.22) will on emerging from the $(s+1)$th layer meet a 'rarer' medium while the other will meet a 'denser' medium. One component therefore gets reflected without any change of phase and the other with a phase change of π. Thus, in contrast to reflexion from a normal dielectric, the sense of circular polarization remains the same after reflexion. Applying the standard formulae for reflexion at normal incidence from the surface of a non-absorbing anisotropic crystal, the reflected components ξ', η', referred to the principal axes of the $(s+2)$th layer, are

$$\begin{bmatrix}\xi'\\ \eta'\end{bmatrix} = -\frac{\beta\delta n}{2n}\begin{bmatrix}i\\ 1\end{bmatrix} \exp[i\{(s+1)\beta - \varphi_{s+1}\}]$$

$$= -iq\begin{bmatrix}1\\ -i\end{bmatrix} \exp[i\{(s+1)\beta - \varphi_{s+1}\}],$$

where $q = \beta\,\delta n/2n$. We make the approximation here that $\sin\beta \simeq \beta$, since β is assumed to be very small ($\sim 10^{-2}$ radian). On reflexion a very slight ellipticity is introduced in the transmitted beam but we shall ignore it in the present discussion. Transforming back to x, y the reflected wave on reaching the surface of the liquid crystal will be

$$\begin{bmatrix} X \\ Y \end{bmatrix} = -iq \begin{bmatrix} 1 \\ -i \end{bmatrix} \exp[i\{(2s+2)\beta - 2\varphi_{s+1}\}],$$

which represents a *right-circular* vibration travelling in the negative direction of z. Clearly the phase difference between this wave and that reflected at the boundary between the first and second layers is $2(s\beta - \varphi_s)$. When $\lambda = n_R P$, we have $2\pi n_R p/\lambda = \beta$ and $\varphi_s = s\beta$ (since $mp = P$ and $m\beta = 2\pi$, where m is the number of layers per turn of the helix). Hence the phase factor $\exp[2i(s\beta - \varphi_s)]$ becomes unity irrespective of the value of s, and there results a strong interference maximum. For a left-handed structure, β is negative and $(s\beta - \varphi_s)$ does not vanish; therefore, the waves from the different layers will not be in phase and the vibration will be transmitted practically unchanged.

Using the kinematical approximation, i.e., neglecting multiplc reflexions within the m layers, the reflexion coefficient per turn of the helix is

$$|Q| = m|q| = \pi\delta n/n. \tag{4.1.23}$$

Dynamical theory of reflexion The complete solution of the problem has to take into account the effect of multiple reflexions. This can be done by setting up difference equations closely similar to those formulated by Darwin[27] in his dynamical theory of X-ray diffraction. For the purposes of this theory we shall regard the liquid crystal as consisting of a set of parallel planes spaced P apart. Each plane therefore replaces the m layers per turn of the helix. We ascribe a reflexion coefficient $-iQ$ per plane for right-circular light at normal incidence. Assuming a kinematical approximation for the m layers, Q is given by (4.1.23). We can then write the difference equations in a simple manner because, as stated earlier, circularly polarized waves travel practically without change of form, so that the interference of the multiply reflected waves with one another and with the primary wave can be evaluated directly.

We shall suppose, as before, that structure is right-handed and that right-circular light is incident normally. Let T_r and S_r be the complex amplitudes of the primary and reflected waves at a point just above the rth plane, the topmost plane being designated by the serial number zero (fig. 4.1.8). Neglecting absorption, the difference equations may be written as

$$S_r = -iQT_r + \exp(-i\varphi)S_{r+1}, \tag{4.1.24}$$

$$T_{r+1} = \exp(-i\varphi)\, T_r - iQ \exp(-2i\varphi)S_{r+1}, \tag{4.1.25}$$

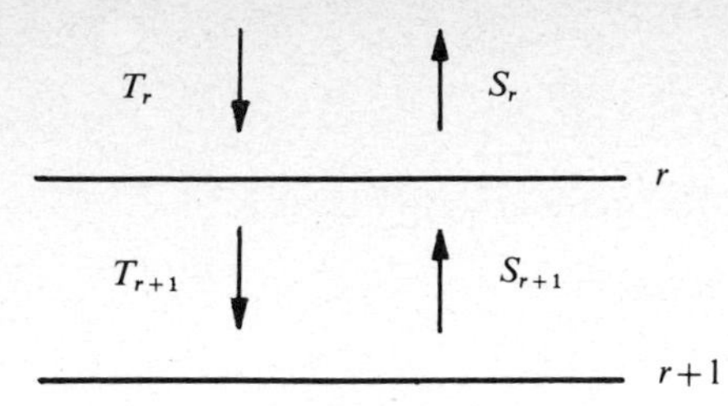

Fig. 4.1.8. Notation for the primary (T) and reflected (S) waves in the dynamical theory.

where $\varphi = 2\pi n_R P/\lambda$. The reflexion coefficient is here taken to be the same on both sides of the plane. Replacing r by $r-1$ in (4.1.24) and (4.1.25), substituting and simplifying, we obtain

$$T_{r+1} + T_{r-1} = yT_r, \tag{4.1.26}$$

$$S_{r+1} + S_{r-1} = yS_r, \tag{4.1.27}$$

where

$$y = \exp(i\varphi) + \exp(-i\varphi) + Q^2 \exp(-i\varphi). \tag{4.1.28}$$

Suppose that the liquid crystal is a film consisting of ν planes. Putting $S_\nu = 0$, we have from (4.1.27)

$$S_{\nu-2} = yS_{\nu-1},$$

$$S_{\nu-3} = yS_{\nu-2} - S_{\nu-1} = (y^2 - 1)S_{\nu-1},$$

$$S_{\nu-4} = (y^3 - 2y)S_{\nu-1}, \text{ etc.,}$$

and

$$S_0 = \left(y^{\nu-1} - \frac{\nu-2}{1!}y^{\nu-3} + \frac{(\nu-4)(\nu-3)}{2!}y^{\nu-5} - \ldots\right)S_{\nu-1}$$

$$= f_\nu(y)S_{\nu-1} \text{ (say).} \tag{4.1.29}$$

Similarly from (4.1.25), (4.1.26) and (4.1.28)

$$T_{\nu-1} = \exp(i\varphi)T_\nu$$

$$T_{\nu-2} = (y \exp(i\varphi) - 1)T_\nu$$

$$T_{\nu-3} = [(y^2 - 1)\exp(i\varphi) - y]T_\nu, \text{ etc.}$$

and

$$T_0 = (f_\nu(y)\exp(i\varphi) - f_{\nu-1}(y))T_\nu. \tag{4.1.30}$$

Since, from (4.1.24)

$$S_{\nu-1}=-\mathrm{i}QT_{\nu-1}=-\mathrm{i}Q\exp(\mathrm{i}\varphi)\,T_\nu,$$

the ratio of the reflected to the incident amplitudes is

$$\frac{S_0}{T_0}=-\frac{\mathrm{i}Qf_\nu(y)\exp(\mathrm{i}\varphi)}{f_\nu(y)\exp(\mathrm{i}\varphi)-f_{\nu-1}(y)}. \tag{4.1.31}$$

Let us assume a solution in the form

$$T_{r+1}=xT_r, \tag{4.1.32}$$

where x is independent of r. Hence x must satisfy

$$x+(1/x)=y=\exp(\mathrm{i}\varphi)+\exp(-\mathrm{i}\varphi)+Q^2\exp(-\mathrm{i}\varphi).$$

We have seen that the reflexion condition is $n_\mathrm{R}P=\lambda_0$ or $\varphi_0=2\pi$. Accordingly, we may write

$$\varphi=2\pi\lambda_0/\lambda=\varphi_0+e,$$

where

$$e=-2\pi(\lambda-\lambda_0)/\lambda,$$

which is a small quantity in the neighbourhood of the reflexion. Therefore

$$x+(1/x)=\exp(\mathrm{i}e)+\exp(-\mathrm{i}e)+Q^2\exp(-\mathrm{i}e). \tag{4.1.33}$$

This suggests that in the neighbourhood of the reflexion we may put

$$x=\exp(-\xi)\exp(-\mathrm{i}\varphi_0)=\exp(-\xi), \tag{4.1.34}$$

where ξ is small and may be complex. From (4.1.33) and (4.1.34)

$$\xi=\pm(Q^2-e^2)^{1/2}.$$

When

$$y=\exp(\xi)+\exp(-\xi)=2\cosh\xi,$$

the series in (4.1.29) is given by[28]

$$f(y)=\frac{\sinh\nu\xi}{\sinh\xi}. \tag{4.1.35}$$

Substituting in (4.1.31) and simplifying

$$\frac{S_0}{T_0}\simeq\frac{-\mathrm{i}Q\exp(\mathrm{i}e)}{\mathrm{i}e+\xi\coth\nu\xi}, \tag{4.1.36}$$

or

$$\mathscr{R} = \left|\frac{S_0}{T_0}\right|^2 = \frac{Q^2}{e^2 + \xi^2 \coth^2 \nu\xi}.$$

For the semi-infinite medium, $\nu = \infty$ in (4.1.36) and

$$\frac{S_0}{T_0} = -\frac{Q}{e \pm \mathrm{i}(Q^2 - e^2)^{1/2}}. \tag{4.1.37}$$

When $-Q < e < Q$, ξ is real and

$$\mathscr{R} = \left|\frac{S_0}{T_0}\right|^2 = 1.$$

The reflexion is total within this range. The spectral width of total reflexion is therefore $\Delta\lambda = Q\lambda_0/\pi$. Using (4.1.23),

$$\Delta\lambda = P\delta n, \tag{4.1.38}$$

a result first derived by de Vries.[(3)] Outside this range, the reflexion decreases rapidly on either side. When $\lambda > \lambda_0$, e is negative and hence the negative value of the square root in the denominator of (4.1.37) has to be taken because $\mathscr{R}$ can never exceed unity; when $\lambda < \lambda_0$ the positive root has to be taken.

Illustrative curves of $\mathscr{R}$ as a function of wavelength are shown in fig. 4.1.9. The semi-infinite medium gives the familiar flat topped curve of the dynamical theory, while the thin film gives a principal maximum accompanied by subsidiary fringes. The fringes are somewhat difficult to observe as even slight inhomogeneities in the specimen and small variations in its thickness tend to obliterate them, but careful experiments by Dreher, Meier and Saupe[(29)] have confirmed that they do occur (fig. 4.1.10).

Primary extinction and anomalous rotatory dispersion If reflexions are neglected, the optical rotation per thickness P of the liquid crystal is $\frac{1}{2}(\varphi_R - \varphi_L)$ and the rotatory power is given by (4.1.21). Near the region of reflexion, the dynamical theory shows that the right-circular component suffers anomalous phase retardation and, under certain circumstances, attenuation as it travels through the medium. Left-circular light, on the other hand, exhibits normal behaviour throughout, and as a consequence the rotatory dispersion is anomalous around the reflecting region.

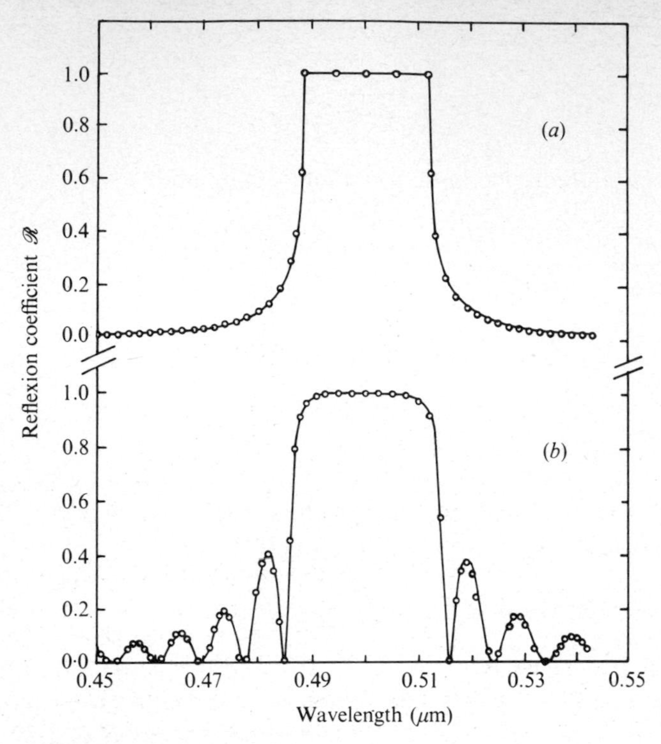

Fig. 4.1.9. Reflexion coefficient $\mathscr{R}$ at normal incidence versus wavelength for a non-absorbing cholesteric: (*a*) semi-infinite medium, (*b*) film of thickness $25P$, where P is the pitch. Curves are derived from the dynamical theory; circles represent values computed from the exact theory (§ 4.1.3) assuming that the medium external to the cholesteric (e.g., glass) has a refractive index 1.5. The parameters used in the calculations are $n = 1.5$, $\delta n = 0.07$, $\lambda_0 = nP = 0.5$ μm. (After reference 25.)

(i) We shall consider first the semi-infinite case. According to (4.1.32), the amplitude of the right-circular wave as it passes from one plane to the next is given by

$$T_{r+1} = xT_r$$

where

$$x = \exp(-\xi)\exp(-i\varphi_0),$$

$$\xi = \pm(Q^2 - e^2)^{1/2},$$

$$\varphi_0 = \varphi_R - e = 2\pi.$$

Inside the totally reflecting range, ξ is real and the wave is strongly attenuated. In X-ray diffraction theory, this phenomenon is referred to as

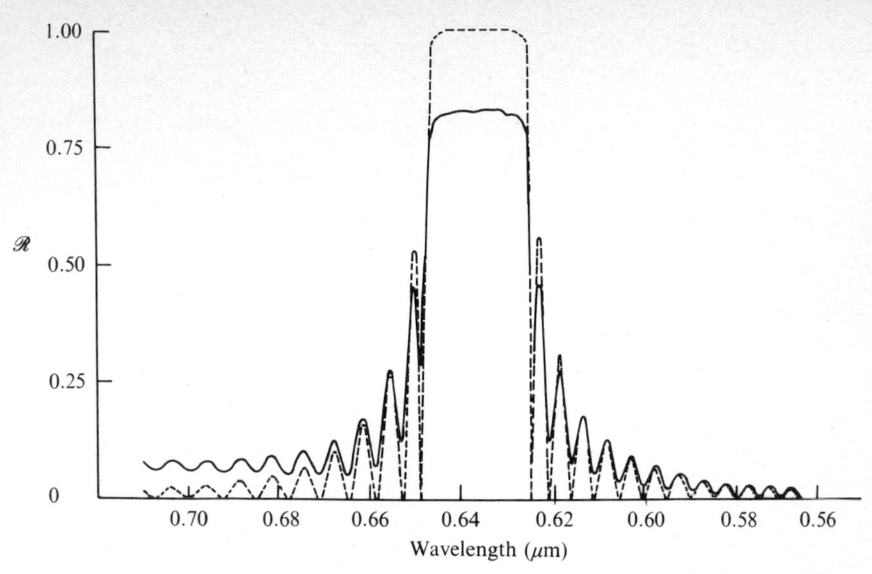

Fig. 4.1.10. Reflexion spectrum from a monodomain cholesteric film at normal incidence. Full curve: experimental spectrum for a mixture of cholesteryl nonanoate, cholesteryl chloride and cholesteryl acetate in weight ratios 21:15:6 at 24 °C (intensity in arbitrary units). Broken curve: spectrum computed from the exact theory for a film thickness of 21.0 μm and pitch 0.4273 μm. (After Dreher, Meier and Saupe.[29])

primary extinction. The extinction length, defined as the distance over which the amplitude of the incident wave decreases to 1/e of its value, is P/Q at the centre of the reflexion band.

Outside the range of total reflexion, ξ is imaginary and primary extinction vanishes. Fig. 4.1.11 gives a plot of the wave-vectors $K_R(=2\pi n_R/\lambda)$ and $K_L(=2\pi n_L/\lambda)$ for right- and left-circular polarizations respectively, as functions of the wavelength. The real part of K_R shows a gap within the reflexion band – analogous to the familiar band gap in solid state physics – while the imaginary part grows rapidly in the same region.

When $e^2>Q^2$, $\xi=\mathrm{i}(e^2-Q^2)^{1/2}$ and may be positive or negative. The optical rotation per pitch is clearly

$$\tfrac{1}{2}[(e^2-Q^2)^{1/2}+\varphi_0-\varphi_L]=\tfrac{1}{2}(\varphi_R-\varphi_L)-\tfrac{1}{2}e\left[1-\left(1-\frac{Q^2}{e^2}\right)^{1/2}\right]$$

and hence the rotatory power in radians per unit length

$$\rho=-\frac{\pi(\delta n)^2P}{4\lambda^2}+\frac{\pi(\lambda-\lambda_0)}{P}\left[1-\left(1-\frac{Q^2}{e^2}\right)^{1/2}\right]. \qquad (4.1.39)$$

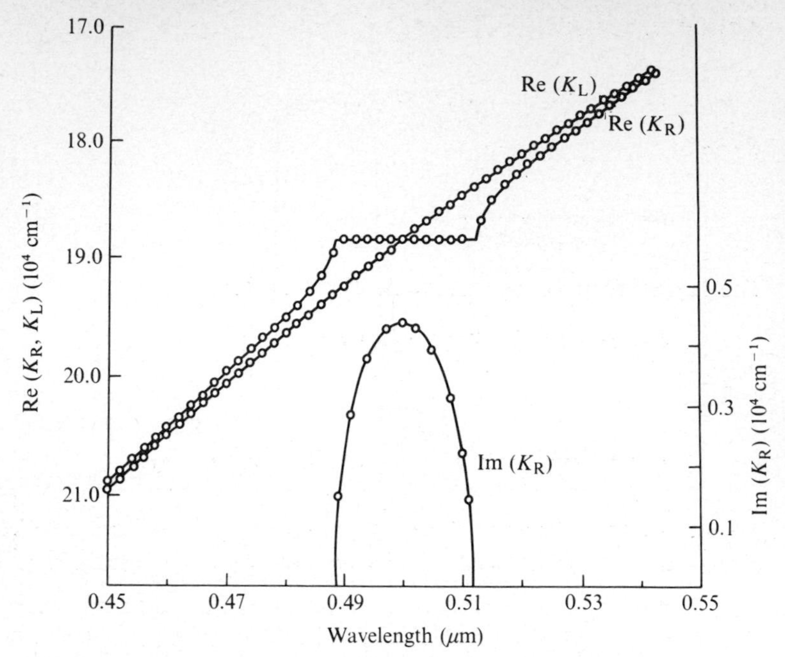

Fig. 4.1.11. The wave-vectors K_R and K_L of the normal waves as functions of wavelength in a semi-infinite non-absorbing medium. Curves are derived from the dynamical theory; circles represent values computed from the exact theory. n, δn and λ_0 same as in fig. 4.1.9.[25]

When $Q^2 > e^2$, i.e., within the region of total reflexion, ρ given by (4.1.39) becomes complex, showing that the medium is now circularly dichroic. The real part which represents the rotatory power is

$$\rho = -\frac{\pi(\delta n)^2 P}{4\lambda^2} + \frac{\pi(\lambda - \lambda_0)}{P\lambda}. \tag{4.1.40}$$

(ii) For a thin film, we have from (4.1.30) and (4.1.35)

$$\frac{T_\nu}{T_0} = \left[\exp(ie)\frac{\sinh \nu\xi}{\sinh \xi} - \frac{\sinh(\nu - 1)\xi}{\sinh \xi}\right]^{-1} \approx \frac{\xi \operatorname{cosech} \nu\xi}{ie + \xi \coth \nu\xi} \tag{4.1.41}$$

and

$$\left|\frac{T_\nu}{T_0}\right|^2 + \left|\frac{S_0}{T_0}\right|^2 = 1.$$

Consequently the oscillations which appear in the reflexion curve should also be seen in transmission and in circular dichroism. Equation (4.1.41) may be expressed as

$$\frac{T_\nu}{T_0} = A \exp[-i\nu(\varphi_0 + \psi)],$$

where

$$\tan \nu\psi = \frac{e}{\xi \coth \nu\xi}.$$

The optical rotation per thickness P is therefore

$$\tfrac{1}{2}(\varphi_0 + \psi - \varphi_L) = \tfrac{1}{2}[(\varphi_R - \varphi_L) + (\psi - e)]$$

and

$$\rho = -\frac{\pi(\delta n)^2 P}{4\lambda^2} + \frac{\psi - e}{2P}.$$

The theoretical variation of ρ with λ is shown in fig. 4.1.12. As observed already in § 4.1.1 (see fig. 4.1.3) the rotatory power of a cholesteric liquid

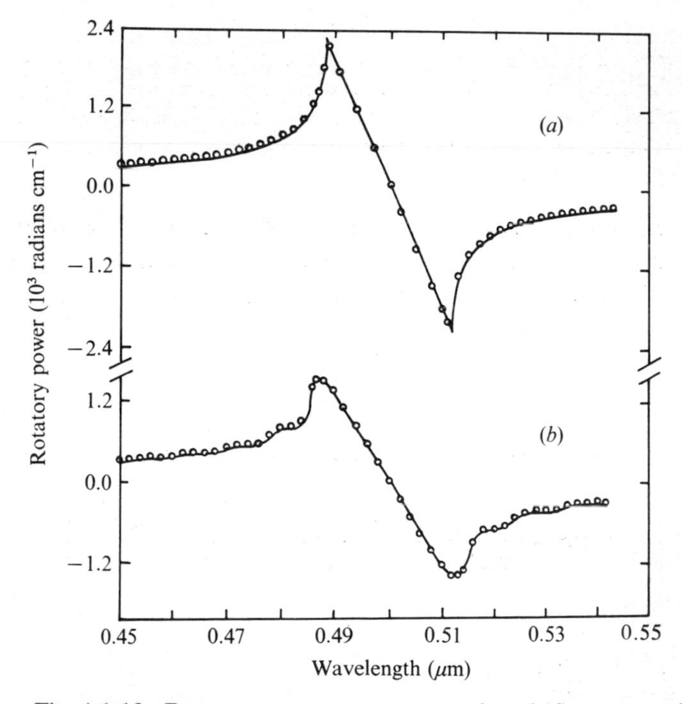

Fig. 4.1.12. Rotatory power versus wavelength for a non-absorbing cholesteric: (*a*) semi-infinite medium, (*b*) film of thickness 25*P*. Curves are derived from the dynamical theory; circles represent values computed from the exact theory. n, δn and λ_0 same as in fig. 4.1.9.[25]

crystal, unlike that of an ordinary optically active substance, is a function of the sample thickness. Some recent measurements[30] of the optical rotation right through the reflexion band using thin films are presented in fig. 4.1.13. The oscillations in the theoretical curve for ρ appear to be smeared out, probably owing to slight imperfections in the sample, but the trends are in agreement with theory. There is also some evidence of subsidiary maxima in the circular dichroism which again is to be expected theoretically.

Absorbing systems: the Borrmann effect The Borrmann effect is the anomalous increase in the transmitted X-ray intensity when a crystal is set for Bragg reflexion.[31] An analogous optical effect in absorbing cholesteric media in the vicinity of the reflexion band has been predicted and confirmed experimentally.[26,32] The origin of the effect can be readily understood by extending the dynamical theory to include absorption. However, in contrast to the X-ray case, the polarization of the wave field and the linear dichroism play an essential part.

Suppose that the birefringent layers are also linearly dichroic and that the principal axes of linear birefringence and linear dichroism are the same. All the equations obtained for the non-absorbing medium hold

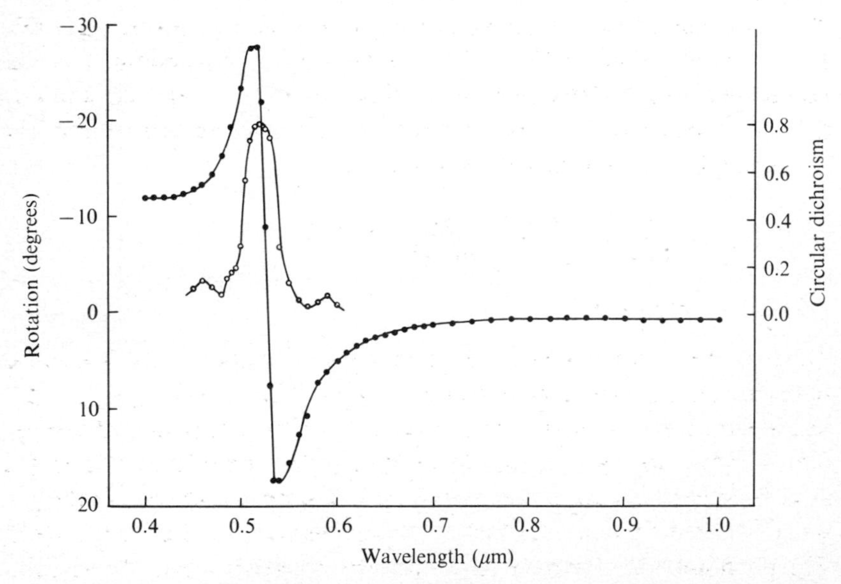

Fig. 4.1.13. Experimental circular dichroism (open circles) and rotatory dispersion (closed circles) of cholesteric cinnamate at 177 °C. Sample thickness ~3 μm. (After reference 30).

good in this case also except that Q, φ_R, φ_L etc. have to be replaced by the corresponding complex quantities:

$$\hat{Q} = \pi\delta\hat{n}/\hat{n},$$

$$\hat{\varphi}_R = \frac{2\pi\hat{n}_R P}{\lambda} = \frac{2\pi\hat{n}P}{\lambda} - \frac{\pi(\delta\hat{n})^2P^2}{4\lambda^2},$$

$$\hat{\varphi}_L = \frac{2\pi\hat{n}_L P}{\lambda} = \frac{2\pi\hat{n}P}{\lambda} + \frac{\pi(\delta\hat{n})^2P^2}{4\lambda^2},$$

$$\hat{e} = \frac{2\pi}{\lambda}(\hat{n}_R P - \lambda), \qquad \hat{\xi} = \pm(\hat{Q}^2 - \hat{e}^2)^{1/2},$$

$$\hat{K}_R = 2\pi\hat{n}_R/\lambda, \qquad \hat{K}_L = 2\pi\hat{n}_L/\lambda, \text{ etc.},$$

with

$$\delta\hat{n} = \hat{n}_a - \hat{n}_b, \qquad \hat{n} = \tfrac{1}{2}(\hat{n}_a + \hat{n}_b),$$

$$\hat{n}_a = n_a - \mathrm{i}\kappa_a, \qquad \hat{n}_b = n_b - \mathrm{i}\kappa_b,$$

where κ_a, κ_b are the principal absorption coefficients. Fig. 4.1.14 gives the reflexion coefficient $\mathscr{R}$ and the dependence of the real part of $\hat{\rho}$ and the imaginary parts of $\hat{K}_R$ and $\hat{K}_L$ on wavelength. Here δn, n and P are taken to be the same as for the non-absorbing case (see fig. 4.1.12) and in addition it is assumed that $\kappa = \frac{1}{2}(\kappa_a + \kappa_b) = 0.02$ and $\delta\kappa = \kappa_a - \kappa_b = 0.028$. The interesting result is obtained that on the shorter wavelength side $\mathrm{Im}(\hat{K}_R)$ is less than $\mathrm{Im}(\hat{K}_L)$, i.e., the right-circular component is less attenuated than the left component, while on the longer wavelength side the opposite is true. To observe this effect thin films have to be used.

For an absorbing film of thickness νP,

$$\mathscr{T}_R = \left|\frac{T_\nu}{T_0}\right|^2 = \left|\frac{\hat{\xi}\ \mathrm{cosech}\ \nu\hat{\xi}}{\mathrm{i}\hat{e} + \hat{\xi}\coth \nu\hat{\xi}}\right|^2$$

$$\mathscr{T}_L = |\exp(-\mathrm{i}\nu\hat{\varphi}_L)|^2.$$

The theoretical dependence of $\mathscr{T}_R$ and $\mathscr{T}_L$ on λ is shown in fig. 4.1.15 for both the non-absorbing and absorbing cases. The structure being right-handed, the right-circular wave is reflected, and hence in a non-absorbing film ($\kappa = \delta\kappa = 0$) $\mathscr{T}_R$ is always less than $\mathscr{T}_L$. On the other hand, in the absorbing case $\mathscr{T}_R$ shows an enhanced value on the short wavelength side of the reflexion band, which is the analogue of the Borrmann effect. The phenomenon is shown up even more convincingly in the circular dichroism curves (fig. 4.1.16). It can be shown that $\mathscr{T}_L$ will exhibit an anomalous increase for a left-handed structure (i.e., negative β), and also

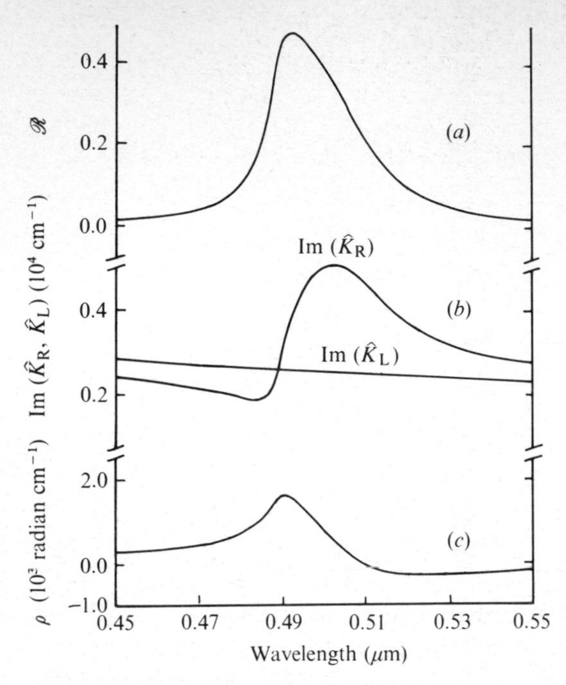

Fig. 4.1.14. (*a*) Reflexion coefficient $\mathcal{R}$ at normal incidence, (*b*) imaginary parts of $\hat{K}_R$ and $\hat{K}_L$, (*c*) rotatory power ρ (the real part of $\hat{\rho}$), plotted against the wavelength for an absorbing semi-infinite medium calculated from the dynamical theory. $\kappa = 0.02$, $\delta\kappa = 0.028$ and other parameters same as in fig. 4.1.9. (After reference 25.)

that the peak transmission will occur on the long wavelength side of the reflexion if $\delta\kappa$ is negative.

Experiments were conducted on cholesteryl nonanoate in which was dissolved small quantities of PAA or n-*p*-methoxybenzylidene-*p*-phenylazoaniline.[26,32] The temperature of the system was adjusted so that the reflexion band overlapped with the strongly linearly dichroic absorption band of the solute molecule. Under these circumstances, the circular dichroism exhibits the features predicted by theory (fig. 4.1.17).

4.1.3. Exact solution of the wave equation for propagation along the optic axis: the Mauguin–Oseen–de Vries model

We next consider the exact solution of the wave equation for propagation along the optic axis. The complete theory is contained in the papers of Mauguin,[1] Oseen[2] and de Vries,[3] and has since been presented in various forms by other authors.[33–36] We shall discuss an elegant treatment of the theory developed recently by Kats[33] and by Nityananda.[34]

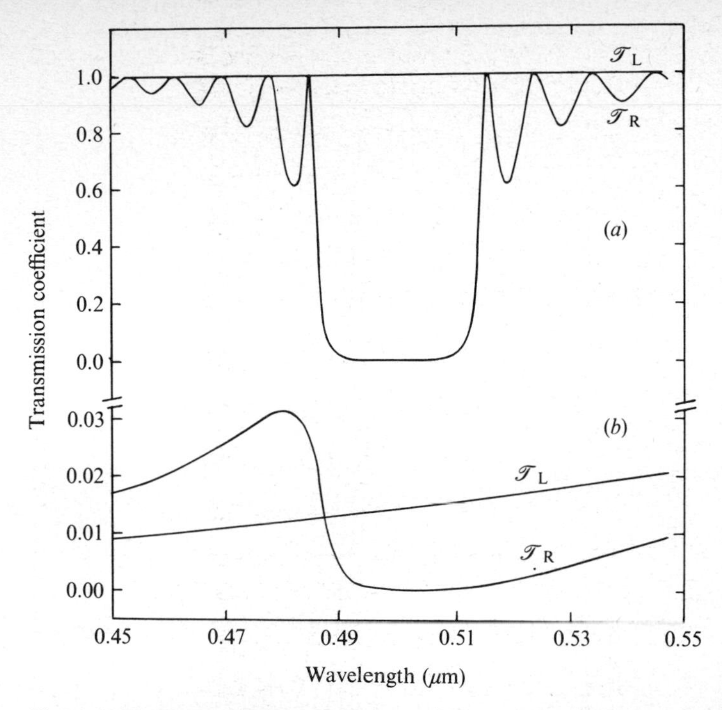

Fig. 4.1.15. Transmission coefficients, $\mathscr{T}_R$ and $\mathscr{T}_L$ for right- and left-circular waves calculated for a film of thickness $25P$; (*a*) non-absorbing and (*b*) absorbing. Parameters same as in fig. 4.1.14. The enhanced transmission for the right-circular component in (*b*) is the analogue of the Borrmann effect. (After reference 25.)

We represent the dielectric tensor by a 'spiralling ellipsoid' whose principal axis Oc is always parallel to z; the other two principal axes Oa and Ob (with principal values ε_a and ε_b) spiral around z with a twist angle $q = 2\pi/P$ per unit length. If Oa, Ob are taken to be along x, y at the origin, the tensor $\boldsymbol{\varepsilon}$ at any point z may be expressed with respect to x, y as

$$\boldsymbol{\varepsilon} = \begin{bmatrix} \cos qz & -\sin qz \\ \sin qz & \cos qz \end{bmatrix} \begin{bmatrix} \varepsilon_a & 0 \\ 0 & \varepsilon_b \end{bmatrix} \begin{bmatrix} \cos qz & \sin qz \\ -\sin qz & \cos qz \end{bmatrix}$$

$$= \begin{bmatrix} \varepsilon + \alpha \cos 2qz & \alpha \sin 2qz \\ \alpha \sin 2qz & \varepsilon - \alpha \cos 2qz \end{bmatrix} \qquad (4.1.42)$$

where $\varepsilon_a = n_a^2$, $\varepsilon_b = n_b^2$, $\varepsilon = \frac{1}{2}(\varepsilon_a + \varepsilon_b)$, $\alpha = \frac{1}{2}(\varepsilon_a - \varepsilon_b) = \frac{1}{2}(n_a + n_b)(n_a - n_b) = n\delta n$. The wave equation for propagation along z is

$$\frac{\partial^2 \mathbf{E}}{\partial z^2} = -\frac{\omega^2}{c^2}\boldsymbol{\varepsilon}\mathbf{E}. \qquad (4.1.43)$$

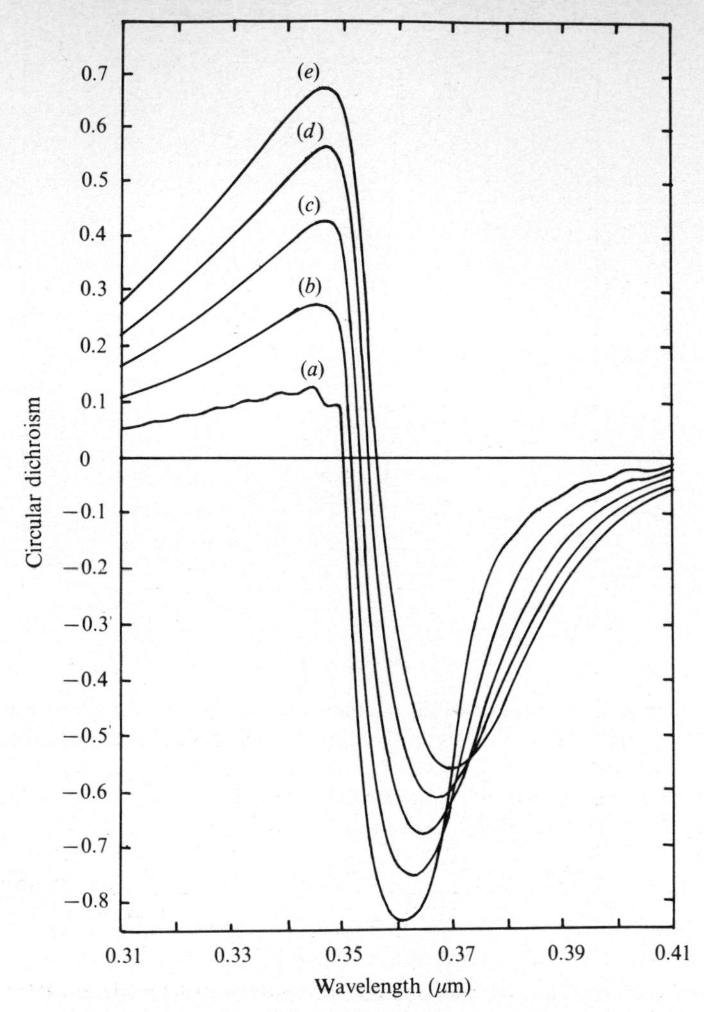

Fig. 4.1.16. Circular dichroism versus wavelength computed for different κ and $\delta\kappa$. Sample thickness $25P$, $n=1.5$, $\delta n=0.07$ and $\lambda_0=0.36$ μm. The absorption coefficients were assumed to be Gaussian curves having a maximum at 0.36 μm and of half-width 0.06 μm. The maximum values of κ and $\delta\kappa$ are respectively as follows: (*a*) 0.0125, 0.0157; (*b*) 0.0250, 0.0314; (*c*) 0.0375, 0.0471; (*d*) 0.0500, 0.0628; (*e*) 0.0625, 0.0785. (After reference 32.)

We introduce the variables

$$E' = 2^{-1/2}(E_x + \mathrm{i}E_y),$$

$$E'' = 2^{-1/2}(E_x - \mathrm{i}E_y).$$

E' is right-circular and E'' left-circular for propagation along $+z$ and vice

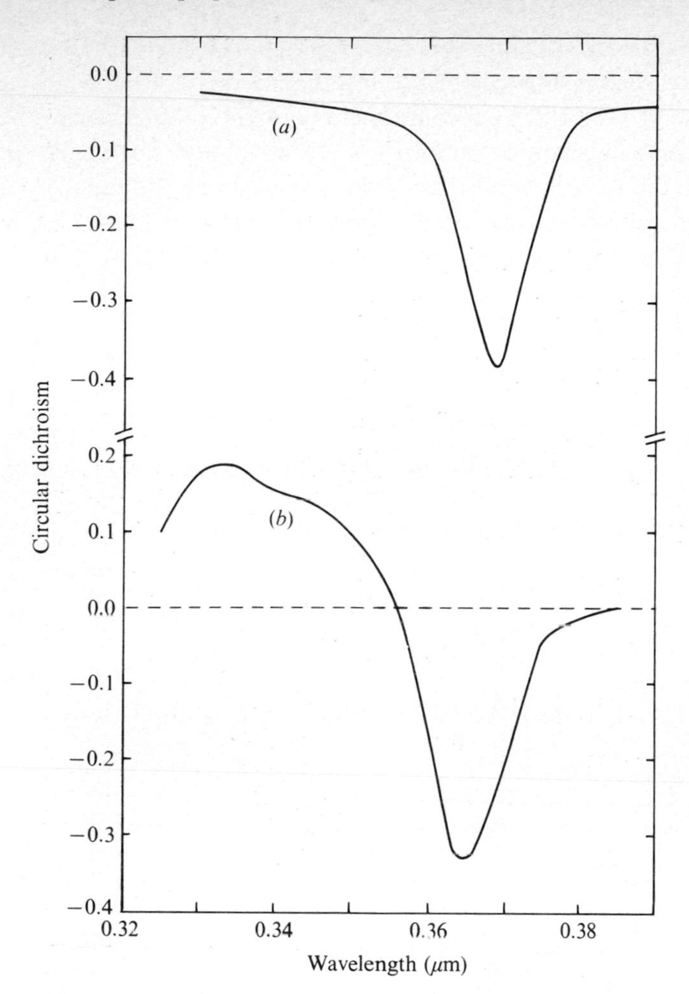

Fig. 4.1.17. Experimental circular dichroism curves versus wavelength. (*a*) Pure cholesteryl nonanoate (CN), (*b*) CN+0.98 per cent by weight of PAA. Sample thickness in both cases 6.5 μm. (After reference 32.)

versa for propagation along $-z$. Substituting in (4.1.43) we obtain

$$\begin{bmatrix} \dfrac{\partial^2 E'}{\partial z^2} \\ \dfrac{\partial^2 E''}{\partial z^2} \end{bmatrix} = -\frac{\omega^2}{c^2} \begin{bmatrix} \varepsilon & \alpha \exp(2iqz) \\ \alpha \exp(-2iqz) & \varepsilon \end{bmatrix} \begin{bmatrix} E' \\ E'' \end{bmatrix}. \tag{4.1.44}$$

The solution of (4.1.44) is of the form

$$\mathbf{u} = \begin{bmatrix} A' \exp[i(k+q)z] \\ A'' \exp[i(k-q)z] \end{bmatrix}, \tag{4.1.45}$$

which is a superposition of two waves of opposite circular polarizations with wave-vectors differing by $2q$. It is readily verified that when substituted into (4.1.44) the $k+q$ component of (4.1.45) suffers a wave-vector shift of $2q$ and is converted into a $k-q$ wave, and vice versa, so that together the two components form a closed set. Equation (4.1.45) therefore represents a true normal wave which can satisfy (4.1.44) with a proper choice of A' and A''. From (4.1.44) and (4.1.45), we have

$$\begin{bmatrix} (k+q)^2-K_m^2 & -\alpha K^2 \\ -\alpha K^2 & (k-q)^2-K_m^2 \end{bmatrix} \begin{bmatrix} A' \\ A'' \end{bmatrix} = 0, \tag{4.1.46}$$

where $K=2\pi/\lambda$ and $K_m = 2\pi n/\lambda$, λ being the wavelength *in vacuo*. The condition for (4.1.46) to yield non-vanishing solutions of A' and A'' is

$$[(k+q)^2-K_m^2][(k-q)^2-K_m^2]-\alpha^2K^4=0,$$

whose roots are

$$k_2, k_1 = [(K_m^2+q^2)\pm(4K_m^2q^2+\alpha^2K^4)^{1/2}]^{1/2}. \tag{4.1.47}$$

Corresponding to the k_1 and k_2 solutions we have respectively

$$\left.\begin{aligned} \frac{A'}{A''} &= \frac{\alpha K^2}{(k_1+q)^2-K_m^2} \\ &\text{and} \\ \frac{A''}{A'} &= \frac{\alpha K^2}{(k_2-q)^2-K_m^2}. \end{aligned}\right\} \tag{4.1.48}$$

When α is small, (4.1.47) and (4.1.48) give for the k_2 solution $A'/A'' \sim \alpha$, and for the k_1 solution $A''/A' \sim \alpha$. In other words, each normal wave is made up of two oppositely polarized circular components with one of the components generally dominating. The mixing of these two components with wave-vectors differing by $2q$ is a consequence of the Bragg reflexion. Equation (4.1.45) may conveniently be rewritten as

$$\mathbf{u}_1 = \begin{bmatrix} \exp(iK_1z) \\ d\exp i(K_1-2q)z \end{bmatrix}, \tag{4.1.49}$$

$$\mathbf{u}_2 = \begin{bmatrix} f\exp i(K_2+2q)z \\ \exp(iK_2z) \end{bmatrix}, \tag{4.1.50}$$

where

$$K_1 = k_1 + q = q + [K_m^2 + q^2 - (4K_m^2 q^2 + \alpha^2 K^4)^{1/2}]^{1/2}, \qquad (4.1.51)$$

$$K_2 = k_2 - q = -q + [K_m^2 + q^2 + (4K_m^2 q^2 + \alpha^2 K^4)^{1/2}]^{1/2}, \qquad (4.1.52)$$

$$d = \frac{K_1^2 - K_m^2}{\alpha K^2}, \qquad f = \frac{K_2^2 - K_m^2}{\alpha K^2}. \qquad (4.1.53)$$

The fact that the wave-vectors K_1 and K_2 are different is responsible for the optical activity of the medium, the optical rotation per unit length being $\rho = \frac{1}{2}(K_1 - K_2)$ radians. However, as emphasized in previous sections the phenomenon is not identical with natural optical activity because the normal waves are not pure circular waves.

The de Vries equation If now we make the approximation that $(K_1 - K_2)/q \ll 1$, or that the rotation per pitch is small compared with π, which is certainly valid in most cholesterics,

$$\rho = \tfrac{1}{2}(K_1 - K_2) = -\frac{x - (x^2 - \alpha^2 K^4)^{1/2}}{4q},$$

where

$$x = K_m^2 - q^2.$$

When $x^2 < \alpha^2 K^4$, ρ becomes a complex quantity. The real part gives the rotatory power and the imaginary part the circular dichroism. Since no dissipative mechanism is built into the model it follows that the imaginary part of ρ is associated with the reflexion of one of the components. The reflexion band is centred at $x = 0$, i.e., at $K_m = q$ or $\lambda_0 = nP$ where λ_0 is the wavelength *in vacuo*. The range of reflexion extends from $x = +\alpha K^2$ to $x = -\alpha K^2$, i.e., from $+q^2(\delta n/n)$ to $-q^2(\delta n/n)$. Since

$$\delta x = \delta(K_m)^2 = 2K_m(\delta K_m) \simeq 2q(\delta K_m),$$

the spectral width of total reflexion is given by

$$\Delta\lambda = P\delta n \qquad (4.1.54)$$

as was first shown by de Vries[3] (see (4.1.38)).

When $x^2 \gg \alpha^2 K^4$, which is not valid close to or inside the reflexion band,

$$\rho = -\frac{\alpha^2 K^4}{8qx}$$

$$= -\frac{\pi(\delta n)^2 P}{4\lambda^2[1-(\lambda^2/\lambda_0^2)]}. \qquad (4.1.55)$$

This is known as the *de Vries equation*. The sign of the rotary power reverses on crossing the reflexion band (λ_0). When $\lambda \ll \lambda_0$, (4.1.55) reduces to (4.1.21), and when $\lambda \gg \lambda_0$, ρ tends asymptotically to 0. The behaviour on either side of the reflexion band has been confirmed experimentally[30] (fig. 4.1.13).

Thin films Nityananda and Kini[37] have applied the theory to obtain exact solutions for reflexion and transmission by a plane parallel film bounded on either side by an isotropic medium. The treatment allows for the contribution due to reflexion at the cholesteric–isotropic interface. In general, for each circular polarization at normal incidence the reflected and transmitted waves consist of both circular polarizations. Four coefficients, two for reflexion and two for transmission, are required to describe the problem fully and the solution consists of matching the incoming and reflected waves on one side of the slab with four waves within the slab (two in the forward direction and two in the backward direction) and the transmitted wave on the other side. An extension of the treatment to absorbing media yields the theory of the Borrmann effect.[26]

Some calculations for the semi-infinite medium and for the thin film are shown in figs. 4.1.9, 4.1.11 and 4.1.12. In these calculations, the isotropic medium external to the liquid crystal is assumed to have a refractive index equal to $n = \frac{1}{2}(n_a + n_b)$ so that the contribution of the ordinary Fresnel reflexion coefficient at the cholesteric–isotropic interface is eliminated. It is clear from these figures that the results of the exact theory differ only slightly from those of the dynamical model, indicating that the latter is probably adequate for most practical calculations. However, the simple formulation of the dynamical model presented in § 4.1.2 does have some inherent limitations: (i) it is valid only for integral values of the pitch, (ii) it is developed for small $e(= -2\pi(\lambda - \lambda_0)/\lambda)$ and therefore does not give exact values for wavelengths away from the reflexion band, and (iii) it fails when the film thickness is very small, or when the extinction length is of the order of a pitch. These limitations arise primarily because of the kinematical approximation made for the reflexion from the *m* layers per turn of the helix. We shall now show that when multiple reflexions within the m layers are also included, the simple difference equations become matrix difference equations and the resulting solutions turn out to be fully equivalent to the exact electromagnetic treatment. Proofs of this result have been given by Joly[35] and by Nityananda.[34,38] We shall follow the latter's treatment[38] which is simpler.

4.1.4. Equivalence of the continuum and the dynamical theories

We go back to the model discussed in § 4.1.1, viz, a twisted stack of thin birefringent layers. The principal refractive indices of each are n_a and n_b, and the angle of twist between the successive layers is β. Let r_a, t_a, r_b and t_b be the reflexion and transmission coefficients for a single layer for light linearly polarized along its principal axes at normal incidence. For polarization along the a axis,

$$r_a = -\frac{(n_a^2-1)(1-\tau_a^2)}{(n_a+1)^2-(n_a-1)^2\tau_a^2}$$

$$t_a = \frac{4n_a\tau_a}{(n_a+1)^2-(n_a-1)^2\tau_a^2}$$

where $\tau_a = \exp(in_aKp)$, K is the wave-vector *in vacuo* and p the layer thickness.[39] Exactly similar expressions may be written for the other polarization also.

We define I_s and $\mathcal{I}_s$ as the amplitudes incident on the sth layer in $+z$ and $-z$ directions respectively, and similarly E_s and $\mathcal{E}_s$ as the amplitudes emerging from the sth layer in the $+z$ and $-z$ directions respectively (fig. 4.1.18). Therefore, for the sth layer we have

$$\left.\begin{aligned} E_s^a &= t_aI_s^a + r_a\mathcal{I}_s^a \\ E_s^b &= t_bI_s^b + r_b\mathcal{I}_s^b \\ \mathcal{E}_s^a &= r_aI_s^a + t_a\mathcal{I}_s^a \\ \mathcal{E}_s^b &= r_bI_s^b + t_b\mathcal{I}_s^b. \end{aligned}\right\} \tag{4.1.56}$$

The first two equations of (4.1.56) can be combined and written as

$$\begin{bmatrix} E_s^a \\ E_s^b \end{bmatrix} = \begin{bmatrix} t_a & 0 \\ 0 & t_b \end{bmatrix}\begin{bmatrix} I_s^a \\ I_s^b \end{bmatrix} + \begin{bmatrix} r_a & 0 \\ 0 & r_b \end{bmatrix}\begin{bmatrix} \mathcal{I}_s^a \\ \mathcal{I}_s^b \end{bmatrix}.$$

The second two lines can also be similarly combined. However, in what follows we shall write them in the form

$$\left.\begin{aligned} \mathbf{E}_s &= \mathbf{t}\mathbf{I}_s + \mathbf{r}\boldsymbol{\mathcal{I}}_s \\ \boldsymbol{\mathcal{E}}_s &= \mathbf{r}\mathbf{I}_s + \mathbf{t}\boldsymbol{\mathcal{I}}_s \end{aligned}\right\} \tag{4.1.57}$$

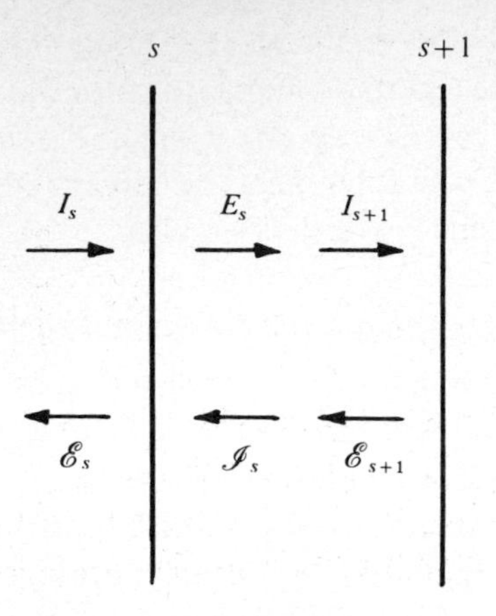

Fig. 4.1.18. Notation for the incident and reflected waves. I_s and $\mathscr{I}_s$ are the amplitudes of the waves incident on the sth layer in the positive and negative directions respectively, and E_s and $\mathscr{E}_s$ the amplitudes emerging from the sth layer in these two directions.

with the understanding that $\mathbf{E}$, $\mathbf{I}$, $\boldsymbol{\mathscr{E}}$ and $\boldsymbol{\mathscr{I}}$ are each column vectors (2×1 matrices), and that $\mathbf{r}$ and $\mathbf{t}$ are each 2×2 diagonal matrices.

Now, the emergent wave $\mathbf{E}_s$ in the $+z$ direction from the sth layer is physically the same as $\mathbf{I}_{s+1}$, the wave incident on the $(s+1)$th layer, but (4.1.57) will apply to the $(s+1)$th layer only if $\mathbf{I}_{s+1}$ is referred to its own principal axes which are rotated through β with respect to those of the sth layer. We therefore write

$$\mathbf{I}_{s+1}=\mathbf{SE}_s \tag{4.1.58}$$

and similarly

$$\boldsymbol{\mathscr{I}}_s=\mathbf{S}^{-1}\boldsymbol{\mathscr{E}}_{s+1}, \tag{4.1.59}$$

where

$$\mathbf{S}\equiv\mathbf{S}(\beta)=\begin{bmatrix}\cos\beta & \sin\beta\\ -\sin\beta & \cos\beta\end{bmatrix}.$$

Strictly these equations should include a phase factor allowing for the air gaps between the layers, but the gap may be taken to be infinitesimally small compared with the thickness of the layer, which itself tends to zero. Using (4.1.58), (4.1.59) and (4.1.57)

$$\mathbf{E}_s=\mathbf{tSE}_{s-1}+\mathbf{r}\,\boldsymbol{\mathscr{I}}_s,$$

$$\mathbf{S}\boldsymbol{\mathscr{I}}_{s-1}=\mathbf{rSE}_{s-1}+\mathbf{t}\,\boldsymbol{\mathscr{I}}_s.$$

Because of the periodicity with respect to s, the difference equations can be solved in the form

$$\mathbf{E}_s = \mathbf{E}\exp(\mathrm{i}s\varphi).$$

This yields

$$\mathbf{E} = \mathbf{tS}\exp(-\mathrm{i}\varphi)\,\mathbf{E} + \mathbf{r}\mathscr{I}$$

$$\mathscr{I} = \mathbf{S}^{-1}\mathbf{rSE} + \mathbf{S}^{-1}\mathbf{t}\exp(\mathrm{i}\varphi)\mathscr{I}.$$

Now the total field in the gap is $\mathbf{F} = \mathbf{E} + \mathscr{I}$, so that

$$\mathbf{F} - \mathscr{I} = \mathbf{tS}\exp(-\mathrm{i}\varphi)(\mathbf{F} - \mathscr{I}) + \mathbf{r}\mathscr{I} \tag{4.1.60}$$

$$\mathscr{I} = \mathbf{S}^{-1}\mathbf{rS}(\mathbf{F} - \mathscr{I}) + \mathbf{S}^{-1}\mathbf{t}\exp(\mathrm{i}\varphi)\mathscr{I}. \tag{4.1.61}$$

From (4.1.60) we get

$$\mathscr{I} = (1 + \mathbf{r} - \mathbf{tS}\exp(-\mathrm{i}\varphi))^{-1}(1 - \mathbf{tS}\exp(-\mathrm{i}\varphi))\mathbf{F}.$$

Using this to eliminate $\mathscr{I}$ from (4.1.61)

$$[(1 + \mathbf{S}^{-1}\mathbf{rS} - \mathbf{S}^{-1}\mathbf{t}\exp(\mathrm{i}\varphi))(1 + \mathbf{r} - \mathbf{tS}\exp(-\mathrm{i}\varphi))^{-1}$$

$$\times(1 - \mathbf{tS}\exp(-\mathrm{i}\varphi)) - \mathbf{S}^{-1}\mathbf{rS}]\mathbf{F} = 0. \tag{4.1.62}$$

As before we effect a change of variables:

$$F_{\pm} = 2^{-1/2}(F_a \pm \mathrm{i}F_b),$$

i.e.,

$$\begin{bmatrix} F_+ \\ F_- \end{bmatrix} = 2^{-1/2}\begin{bmatrix} 1 & \mathrm{i} \\ 1 & -\mathrm{i} \end{bmatrix}\begin{bmatrix} F_a \\ F_b \end{bmatrix} = \mathbf{A}\begin{bmatrix} F_a \\ F_b \end{bmatrix}, \text{ say.}$$

The matrices in the difference equations should also be transformed to the new variables. For example $\mathbf{r}$ should be replaced by

$$\mathbf{ArA}^{-1} \equiv \begin{bmatrix} \frac{1}{2}(r_a + r_b) & \frac{1}{2}(r_a - r_b) \\ \frac{1}{2}(r_a - r_b) & \frac{1}{2}(r_a + r_b) \end{bmatrix} = \begin{bmatrix} \bar{r} & \delta r/2 \\ \delta r/2 & \bar{r} \end{bmatrix},$$

where $\bar{r} = \frac{1}{2}(r_a + r_b)$ and $\delta r = r_a - r_b$, and $\mathbf{S}$ by

$$\mathbf{ASA}^{-1} \equiv \begin{bmatrix} \exp(-\mathrm{i}\beta) & 0 \\ 0 & \exp(\mathrm{i}\beta) \end{bmatrix}.$$

Since $\mathbf{r}$ and $\mathbf{t}$ are functions of the thickness p of the layer, we expand them in powers of p. It is sufficient to retain the first power in each case:

$$\bar{r} = \tfrac{1}{2}(\varepsilon - 1)\mathrm{i}Kp,$$

$$\bar{t} = \tfrac{1}{2}(t_a + t_b) = 1 + \tfrac{1}{2}(\varepsilon + 1)\mathrm{i}Kp,$$

$$\delta r/2 = \tfrac{1}{2}\mathrm{i}\alpha Kp, \qquad \delta t/2 = (t_a - t_b)/2 = \tfrac{1}{2}\mathrm{i}\alpha Kp,$$

where as before $\varepsilon=\frac{1}{2}(\varepsilon_a+\varepsilon_b)$ and $\alpha=\frac{1}{2}(\varepsilon_a-\varepsilon_b)$. Writing $\beta=qp$ and $\varphi=kp$ (and remembering that since $\mathbf{r}$ is of order p, $\mathbf{rS}\sim\mathbf{r}$ to this order) (4.1.62) may be expressed as

$$\mathbf{R}\begin{bmatrix}F_+\\F_-\end{bmatrix}=0, \tag{4.1.63}$$

where $\mathbf{R}$ simplifies to (4.1.64) below:

$$\mathbf{R}=\frac{1}{K^2}\begin{bmatrix}\left(\frac{k+q}{K}-1\right)^{-1} & 0\\ 0 & \left(\frac{k-q}{K}-1\right)^{-1}\end{bmatrix}\begin{bmatrix}(k+q)^2-\varepsilon K^2 & -\alpha K^2\\ -\alpha K^2 & (k-q)^2-\varepsilon K^2\end{bmatrix} \tag{4.1.64}$$

Since the first matrix on the right-hand side is non-singular we may premultiply by its inverse to obtain (4.1.63) in the form

$$\begin{bmatrix}(k+q)^2-\varepsilon K^2 & -\alpha K^2\\ -\alpha K^2 & (k-q)^2-\varepsilon K^2\end{bmatrix}\begin{bmatrix}F_+\\F_-\end{bmatrix}=0,$$

which is precisely the same as (4.1.46). The dynamical theory applied in this manner to a twisted pile of birefringent layers is then exactly valid for any arbitrary thickness of the sample and for the entire range of wavelengths.

4.1.5. Oblique incidence

The theory of propagation inclined to the optic axis is, of course, very much more complicated, and analytical solutions have not so far been found.[40] The first attempt at solving the problem numerically was by Taupin,[41] but the most complete calculations are those of Berreman and Scheffer[42] who also carried out a precise experimental study of reflexion from monodomain samples at oblique incidence. Fig. 4.1.19 presents their observed reflexion spectra for two polarizations.

Berreman used a 4×4 matrix multiplication method. Assuming the incident and reflected wave-vectors to be in the xz plane, z being along the helical axis of the cholesteric, the dependence on the y coordinate may be ignored altogether. Writing $E_x=E_x^0\exp[\mathrm{i}(\omega t-kx)]$ etc., it is

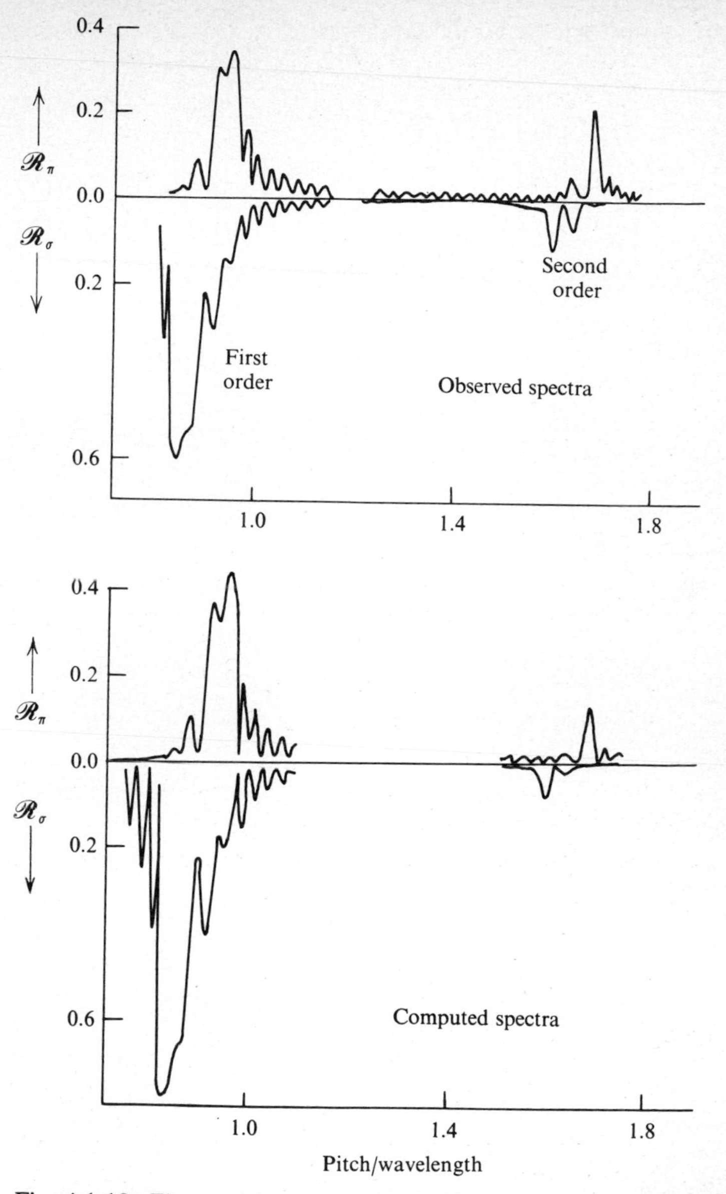

Fig. 4.1.19. First and second order reflexion spectra of a cholesteric liquid crystal film (0.45:0.55 mole fraction mixture of 4′-bis(2-methylbutoxy)-azoxybenzene and 4,4′-di-n-hexoxyazoxybenzene) 15 pitch lengths or 11.47 μm thick. Angle of incidence 45°. Polarizer and analyser are parallel to the plane of reflexion for $\mathscr{R}_\pi$ and normal to it for $\mathscr{R}_\sigma$ measurements. The small oscillations are interference fringes from the two cholesteric–glass interfaces. (After Berreman and Scheffer.[42])

easily verified that Maxwell's equations can be reduced to the matrix form

$$\frac{\partial}{\partial z}\begin{bmatrix} E_x \\ \mathrm{i}H_y \\ E_y \\ -\mathrm{i}H_x \end{bmatrix} = \frac{\omega}{c}\begin{bmatrix} -\mathrm{i}\dfrac{kc\varepsilon_{xz}}{\omega\varepsilon_{zz}} & 1-\dfrac{1}{\varepsilon_{zz}}\left(\dfrac{kc}{\omega}\right)^2 & -\mathrm{i}\dfrac{kc}{\omega}\dfrac{\varepsilon_{yz}}{\varepsilon_{zz}} & 0 \\ -\varepsilon_{xx}+\dfrac{\varepsilon_{xz}^2}{\varepsilon_{zz}} & -\mathrm{i}\dfrac{kc}{\omega}\dfrac{\varepsilon_{xz}}{\varepsilon_{zz}} & \dfrac{\varepsilon_{xz}\varepsilon_{yz}}{\varepsilon_{zz}}-\varepsilon_{xy} & 0 \\ 0 & 0 & 0 & 1 \\ \dfrac{\varepsilon_{xz}\varepsilon_{yz}}{\varepsilon_{zz}}-\varepsilon_{xy} & -\mathrm{i}\dfrac{kc}{\omega}\dfrac{\varepsilon_{yz}}{\varepsilon_{zz}} & \dfrac{\varepsilon_{yz}^2}{\varepsilon_{zz}}-\varepsilon_{yy}+\left(\dfrac{kc}{\omega}\right)^2 & 0 \end{bmatrix}\begin{bmatrix} E_x \\ \mathrm{i}H_y \\ E_y \\ -\mathrm{i}H_x \end{bmatrix}$$

or

$$\frac{\partial \boldsymbol{\psi}}{\partial z} = \frac{\omega}{c}\mathscr{D}\boldsymbol{\psi}$$

where, assuming a spiralling ellipsoid model, the components of the dielectric tensor are given by

$$\varepsilon_{xx} = \varepsilon + \alpha \cos 2qz$$

$$\varepsilon_{yy} = \varepsilon - \alpha \cos 2qz$$

$$\varepsilon_{xy} = \varepsilon_{yx} = \alpha \sin 2qz$$

$$\varepsilon_{zz} = \varepsilon_c$$

and all other components are zero (see (4.1.42)).

To a first order of approximation

$$\boldsymbol{\psi}(z+\delta z) \equiv \mathscr{P}(z, \delta z)\boldsymbol{\psi}(z) = \left(\mathbf{E} + \frac{\omega}{c}\delta z\,\mathscr{D}(z)\right)\boldsymbol{\psi}(z)$$

where $\mathscr{P}$ is a 4×4 propagation matrix and $\mathbf{E}$ is the unit 4×4 matrix. Repeated matrix multiplication in very small steps δz gives the matrix for the total film, and by taking into account the appropriate boundary conditions on either side (glass) one can work out the transmitted and reflected waves. Of course, cyclic and other symmetry properties of $\mathscr{P}(z)$ reduce the number of matrix multiplications to a reasonable number in practical cases. In the actual calculation, Berreman and Scheffer included a second order term subject to the symmetry property

$$\mathscr{P}(z, \delta z) = \mathscr{P}^{-1}(z, -\delta z).$$

The results of their computations are shown in fig. 4.1.19. The agreement with the experimental spectra can be seen to be good. There is a difference in the intensities, which may conceivably be due to thin regions near the surface with anomalous dielectric properties or due to the neglect of absorption.

An important fact that emerged from this study is that the observed features were best reproduced when the local dielectric ellipsoid was taken to be a prolate spheroid, with the principal axis Oc parallel to z and $\varepsilon_c = \varepsilon - \alpha$. Thus the assumption that is generally made that the local dielectric ellipsoid is uniaxial would appear to be valid to a very good approximation as far as optical calculations are concerned (see, however, § 4.8.4).

4.1.6. Propagation normal to the optic axis

When light is incident normal to the optic axis polarized diffraction maxima are seen in transmission.[15] For the electric vector polarized parallel to z the refractive index is independent of the z coordinate, whereas for the vector polarized perpendicular to it the refractive index varies periodically from n_a to n_b. The wavefront having the latter polarization therefore suffers changes of phase which vary along z with a periodicity equal to half the pitch. There can also be changes of amplitude as parallel rays tend to acquire a slight curvature when travelling in a medium in which the gradient of refractive index is normal to the direction of propagation.[43] This periodicity in the phase and amplitude gives rise to polarized diffraction effects. Approximate expressions for the intensity of the maxima may be derived by applying the Raman-Nath theory of the diffraction of light by ultrasonic waves.[44]

Sackmann *et al.*[15] have investigated the temperature variation of the pitch of a mixture of right-handed cholesteryl chloride and left-handed cholesteryl myristate by this method. At a certain temperature (T_N) there is an exact compensation of the two opposite helical structures and the sample becomes nematic. At this temperature only the central spot (zero order) is observed, while at the other temperatures, polarized diffraction maxima of higher order make their appearance. The inverse pitch varies almost linearly with temperature passing through zero at T_N (fig. 4.1.20).

4.2. Disclinations

4.2.1. Grandjean–Cano walls

Because of the spontaneous twist in the director field, it is to be expected that the properties of disclinations in cholesteric liquid crystals are rather different from those in nematics. We shall consider first the disclinations that give rise to what have come to be known as the *Grandjean–Cano*

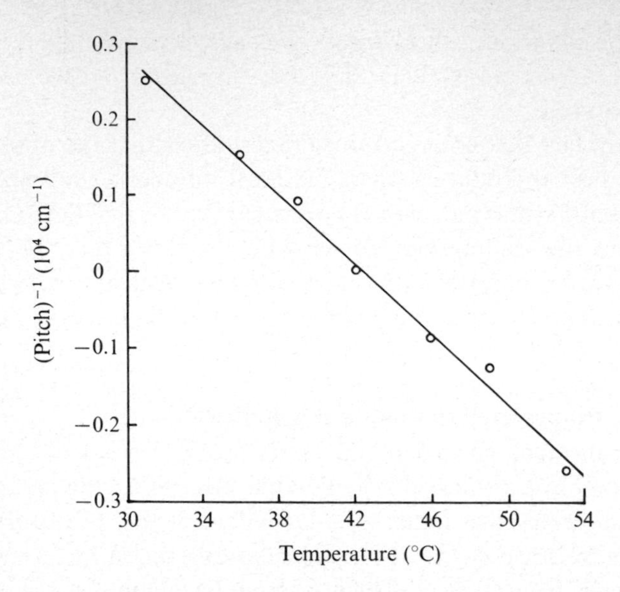

Fig. 4.1.20. Variation of inverse pitch with temperature in a 1.75:1 weight mixture of right-handed cholesteryl chloride and left-handed cholesteryl myristate as determined by laser diffraction. The mixture becomes nematic at 42 °C. (Sackmann *et al.*[15])

walls.[45,46] These walls appear when the cholesteric is prepared in the form of a thin wedge between two glass plates inclined at a small angle. The surfaces of both plates are rubbed along the same direction so that the director has the same orientation at either end, the helical axis being roughly normal to the plates. Viewed under a polarizing microscope, regular striations are seen running across the film (plate 11(a)). From optical studies, Cano[46] established that the director changes by half a turn on going from one strip to the next ($s = \pm\frac{1}{2}$). The mechanism of this transition was explained by de Gennes[47] who postulated a line discontinuity parallel to the plates located in the midplane of the film (fig. 4.2.1). He also calculated the elastic energy of deformation around the disclination using the one-constant approximation. A slightly more elaborate calculation has since been made by Scheffer[48] whose treatment we outline below.

We choose a cartesian coordinate system with z normal to the film (supposed for the present to be plane parallel). We assume the director to be throughout normal to z and invariant along y so that

$$\mathbf{n} = (\cos\varphi(x, z), \sin\varphi(x, z), 0).$$

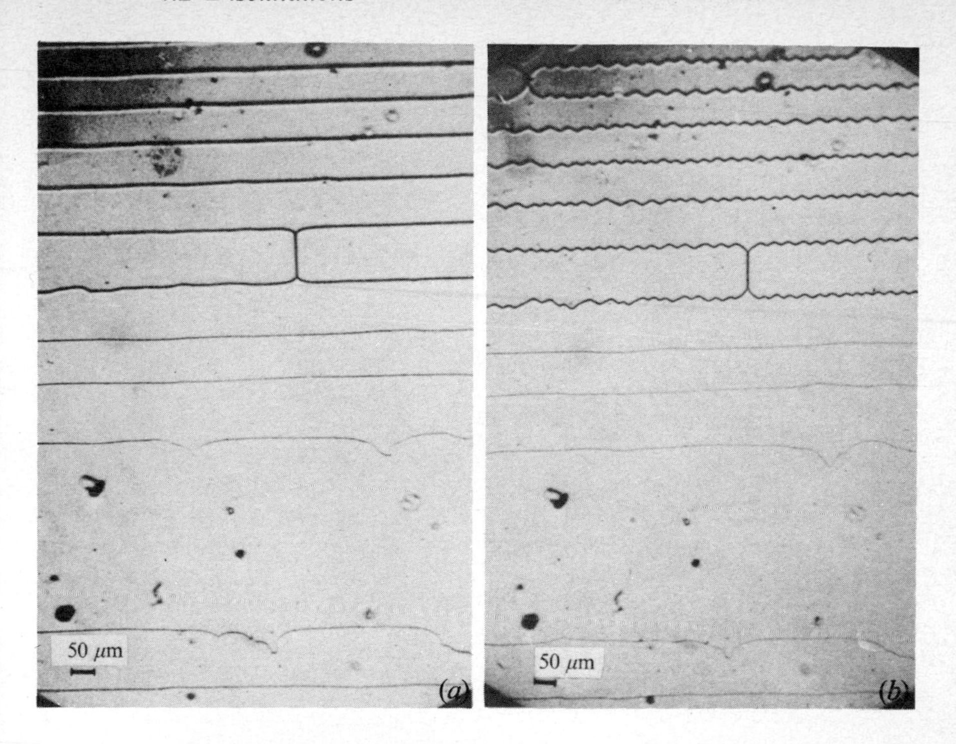

Plate 11. Grandjean–Cano walls in a small-angled wedge of cholesteric liquid crystal of large pitch: (*a*) without a magnetic field and (*b*) with a magnetic field normal to the helical axis. The double disclinations – the darker lines – become zigzag in the presence of a magnetic field while the single disclinations remain stable. (Orsay Liquid Crystals Group.[49])

The elastic free energy density for a structure having an intrinsic pitch $2\pi/q_0$ is conveniently written as

$$F=\tfrac{1}{2}k_{11}(\nabla\cdot\mathbf{n})^2+\tfrac{1}{2}k_{22}(\mathbf{n}\cdot\nabla\times\mathbf{n}+q_0)^2+\tfrac{1}{2}k_{33}[\mathbf{n}\cdot(\nabla\mathbf{n})]^2 \qquad (4.2.1)$$

which is identical with (3.2.7) except for a constant term $\tfrac{1}{2}k_{22}q_0^2$. On making the simplifying assumption that $k_{11}=k_{33}$, (4.2.1) reduces to

$$F=\frac{1}{2}\left[k_{11}\left(\frac{\partial\varphi}{\partial x}\right)^2+k_{22}\left(q_0-\frac{\partial\varphi}{\partial z}\right)^2\right].$$

The total energy is

$$\mathscr{F}=\int\int F\,\mathrm{d}x\,\mathrm{d}z$$

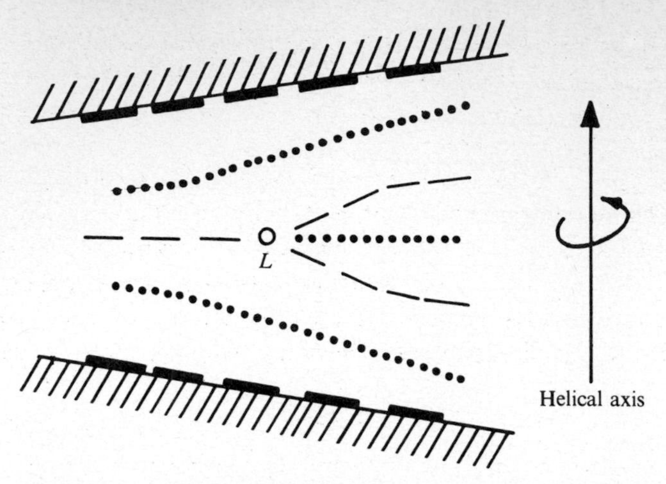

Fig. 4.2.1. Grandjean–Cano walls in a wedge-shaped cholesteric sample. L is a line singularity in the midplane of the sample. The director orientation changes by half a turn on going from one side of the singularity to the other. Dots signify that the director is normal to the paper and dashes that it is parallel to the plane of the paper.

and the equation of equilibrium is

$$k_{11}\frac{\partial^2\varphi}{\partial x^2}+k_{22}\frac{\partial^2\varphi}{\partial z^2}=0$$

or, putting $\xi=x(k_{22}/k_{11})^{1/2}$,

$$\frac{\partial^2\varphi}{\partial\xi^2}+\frac{\partial^2\varphi}{\partial z^2}=0, \tag{4.2.2}$$

and

$$\mathscr{F}=\tfrac{1}{2}k_{22}\iint\left[\left(\frac{\partial\varphi}{\partial\xi}\right)^2+\left(q_0-\frac{\partial\varphi}{\partial z}\right)^2\right]\mathrm{d}\xi\,\mathrm{d}z. \tag{4.2.3}$$

Let the origin of the coordinate system be at the disclination line itself. The solution of (4.2.2) is then

$$\varphi=\tfrac{1}{2}\tan^{-1}\left\{\frac{\tan(\pi z/d)}{\tanh(\pi\xi/d)}\right\}+q_0z+\varphi_0, \tag{4.2.4}$$

where d is the film thickness and

$$\varphi_0=\begin{cases}-\pi/2 & \text{for } \xi>0,\, z>0\\ 0 & \text{for } \xi<0\\ +\pi/2 & \text{for } \xi>0,\, z<0.\end{cases}$$

Contours of constant director orientation calculated from (4.2.4) are shown in fig. 4.2.2. The perturbation in the cholesteric structure is confined largely to a circular region around the singularity extending to about half the film thickness.

The energy of a disclination can be derived by substituting (4.2.4) in (4.2.3) and integrating. Scheffer has made some calculations, taking into account the fact that in reality the plates are not parallel but form a wedge. In the absence of any knowledge of the core structure, the calculations are necessarily incomplete, but nevertheless they do indicate that the Grandjean disclination stabilizes a disclination-free wedge.

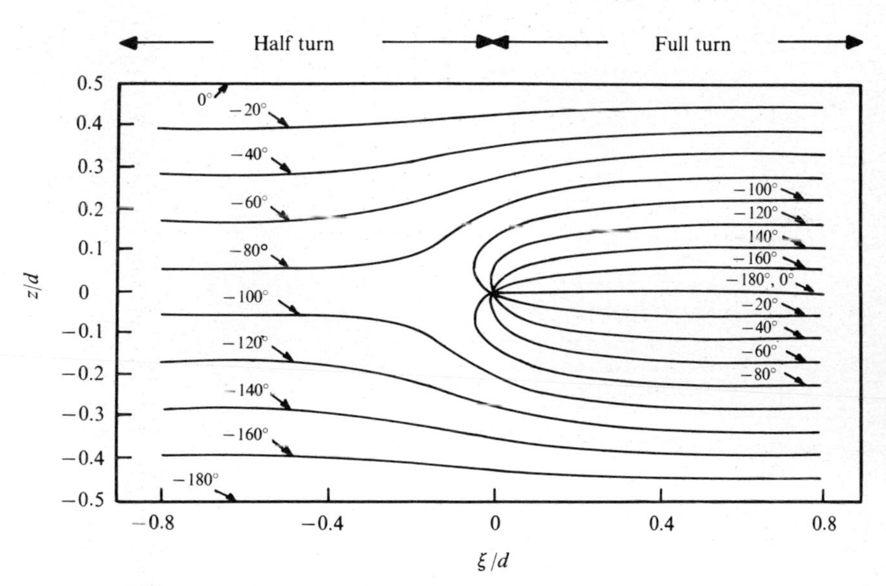

Fig. 4.2.2. Contours of constant director orientation calculated from (4.2.4) for $q_0 = 3\pi/2d$. (After Scheffer.[48])

Double disclinations with the director changing orientation by a full turn around the singularity ($s = \pm 1$) have also been observed.[49] These are the more strongly contrasted lines in plate 11(a).

4.2.2. λ, τ and χ disclinations

The topological features of disclinations can be visualized easily by applying certain concepts of dislocation theory, as was first shown by Kléman and Friedel.[50] We define the disclination line L as the limit of a cut made into a cholesteric (fig. 4.2.3(a)). We assume that the molecules are not disturbed by the cut and that they are anchored firmly to S_1 and S_2, the two surfaces created by the cut. We rotate S_1 by $\pi/2$ and S_2 by

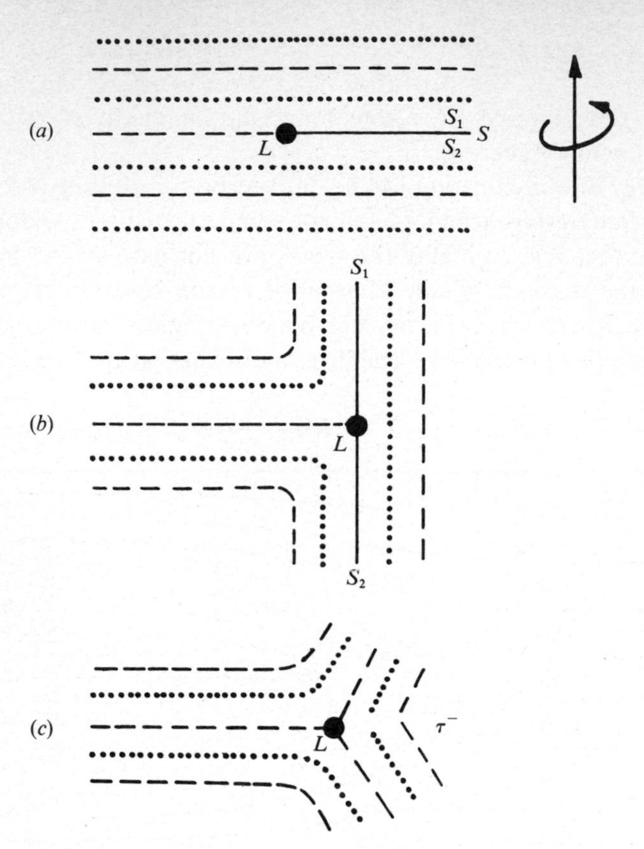

Fig. 4.2.3. Creation of a τ^- disclination (the Volterra process). Dots signify that the director axis is normal to the paper, dashes that it is parallel to plane of the paper.

$-\pi/2$ (fig. 4.2.3(b)), add some material to fill the empty space on the right-hand side, and then allow the system to relax viscously (fig. 4.2.3(c)). Since in this case the molecules in the cut plane are normal to the line L, we refer to the configuration of fig. 4.2.3(c) as a τ^- disclination, τ denoting 'transverse' and the minus sign indicating that material has been *added* to arrive at the final configuration. If on the other hand, the cut is such that the molecules are parallel to the line L, we have the λ^- disclination, λ for 'longitudinal' (fig. 4.2.4(a)).

To arrive at positive disclinations, we rotate S_1 by $-\pi/2$ and S_2 by $+\pi/2$, and *remove* the overlapping matter. Again two cases result, viz, τ^+ and λ^+ (fig. 4.2.4(b) and (c)). Isolated λ and τ are necessarily straight.

For both λ and τ, the axis of rotation is perpendicular to the spiral axis. When the axis of rotation is parallel to the spiral axis, we have χ

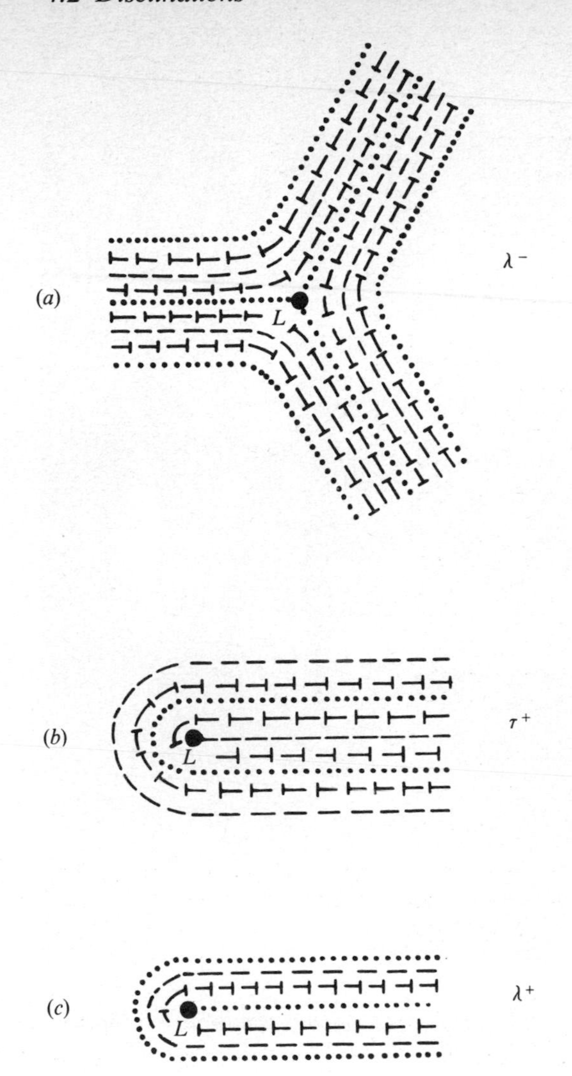

Fig. 4.2.4. The configurations for (*a*) λ^-, (*b*) τ^+ and (*c*) λ^+ disclinations. Dots signify that the director axis is normal to the paper, dashes that it is parallel to the plane of the paper, and nails that it is tilted.

disclinations. These may take any shape since the viscosity in the cholesteric planes (i.e., in the planes normal to the spiral axis) is comparable to that of a nematic. The simplest example of a χ disclination is that proposed by de Gennes to explain the Grandjean–Cano walls (fig. 4.2.1). A remarkable instance of a helicoidal χ disclination ($s=-1$) wound round a straight χ disclination ($s=+1$) is shown in plate 12.[51]

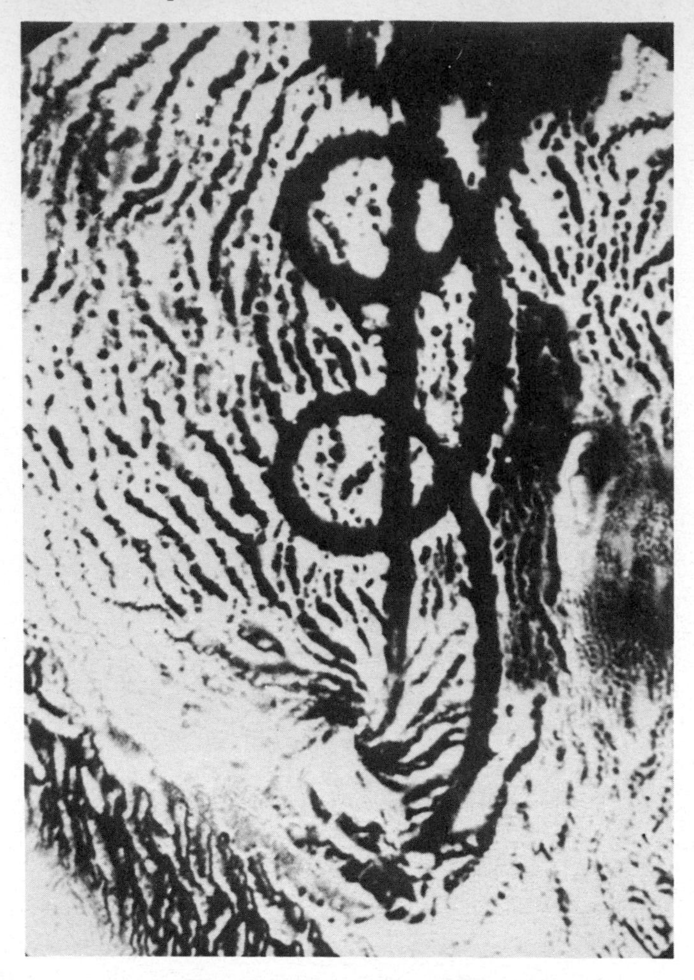

Plate 12. A helical $\chi(s=-1)$ disclination wound round a straight $\chi(s=+1)$ disclination in a cholesteric liquid crystal. (Rault.[51])

λ or τ of opposite signs tend to occur in pairs[52] (fig. 4.2.5). Examples of such pairing are shown in plate 13 and fig. 4.2.6. From a distance such pairs are equivalent to the χ disclinations of de Gennes, but the core is, of course, different. Kléman and Friedel[50] have suggested that the double disclination in the Grandjean steps may in fact be a pair of straight disclinations. When a magnetic field is applied normal to the spiral axis the pair tends to rotate by $\pi/2$, the effect being greater the larger the core. A simple calculation shows that the field required to distort a double line is about half that for a single line, which could either be a χ or a pair. Thus

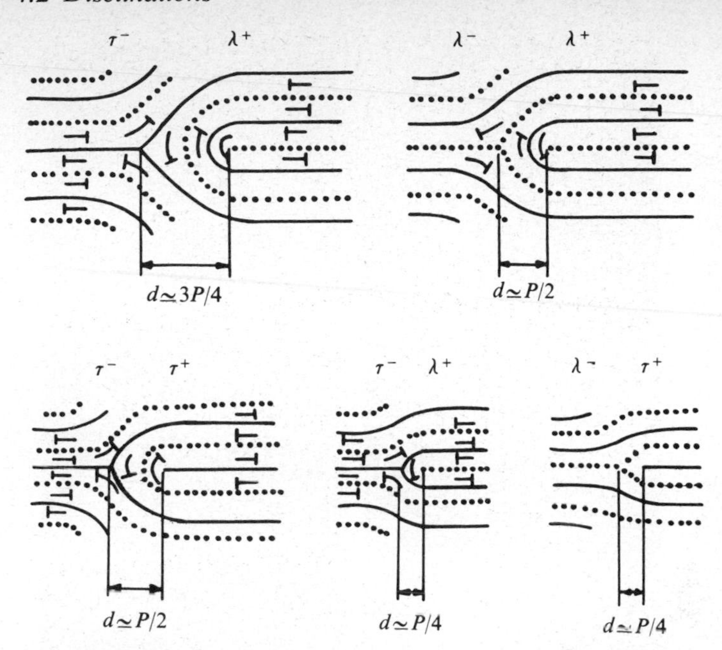

Fig. 4.2.5. Examples of pairing of τ and λ disclinations of opposite signs. From a distance these are equivalent to a χ disclination (see fig. 4.2.1). (After Bouligand and Kléman.[52])

in the presence of a magnetic field, the double disclinations become zigzag whereas the single disclination remains stable [49] (plate 11(*b*)).

The stratification in the cholesteric structure also gives rise to focal conic textures, similar to those found in smectic liquid crystals (see § 5.3.1). More complex networks are sometimes formed.[53] Perturbation by external magnetic fields has proved to be a useful method of elucidating their structures.[54]

4.3. Leslie's theory of thermomechanical coupling

The fact that the properties of the cholesteric liquid crystal are not invariant with respect to reflexion introduces some additional complexity into the equations of the continuum theory. The possibility now exists of a coupling between thermal and mechanical effects which symmetry considerations rule out automatically from the theory of the nematic state. Thus translational or rotational motion of the fluid can, in principle, cause heat transfer, and equally a thermal gradient can create motion.[55]

Consider an incompressible cholesteric fluid with a non-polar director of constant magnitude. The basic equations developed in § 3.1 need to be modified slightly because of the absence of a plane of symmetry. Allowing

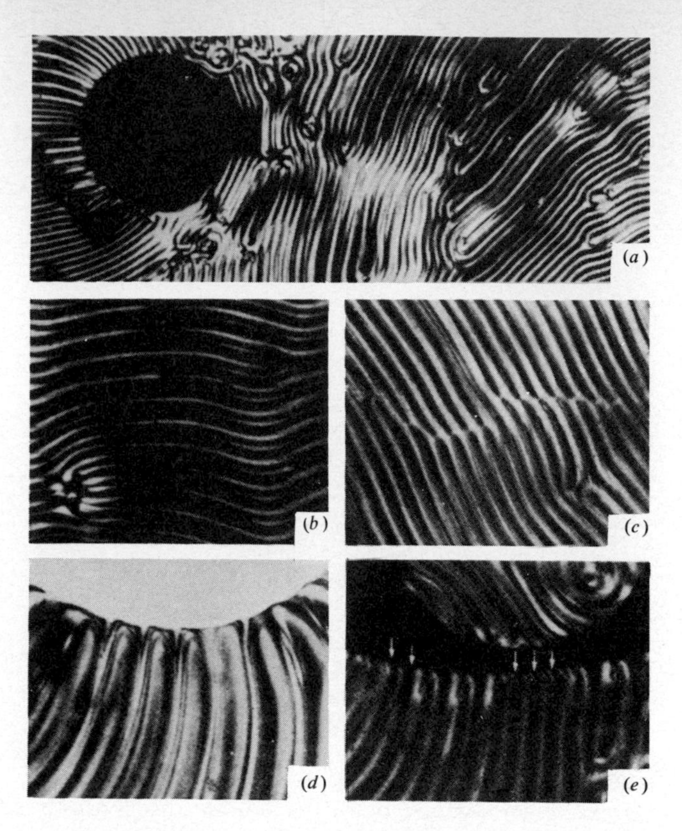

Plate 13. Disposition of layers and of disclinations in a cholesteric liquid crystal of large pitch. (Bouligand and Kléman.[52])

for heat flux and thermal gradients the conservation laws are

$$\dot{\rho} = 0, \tag{4.3.1}$$

$$\rho \dot{v}_i = f_i + t_{ji,j}, \tag{4.3.2}$$

$$\dot{U} = Q - q_{i,i} + t_{ji} d_{ij} + \pi_{ji} N_{ij} - g_i N_i, \tag{4.3.3}$$

$$\rho_1 \ddot{n}_i = G_i + g_i + \pi_{ji,j}, \tag{4.3.4}$$

where Q is the heat supply per unit volume, q_i the heat flux vector per unit area per unit time and the other symbols have the usual meanings (see § 3.3). If p_i is the entropy flux vector per unit area per unit time, it is convenient to introduce a vector φ_i such that

$$\varphi_i = q_i - Tp_i.$$

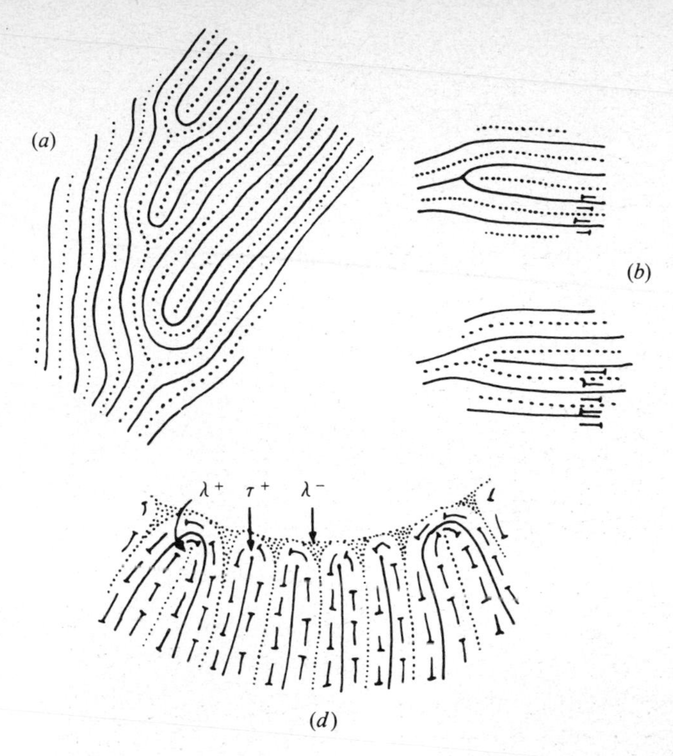

Fig. 4.2.6. Interpretation of (*a*), (*b*) and (*d*) of plate 13. (Bouligand and Kléman.[52])

The entropy inequality may then be written as

$$t_{ji}d_{ij}+\pi_{ji}N_{ij}-g_iN_i-p_iT_{,i}-\dot{F}-S\dot{T}-\varphi_{i,i}\geqslant 0$$

where $F=U-TS$ is the free energy function. Making use of the fact that $T_{,i}$, d_{ij} and $n_{i,j}$ can be chosen arbitrarily and independently, we obtain the following equations:

$$\frac{\partial\varphi_i}{\partial n_{k,j}}+\frac{\partial\varphi_j}{\partial n_{k,i}}=0,$$

$$\frac{\partial\varphi_i}{\partial T_{,j}}+\frac{\partial\varphi_j}{\partial T_{,i}}=0,$$

$$\frac{\partial\varphi_i}{\partial d_{jk}}+\frac{\partial\varphi_i}{\partial d_{kj}}+n_k\frac{\partial\varphi_j}{\partial N_i}+n_j\frac{\partial\varphi_k}{\partial N_i}-n_i\frac{\partial\varphi_j}{\partial N_k}-n_i\frac{\partial\varphi_k}{\partial N_j}=0,$$

with

$$\varphi_i=\alpha e_{ijk}n_j(N_k+d_{kp}n_p),$$

where α is a constant. The inequality (4.3.5) then becomes

$$\left(t_{ji}+\frac{\partial F}{\partial n_{k,j}}n_{k,i}\right)d_{ij}+\left(\pi_{ji}-\frac{\partial F}{\partial n_{i,j}}-\frac{\partial \varphi_j}{\partial N_i}\right)N_{ij}$$

$$-\left(g_i+\frac{\partial F}{\partial n_i}\right)N_i-\frac{\partial \varphi_i}{\partial n_j}n_{j,i}-\left(p_i+\frac{\partial \varphi_i}{\partial T}\right)T_{,i}\geqslant 0. \qquad (4.3.6)$$

Resolving t_{ij}, g_i and p_i into their static and hydrodynamic parts,

$$t_{ij}=t^0_{ij}+t'_{ij},$$

$$g_i=g^0_i+g'_i,$$

$$p_i=p^0_i+p'_i,$$

we have

$$t^0_{ji}=-p\delta_{ji}-\frac{\partial F}{\partial n_{k,j}}n_{k,i}+\alpha e_{jkp}(n_p n_i)_{,k}, \qquad (4.3.7)$$

$$g^0_i=\gamma n_i-\beta_j n_{i,j}-\frac{\partial F}{\partial n_i}-\alpha e_{ijk}n_{k,j}, \qquad (4.3.8)$$

$$\pi_{ji}=\beta_j n_i+\frac{\partial F}{\partial n_{i,j}}+\alpha e_{ijk}n_k, \qquad (4.3.9)$$

$$p^0_i=0,$$

and (4.3.6) reduces to

$$t'_{ji}d_{ij}-g'_iN_i-\left(p'_i+\frac{\partial \varphi_i}{\partial T}\right)T_{,i}\geqslant 0. \qquad (4.3.10)$$

Also

$$t'_{ij}+g'_in_j=t'_{ji}+g'_jn_i, \qquad (4.3.11)$$

and the entropy generation per unit volume

$$T\dot{S}=Q-q_{i,i}+\alpha e_{ijk}[n_j(N_k+d_{kp}n_p)]_{,i}+t'_{ji}d_{ij}-g'_iN_i. \qquad (4.3.12)$$

The hydrodynamic components of the stress tensor etc., are

$$t'_{ji}=\mu_1 n_k n_p d_{kp} n_i n_j+\mu_2 N_i n_j+\mu_3 N_j n_i+\mu_4 d_{ji}$$

$$+\mu_5 d_{ik} n_k n_j+\mu_6 d_{jk} n_k n_i+\mu_7 e_{ipq} T_{,q} n_j n_p$$

$$+\mu_8 e_{jpq} T_{,q} n_i n_p, \qquad (4.3.13)$$

$$g'_i=\lambda_1 N_i+\lambda_2 d_{ij} n_j+\lambda_3 e_{ipq} n_p T_{,q}, \qquad (4.3.14)$$

$$q_i=K_1 T_{,i}+K_2 n_k T_{,k} n_i+K_3 e_{ipq} n_p N_q+K_4 e_{ipq} n_p d_{qr} n_r, \qquad (4.3.15)$$

$$\left.\begin{aligned} \lambda_1 &= \mu_2 - \mu_3, \\ \lambda_2 &= \mu_5 - \mu_6, \\ \lambda_3 &= \mu_7 - \mu_8, \end{aligned}\right\} \tag{4.3.16}$$

where $\mu_1 \ldots \mu_6$ are the six viscosity coefficients already defined in continuum theory of the nematic state; μ_7 and μ_8 are two additional viscosity coefficients coupling thermal and mechanical effects; $K_1 \ldots K_4$ are the coefficients of thermal conductivity.

The free energy of elastic deformation per unit volume is given by (3.2.7).

4.4. The Lehmann rotation phenomenon

An example of this type of thermomechanical coupling appears to have been observed by Lehmann[56] in cholesteric liquid crystals very soon after their discovery. He found that the material spread between two glass surfaces was set into motion by a flow of heat coming from below. The different drops of liquid seemed to be rotating violently, but from optical studies Lehmann concluded that it was not the drops themselves but the structure that was rotating (plate 14). Leslie's theory[55] offers a simple explanation of this phenomenon.

Let the cholesteric film be bounded between the planes $z = 0$ and $z = h$, and let there be a temperature gradient along the screw axis z. The components of the director in a right-handed cartesian coordinate system are $\{\cos\theta(z, t), \sin\theta(z, t), 0\}$. We assume that there are no heat sources within the liquid crystal, no external body forces and that the velocity vector is zero. Hence $T = T(z)$, $f_i = G_i = d_{ij} = w_{ij} = 0$. Thus from (4.3.4)

$$\rho_1 \frac{\partial^2\theta}{\partial t^2} = \lambda_1 \frac{\partial\theta}{\partial t} - \lambda_3 \frac{\partial T}{\partial z} + \frac{\partial}{\partial z}\left(\alpha - k_2 + k_{22}\frac{\partial\theta}{\partial z}\right) \tag{4.4.1}$$

and from (4.3.12)

$$T\frac{\partial S}{\partial t} = \alpha \frac{\partial^2\theta}{\partial z\,\partial t} - \frac{\partial}{\partial z}\left(K_1 \frac{\partial T}{\partial z} + K_3 \frac{\partial\theta}{\partial t}\right) - \left(\lambda_1 \frac{\partial\theta}{\partial t} - \lambda_3 \frac{\partial T}{\partial z}\right)\frac{\partial\theta}{\partial t}, \tag{4.4.2}$$

where

$$S = -\frac{\partial}{\partial T}\left[F_0 - k_2 \frac{\partial\theta}{\partial z} + \frac{1}{2}k_{22}\left(\frac{\partial\theta}{\partial z}\right)^2\right]$$

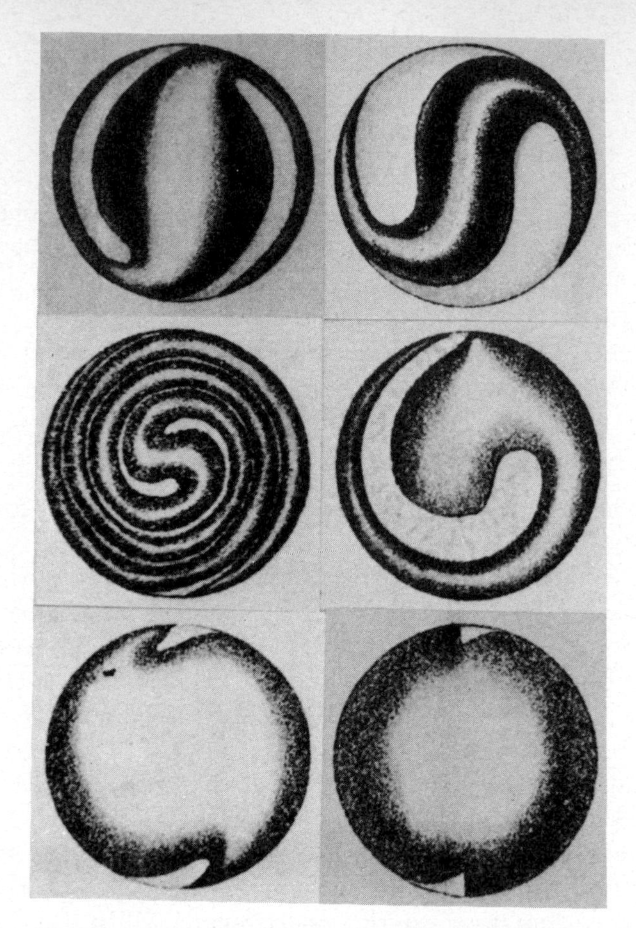

Plate 14. Lehmann's diagrams depicting the rotation phenomenon in open cholesteric droplets heated from below. (After Lehmann.[56])

and F_0 is the free energy in the absence of elastic deformation. In static situations, $\partial^2\theta/\partial t^2 = \partial\theta/\partial t = 0$, and (4.4.1) and (4.4.2) yield

$$\frac{\partial}{\partial z}\left(\alpha - k_2 + k_{22}\frac{\partial\theta}{\partial z}\right) - \lambda_3\frac{\partial T}{\partial z} = 0,$$

$$-\frac{\partial}{\partial z}\left(K_1\frac{\partial T}{\partial z}\right) = 0,$$

giving

$$\int_0^T K_1\,\mathrm{d}\xi = Az + B,$$

and

$$\theta = \frac{1}{A}\int_0^T \frac{K_1}{k_{22}}\left(\int_0^{\xi} \lambda_3 d\eta - \alpha + k_2\right) d\xi + C\int_0^T \frac{K_1}{k_{22}} d\xi + D,$$

where A, B, C, D are constants which can be determined from the boundary conditions.

If $T(0) = T_0$, $T(h) = T_1$ and the torques $\tau_{il} = e_{ijk}n_j\pi_{lk}$ at the boundary are zero,

$$\left(k_{22}\frac{\partial\theta}{\partial z} - k_2 + \alpha\right)_{z=0} = \left(k_{22}\frac{\partial\theta}{\partial z} - k_2 + \alpha\right)_{z=h} = 0.$$

One can then work out a simple solution if the material constants are assumed to be independent of temperature. Let $\theta = \omega t + f(z)$ and $T = g(z)$; (4.4.1) and (4.4.2) then reduce to

$$k_{22}\frac{\partial^2 f}{\partial z^2} - \lambda_3\frac{\partial g}{\partial z} + \lambda_1\omega = 0, \tag{4.4.3}$$

$$K_1\frac{\partial^2 g}{\partial z^2} - \lambda_3\omega\frac{\partial g}{\partial z} + \lambda_1\omega^2 = 0. \tag{4.4.4}$$

Equation (4.4.4) gives

$$g = (\lambda_1/\lambda_3)\omega z + A\exp(\lambda_3\omega z/K_1) + B,$$

and (4.4.3) then gives

$$f = (K_1 A/\omega k_{22})\exp(\lambda_3\omega z/K_1) + Cz + D.$$

From the boundary conditions, one obtains finally

$$T = T_0 + \{(T_1 - T_0)z/h\}, \tag{4.4.5}$$

$$\theta = \theta_0 + \omega t + \{(k_2 - \alpha)z/k_{22}\}, \tag{4.4.6}$$

where θ_0 is the orientation at $z = 0$, and

$$\omega = \lambda_3(T_1 - T_0)/\lambda_1 h. \tag{4.4.7}$$

Thus the director rotates about Oz with an angular velocity ω, which explains Lehmann's observations.

In the absence of a temperature gradient (4.4.6) reduces to

$$\theta = \theta_0 + \{(k_2 - \alpha)z/k_{22}\}$$

which describes the normal cholesteric structure with a pitch $P = 2\pi k_{22}/(k_2 - \alpha)$. The coefficient α has not yet been determined experimentally and it is not known whether its contribution is of practical significance. According to the Oseen–Zöcher–Frank elasticity equations $\alpha = 0$, and in the absence of any evidence to the contrary it is generally neglected in most discussions.

4.5. Flow properties

The flow properties of a cholesteric liquid crystal are surprisingly different from those of a nematic. Its viscosity increases by about a million times as the shear rate drops to a very low value[57] (fig. 4.5.1). One of the difficulties in interpreting this highly non-newtonian behaviour is the uncertainty in the wall orientation which cannot be controlled as easily as in the nematic case. Recently, some careful measurements of the apparent viscosity η_{app} in Poiseuille flow have been made by Candau, Martinoty and Debeauvais[58] of a nematic-cholesteric mixture whose pitch

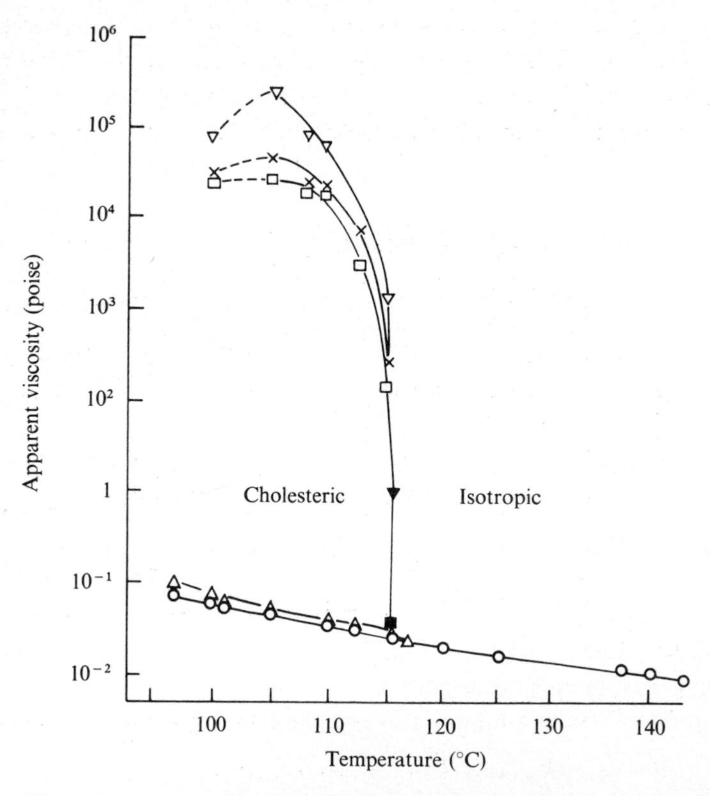

Fig. 4.5.1. Apparent viscosity of cholesteryl acetate versus temperature. Capillary shear rate (s^{-1}): ▽ 10; × 50; □ 100; ▼ 1000; ■ 5000. Rotational shear rate (s^{-1}) △ 10^4; ○ high shear rate and normal liquid behaviour. (After Porter, Barrall and Johnson.[57])

could be varied by changing the composition. Microscopic examination revealed that the helical axes were oriented radially in the capillary so that the cholesteric layers were rolled up in the form of coaxial cylinders. The flow direction was therefore normal to the helical axis at every point. In this geometry there is a slight dependence of the viscosity on the shear rate and on the pitch, but the significant fact emerges that η_{app} is approximately of the same order of magnitude as that for a nematic even at low shear rates (figs. 4.5.2 and 4.5.3).

The application of the Ericksen–Leslie equations to cholesteric flow is less straightforward than in the case of nematics and no detailed solutions have so far been possible even for simple geometries. However, the behaviour in certain limiting situations can be explained qualitatively.

4.5.1. Flow along the helical axis

Helfrich[59] proposed a simple physical mechanism, which he called *permeation*, to account for the very high apparent viscosity at low shear rates. He suggested that flow takes place along the helical axis without the helical structure itself moving owing to the anchoring effects at the walls (fig. 4.5.4) and that the velocity profile is flat rather than parabolic. Under

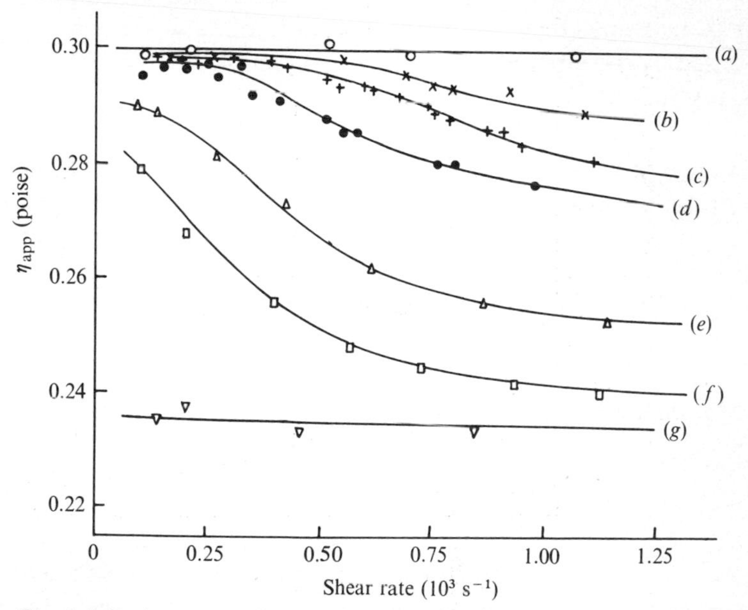

Fig. 4.5.2. Apparent viscosity in Poiseuille flow as a function of shear rate. Flow normal to the helical axis (see text). Pitch (μm) = (*a*) 1.9, (*b*) 2.6, (*c*) 3, (*d*) 3.9, (*e*) 6, (*f*) 9.1, and (*g*) ∞. (After Candau, Martinoty and Debeauvais.[58])

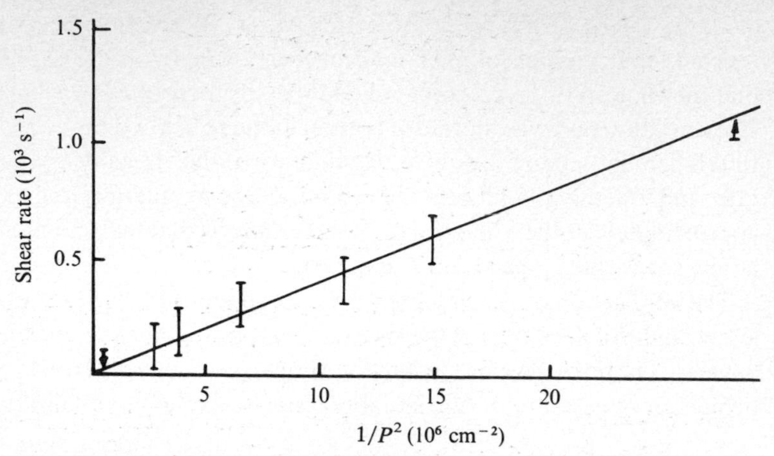

Fig. 4.5.3. Threshold of shear rate above which the fluid becomes non-newtonian plotted against $1/P^2$, where P is the undistorted value of the pitch. The arrows represent the upper and lower limits of the shear rate (see fig. 4.5.2). (After Candau, Martinoty and Debeauvais.[58])

these circumstances, the translational motion of the fluid along the capillary can be directly related to the rotational motion of the director. The energy gained by the motion in the pressure gradient should be equal to that dissipated by the rotational motion. Now the viscous torque exerted by the director

$$\mathbf{\Gamma} = \mathbf{n} \times \mathbf{g}',$$

where $\mathbf{g}'$ is given by (3.3.13). In the absence of velocity gradients this torque is evidently $-\lambda_1 qv$, where v is the linear velocity and $q = 2\pi/P$ is the cholesteric twist per unit length. In nematics, $\lambda_1 < 0$; we shall assume this to be true in the present case also. Thus

$$\lambda_1 (qv)^2 = \frac{\mathrm{d}p}{\mathrm{d}z} v,$$

where $\mathrm{d}p/\mathrm{d}z$ is the pressure gradient. The quantity of fluid flowing per second is

$$Q = -\frac{\pi R^2 (\mathrm{d}p/\mathrm{d}z)}{\lambda_1 q^2}$$

where R is the radius of the capillary. Applying Poiseuille's law,

$$\eta_{\mathrm{app}} = -\frac{\lambda_1 q^2 R^2}{8}. \qquad (4.5.1)$$

Typically, $R \sim 500\,\mu$m and $P = 2\pi/q \sim 1\,\mu$m so that $\eta_{\mathrm{app}} \sim -10^6 \lambda_1$ which explains the very high viscosity at low pressure gradients.

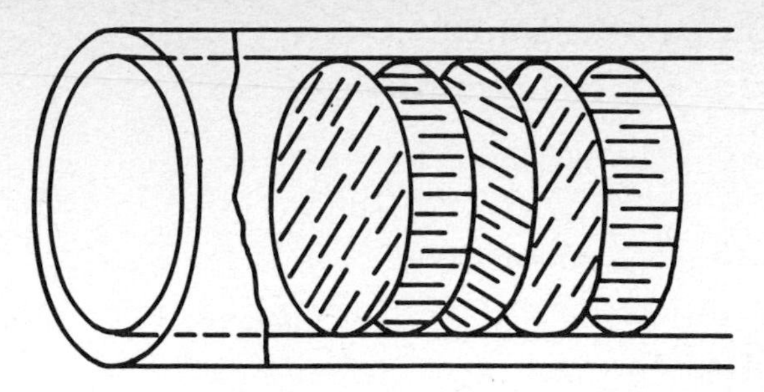

Fig. 4.5.4. Helfrich's model of 'permeation' in a cholesteric liquid crystal. At low shear rates flow takes place along the helical axis without the helical structure itself moving.

We shall now show that the essential features of Helfrich's model can be derived on the basis of the Ericksen–Leslie theory.[60]

Flow between parallel plates We shall consider flow between two parallel plates, caused by a pressure gradient. Choosing a right-handed cartesian system such that the plates occupy $x = \pm h$, we seek solutions of the form

$$n_x = \cos(qz + \varphi) \cos \theta, \qquad n_y = \sin(qz + \varphi) \cos \theta,$$

$$n_z = \sin \theta, \qquad v_x = 0, \qquad v_y = 0, \qquad v_z = w,$$

with $\theta = \theta(x, z)$, $\varphi = \varphi(x, z)$ and $w = w(x)$. This gives a cholesteric of pitch $P = 2\pi/q$ with the helical axis along z for $\theta = \varphi = 0$. Strictly speaking, a general theory should allow for non-zero values of v_x and v_y but as we shall see shortly the present approximation is valid except in a negligibly thin layer very near the boundary.

We consider very low pressure gradients and retain only the first powers of θ, φ and w. Then

$$n_x = C - \varphi S, \qquad n_y = S + \varphi C, \qquad n_z = \theta,$$

where $C = \cos qz$ and $S = \sin qz$. Neglecting director inertia and product terms involving $w\theta$, $w_{,x}\theta_{,i}$, $\theta_{,xx}\theta$, $w_{,x}\theta_{,x}$ etc., (4.3.4) reduces to

$$\begin{aligned} &\theta_{,zx}(k_{11} - k_{22}S^2) - \varphi_{,xx}(k_{11}S + k_{33}SC^2) - \varphi_{,zz}k_{22}S \\ &\quad - \theta_{,x}(k_{33} + 3k_{22})qSC \\ &\quad - 2\varphi_{,z}k_{22}qC - \lambda_1 wqS + \gamma(C - S\varphi) = 0, \end{aligned} \qquad (4.5.2)$$

$$\begin{aligned} &\theta_{,zx}k_{22}SC + \varphi_{,xx}k_{33}C^3 + \varphi_{,zz}k_{22}C + \theta_{,x}[k_{33}qC^2 + k_{22}(C^2 - 2S^2)q] \\ &\quad - 2\varphi_{,z}k_{22}qS + \lambda_1 wqC + \gamma(S + C\varphi) = 0, \end{aligned} \qquad (4.5.3)$$

$$\begin{aligned} &\theta_{,xx}(k_{22}S^2 + k_{33}C^2) + \varphi_{,zx}(k_{22} - k_{11})S + \theta_{,zz}k_{11} - \varphi_{,x}(k_{11} + k_{33})qC \\ &\quad - \theta k_{33}q^2 + \tfrac{1}{2}(\lambda_2 - \lambda_1)w_{,x}C + \gamma\theta = 0. \end{aligned} \qquad (4.5.4)$$

In the above equations $\theta_{,x} = \partial\theta/\partial x$, $\theta_{,xz} = \partial^2\theta/\partial x\,\partial z$ etc. Similarly, under the same approximations, (4.3.2) becomes

$$p_{,x} = -[\tfrac{1}{2}(\mu_6+\mu_3)+\mu_2]qSCw_{,x}, \tag{4.5.5}$$

$$p_{,y} = [\mu_2 C^2 - \tfrac{1}{2}\mu_3 + \tfrac{1}{2}\mu_6(C^2-S^2)]qw_{,x}, \tag{4.5.6}$$

$$p_{,z} = \theta_{,zx}(k_{11}-k_{22})qS - \varphi_{,xx}(k_{11}S^2+k_{33}C^2)q - \varphi_{,zz}k_{22}q$$
$$-\theta_{,x}(k_{22}+k_{33})q^2C + \tfrac{1}{2}w_{,xx}[\mu_4+(\mu_5-\mu_2)C^2]. \tag{4.5.7}$$

From (4.5.2) and (4.5.3) we get

$$\theta_{,zx}(k_{11}-k_{22})S - \varphi_{,xx}(k_{11}S^2+k_{33}C^2) - \varphi_{,zz}k_{22}$$
$$-\theta_{,x}(k_{22}+k_{33})qC - \lambda_1 wq = 0, \tag{4.5.8}$$

and from (4.5.7) and (4.5.8)

$$\tfrac{1}{2}w_{,xx}[\mu_4+(\mu_5-\mu_2)C^2] + w\lambda_1 q^2 - p_{,z} = 0. \tag{4.5.9}$$

We make a 'coarse-grained' approximation and replace $\tfrac{1}{2}[\mu_4 + (\mu_5-\mu_2)C^2]$ by an average value $\bar{\eta}$ and rewrite (4.5.9) as

$$\bar{\eta}w_{,xx} + w\lambda_1 q^2 - p_{,z} = 0. \tag{4.5.10}$$

A solution of (4.5.10) with the boundary conditions $w(\pm h) = 0$ is[60]

$$w(x) = \frac{p_{,z}}{\lambda_1 q^2}\left(1 - \frac{\cosh Kx}{\cosh Kh}\right) \tag{4.5.11}$$

where $K^2 = -\lambda_1 q^2/\bar{\eta}$. The velocity is symmetric about $x = 0$. The amount of fluid flowing per second in the z direction is

$$Q = \int_{-h}^{h} w(x)\,\mathrm{d}x = 2\int_0^h w(x)\,\mathrm{d}x$$
$$= \frac{2p_{,z}h}{\lambda_1 q^2}\left(1 - \frac{\tanh Kh}{Kh}\right). \tag{4.5.12}$$

Hence the apparent viscosity

$$\eta_{\text{app}} = -\frac{2p_{,z}h^3}{3Q} = -\frac{\lambda_1 q^2 h^2}{3(1-\tanh Kh/Kh)}. \tag{4.5.13}$$

Taking $2h = 100\ \mu\text{m}$ and $P = 1\ \mu\text{m}$, the velocity attains 0.99 of the maximum value within a thickness of $0.5\ \mu$m of the boundary. Thus in all practical situations the velocity profile is flat over most of the region between the plates and

$$\eta_{\text{app}} \simeq -\frac{\lambda_1 q^2 h^2}{3} \tag{4.5.14}$$

which is the analogue of (4.5.1).

Poiseuille flow In cylindrical polar coordinates, we seek solutions of the form

$$n_r = \cos(qz - \psi + \varphi)\cos\theta, \qquad n_\psi = \sin(qz - \psi + \varphi),$$

$$n_z = \sin\theta,$$

where θ and φ are functions of r, ψ and z. Considering very small pressure gradients we obtain to a first order in θ and φ,

$$n_r = C - \varphi S, \qquad n_\psi = S + \varphi C, \qquad n_z = \theta,$$

where $C = \cos(qz - \psi)$ and $S = \sin(qz - \psi)$. For the velocity field we assume

$$v_r = 0, \qquad v_\psi = 0, \qquad v_z = w(r).$$

Proceeding as before we obtain

$$p_{,z} = \tfrac{1}{2}w_{,rr}[\mu_4 + (\mu_5 - \mu_2)C^2] + \frac{w_{,r}}{2r}[\mu_4 + (\mu_5 - \mu_2)S^2] + \lambda_1 w q^2. \tag{4.5.15}$$

Again replacing the coefficients of $w_{,rr}$ and $w_{,r}$ by an average value

$$w_{,rr} + \frac{1}{r}w_{,r} + \frac{\lambda_1 q^2 w}{\bar{\eta}} - \frac{p_{,z}}{\bar{\eta}} = 0. \tag{4.5.16}$$

A well behaved solution of (4.5.16) is[60]

$$\frac{\lambda_1 q^2 w}{\bar{\eta}} = \frac{p_{,z}}{\bar{\eta}} + AI_0(Kr),$$

where A is a constant, I_0 is the modified zero order Bessel function of the first kind and $K^2 = -\lambda_1 q^2/\bar{\eta}$. Using the boundary condition $w(R) = 0$,

$$A = -\frac{p_{,z}}{\bar{\eta} I_0(KR)},$$

and

$$w(r) = \frac{p_{,z}}{\lambda_1 q^2}\left(1 - \frac{I_0(Kr)}{I_0(KR)}\right),$$

where R is the radius of the capillary. The quantity of liquid crystal flowing per second

$$Q = \frac{2\pi p_{,z}}{\lambda_1 q^2}\left(\frac{R^2}{2} - \frac{RI_1(KR)}{KI_0(KR)}\right)$$

where I_1 is the modified first order Bessel function of the first kind.

Hence

$$\eta_{\text{app}} = -\frac{\lambda_1 q^2 R^2}{8\{1-[2I_1(KR)/KRI_0(KR)]\}}.$$

Again, in practical situations the velocity profile is flat except very near the boundary and

$$\eta_{\text{app}} \simeq -\frac{\lambda_1 q^2 R^2}{8},$$

which is identical with (4.5.1). Thus Helfrich's idea of permeation along the helical axis of a cholesteric can be justified in terms of the Ericksen–Leslie equations.

4.5.2. Flow normal to the helical axis

The general theory of shear flow normal to the helical axis has been discussed by Leslie.[61] An interesting new feature that comes out of this analysis is that a shear in the xz plane can give rise to secondary flow along y. (In principle, secondary flow should occur in the Helfrich configuration also, but only in a negligibly thin layer very near the boundary where the velocity profile is not flat.) We shall now present a simplified version of Leslie's theory ignoring the thermomechanical coupling.

Consider a cholesteric film between two plane parallel plates, one of which is moving with constant velocity V in its own plane. The plates occupy the planes $z = \pm h$. We examine solutions of the form

$$n_x = \cos\theta\cos\varphi, \qquad n_y = \cos\theta\sin\varphi, \qquad n_z = \sin\theta, \tag{4.5.17}$$

$$v_x = u(z), \qquad v_y = v(z), \qquad v_z = 0, \tag{4.5.18}$$

where $\theta = \theta(z)$ and $\varphi = \varphi(z)$. Then, $t_{zx} = a$ (constant shear), $t_{zy} = 0$ and $t_{zz} = -p$ (an arbitrary constant). Using (4.3.7) and (4.3.13), we get

$$(H_1 + H_2\cos^2\varphi)\xi + H_2\eta\sin\varphi\cos\varphi = a, \tag{4.5.19}$$

$$H_2\xi\sin\varphi\cos\varphi + (H_1 + H_2\sin^2\varphi)\eta = 0, \tag{4.5.20}$$

where

$$2\xi = \frac{du}{dz}, \qquad 2\eta = \frac{dv}{dz},$$

$$H_1 = \mu_4 + (\mu_5 - \mu_2)\sin^2\theta$$

and

$$H_2 = (2\mu_1\sin^2\theta + \mu_3 + \mu_6)\cos^2\theta.$$

From (4.3.4) we obtain

$$F_1\frac{d^2\theta}{dz^2}+\frac{1}{2}\frac{dF_1}{d\theta}\left(\frac{d\theta}{dz}\right)^2-\frac{1}{2}\frac{dF_2}{d\theta}\left(\frac{d\varphi}{dz}\right)^2$$

$$-2k_2\sin\theta\cos\theta\frac{d\varphi}{dz}+(\lambda_1+\lambda_2\cos 2\theta)(\xi\cos\varphi+\eta\sin\varphi)=0, \tag{4.5.21}$$

and

$$F_2\frac{d^2\varphi}{dz^2}+\frac{dF_2}{d\theta}\frac{d\theta}{dz}\frac{d\varphi}{dz}+2k_2\sin\theta\cos\theta\frac{d\theta}{dz}$$

$$+(\lambda_1-\lambda_2)\sin\theta\cos\theta\,(\xi\sin\varphi-\eta\cos\varphi)=0, \tag{4.5.22}$$

where

$$F_1=k_{11}\cos^2\theta+k_{33}\sin^2\theta,$$

$$F_2=(k_{22}\cos^2\theta+k_{33}\sin^2\theta)\cos^2\theta.$$

From the symmetry of the problem it is clear that θ and v should be even functions of z, while $u-\frac{1}{2}V$ and φ are odd functions of z. Equations (4.5.19) and (4.5.20) yield

$$\xi=a[1+(H_2/H_1)\sin^2\varphi]/(H_1+H_2), \tag{4.5.23}$$

$$\eta=-a[(H_2/H_1)\sin\varphi\cos\varphi]/(H_1+H_2), \tag{4.5.24}$$

which immediately give the velocity profiles

$$u=2\int_{-h}^{z}\xi\,dz, \qquad v=-2\int_{-h}^{z}\eta\,dz. \tag{4.5.25}$$

It is seen that $v\neq 0$ even though the shear is confined to the zx plane; in other words, secondary flow occurs.

Using (4.5.23) and (4.5.24), (4.5.21) and (4.5.22) can be simplified to

$$F_1\frac{d^2\theta}{dz^2}+\frac{1}{2}\frac{dF_1}{d\theta}\left(\frac{d\theta}{dz}\right)^2-\frac{1}{2}\frac{dF_2}{d\theta}\left(\frac{d\varphi}{dz}\right)^2$$

$$-2k_2\sin\theta\cos\theta\frac{d\varphi}{dz}+aQ\cos\varphi=0 \tag{4.5.26}$$

and

$$F_2\frac{d^2\varphi}{dz^2}+\frac{dF_2}{d\theta}\frac{d\theta}{dz}\frac{d\varphi}{dz}+2k_2\sin\theta\cos\theta\frac{d\theta}{dz}-aP\sin\varphi=0, \tag{4.5.27}$$

where

$$Q=(\lambda_1+\lambda_2\cos 2\theta)/(H_1+H_2)$$

and

$$P=(\lambda_2-\lambda_1)\sin\theta\cos\theta/H_1.$$

Leslie assumed the following boundary conditions:

$$\left.\begin{aligned}&\theta(h)=\theta(-h)=0,\\&\left(\frac{\mathrm{d}\varphi}{\mathrm{d}z}\right)_h=\left(\frac{\mathrm{d}\varphi}{\mathrm{d}z}\right)_{-h}=\frac{k_2}{k_{22}}.\end{aligned}\right\}\qquad(4.5.28)$$

Now

$$V=4\int_0^h \xi\,\mathrm{d}z$$

and the apparent viscosity

$$\begin{aligned}\eta_{\mathrm{app}}&=\frac{a}{V/2h}\\&=\left[\frac{2}{h}\int_0^h\frac{[1+(H_2/H_1)\sin^2\varphi]\,\mathrm{d}z}{H_1+H_2}\right]^{-1}.\end{aligned}\qquad(4.5.29)$$

For very small values of a[62]

$$\eta_{\mathrm{app}}\simeq\frac{\mu_4(\mu_3+\mu_6)}{2\mu_4+\mu_3+\mu_6-(\mu_3+\mu_6)(\sin qh/qh)}.\qquad(4.5.30)$$

In practical cases, qh is usually so large that $\sin qh/qh$ is negligibly small and η_{app} approaches a maximum limiting value which is independent of the pitch or gap width. This value is several orders of magnitude less than that for flow along the helical axis and is comparable to that for a nematic.

When a is sufficiently large

$$\varphi=0\quad\text{and}\quad\theta=\theta_0=\tfrac{1}{2}\cos^{-1}(\lambda_1/\lambda_2),$$

i.e., the helix is unwound completely except in a layer of thickness of the order of the q^{-1} at the boundaries, and

$$\eta_{\mathrm{app}}\simeq\tfrac{1}{2}(H_1(\theta_0)+H_2(\theta_0))$$

which is again independent of the pitch or gap width. This lower limit is reached when $a\simeq k_{22}q^2$ or more. All these predictions are in qualitative agreement with the observations of Candau *et al.*[58] (see fig. 4.5.2).

4.6. Distortions of the structure by external fields

4.6.1. Magnetic field normal to the helical axis: the cholesteric–nematic transition

When a magnetic field is applied at right angles to the helical axis of an unbounded cholesteric liquid crystal composed of molecules of positive diamagnetic anisotropy ($\chi_a = \chi_\parallel - \chi_\perp > 0$) the structure gets distorted as illustrated schematically in fig. 4.6.1. As the field strength approaches a certain critical value H_c the pitch increases logarithmically; for $H > H_c$ the helix is destroyed completely and the structure becomes nematic.[63] The dependence of the pitch on the field strength was calculated by de Gennes[64] and by Meyer.[65]

Taking the helical axis to be along z, $\mathbf{H} = (H, 0, 0)$ and $\mathbf{n} = (\cos\varphi, \sin\varphi, 0)$, the free energy of the system is

$$\mathscr{F} = \int F\,\mathrm{d}z = \frac{1}{2}\int\left[\left(\frac{\mathrm{d}\varphi}{\mathrm{d}z} - q_0\right)^2 k_{22} - \chi_a H^2 \sin^2\varphi\right]\mathrm{d}\varphi + \text{constant}$$

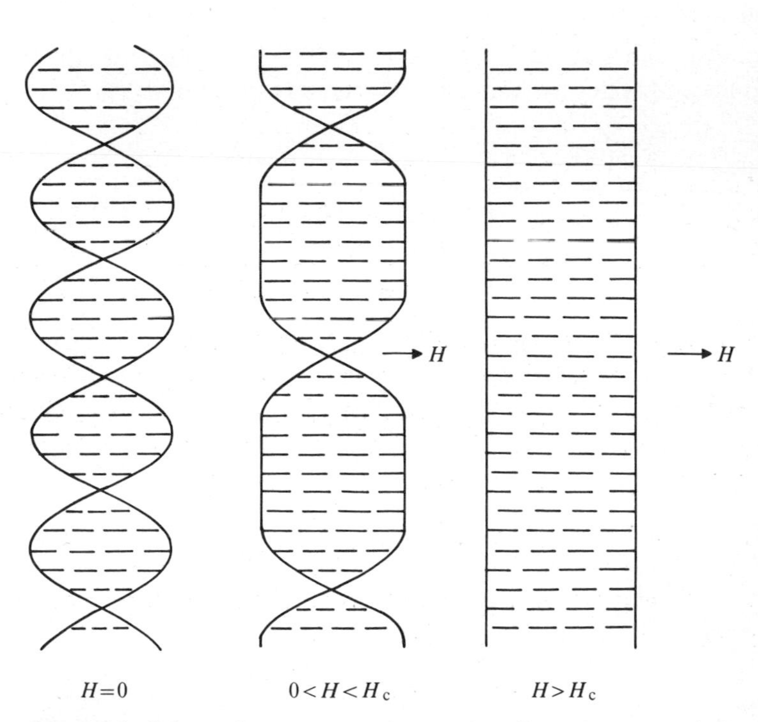

Fig. 4.6.1. Schematic representation of the effect of a magnetic field applied normal to the helical axis of a cholesteric liquid crystal composed of molecules of positive diamagnetic anistropy. For $H > H_c$ the cholesteric is transformed into a nematic.

where $q_0 = 2\pi/P_0$, P_0 being the pitch of the undistorted structure in the absence of a magnetic field. The equation of equilibrium is therefore

$$k_{22}\left(\frac{d^2\varphi}{dz^2}\right) - \chi_a H^2 \sin\varphi \cos\varphi = 0$$

which yields

$$z = 2A\xi K(A)$$

where

$$\xi = \left(\frac{k_{22}}{\chi_a}\right)^{1/2} H^{-1}$$

and

$$K(A) = \int_0^{\pi/2} \frac{\mathrm{d}\varphi}{(1 - A^2 \sin^2\varphi)^{1/2}}$$

is the complete elliptic integral of the first kind; A is a constant which can be determined from the condition that $\mathscr{F}$ should be a minimum; $z = P/2$ the half-pitch of the distorted structure. It is assumed that the sample is sufficiently thick for boundary effects to be neglected. Therefore

$$\mathscr{F} = \left(\frac{1}{2}k_{22}q_0^2 \int \mathrm{d}z\right)\left[1 - \frac{2\pi}{q_0 z} + \frac{2hJ_0}{q_0 z} - h^2(1+B)\right],$$

where

$$1 + B = A^{-2}, \qquad h = (\xi q_0)^{-1},$$

$$J_0 = 2\int_0^{\pi/2} (B + \cos^2\varphi)^{1/2}\,\mathrm{d}\varphi = \frac{4}{A}E(A),$$

and

$$E(A) = \int_0^{\pi/2} (1 - A^2 \sin^2\varphi)^{1/2}\,\mathrm{d}\varphi$$

is the elliptic integral of the second kind. The condition $\partial\mathscr{F}/\partial B = 0$ leads to the relations

$$J_0 = 2\frac{\pi}{h} \quad \text{and} \quad \frac{\partial J}{\partial B} = zq_0 h.$$

Putting $z_0 = \pi/q_0 = \frac{1}{2}P_0$, we have

$$\frac{z}{z_0} = \frac{P}{P_0} = \left(\frac{2}{\pi}\right)^2 K(A)E(A) \tag{4.6.1}$$

and

$$h = \frac{\pi}{2} \frac{A}{E(A)}. \tag{4.6.2}$$

When $z \to \infty$, $A \to 1$, $E(A) \to 1$, $K(A) \to \infty$ and $H \to H_c$, so that $h = \pi/2$ or

$$H_c = \tfrac{1}{2}\pi q_0 \left(\frac{k_{22}}{\chi_a}\right)^{1/2}, \tag{4.6.3}$$

which is the critical field at which the structure becomes nematic.

The variation of pitch with magnetic field strength predicted by (4.6.1) has been verified experimentally[66,67] (fig. 4.6.2). It has also been confirmed that H_c is inversely proportional to P_0 the pitch of the undistorted structure.[67] It turns out that with the usually available magnetic field strengths, the experiment is conveniently performed only with cholesterics of relatively large pitch. For example, in a typical measurement using nematic PAA doped with a small amount of cholesteryl acetate H_c was 8.3 kG for $P_0 = 26\ \mu$m.

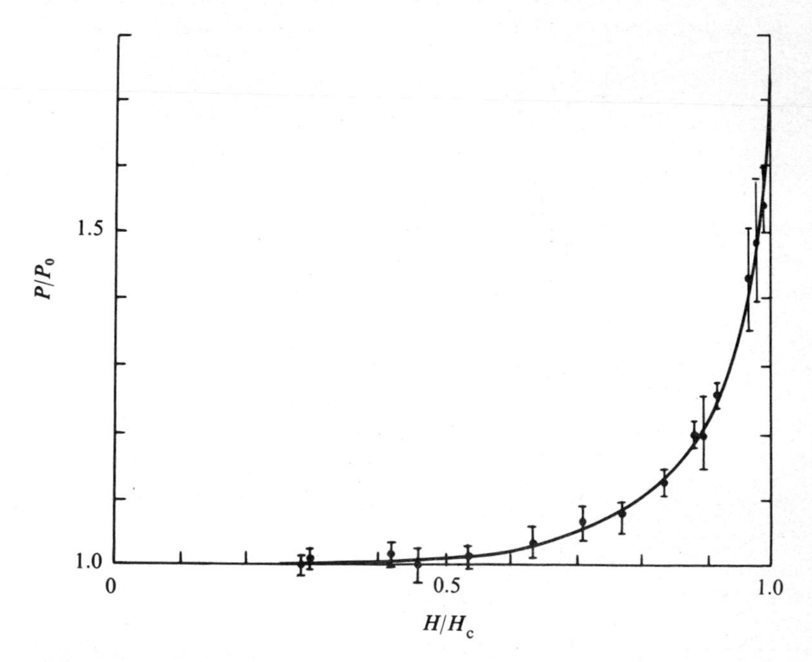

Fig. 4.6.2. Dependence of the pitch P on the magnetic field strength H in PAA mixed with a small quantity of cholesteryl acetate. Curve represents the theoretical variation predicted by de Gennes's equation (4.6.1). (After Meyer.[66])

4.6.2. Magnetic field along the helical axis: the square grid pattern

We next examine the effect of a magnetic field acting along the helical axis of a cholesteric film having a planar texture. If $\chi_a > 0$ and boundary constraints are absent, there is a possibility of a 90° rotation of the helical axis because $\frac{1}{2}(\chi_{\parallel} + \chi_{\perp}) > \chi_{\perp}$. If, on the other hand, boundary effects are such as to maintain the orientation of the helix, an expected type of deformation is for the director at every point to be tilted towards the field, i.e., a conical distortion.[65,68] However, it was pointed out by Helfrich[69] that yet another type of deformation can set in, viz, a corrugation of the layers (fig. 4.6.3). This has since been confirmed experimentally using both magnetic[70] and electric fields,[71,72] and takes place at a much lower threshold. It results in the so-called *square grid pattern* (plate 15), the theory of which was first proposed by Helfrich[69] and subsequently elaborated by Hurault.[74] We shall discuss first the magnetic field case.

For the unperturbed cholesteric

$$\mathbf{n} = (\cos q_0 z, \sin q_0 z, 0),$$

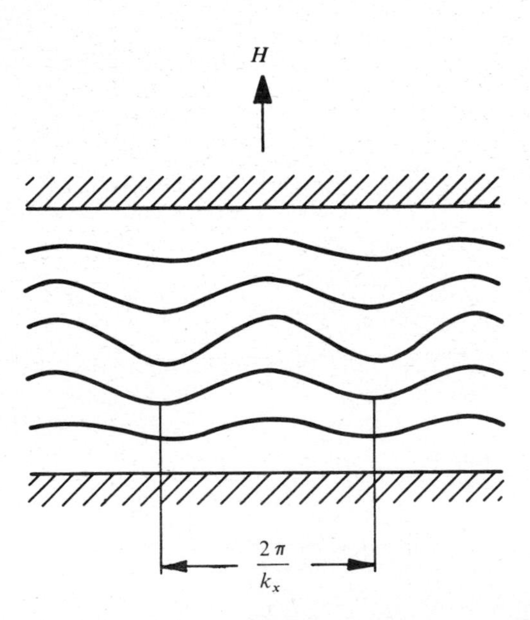

Fig. 4.6.3. Deformation of a planar structure due to a magnetic field acting along the helical axis of cholesteric liquid crystal composed of molecules of positive diamagnetic anisotropy. A similar deformation superposed in an orthogonal direction results in the square-grid pattern (see plate 15). (Helfrich.[69])

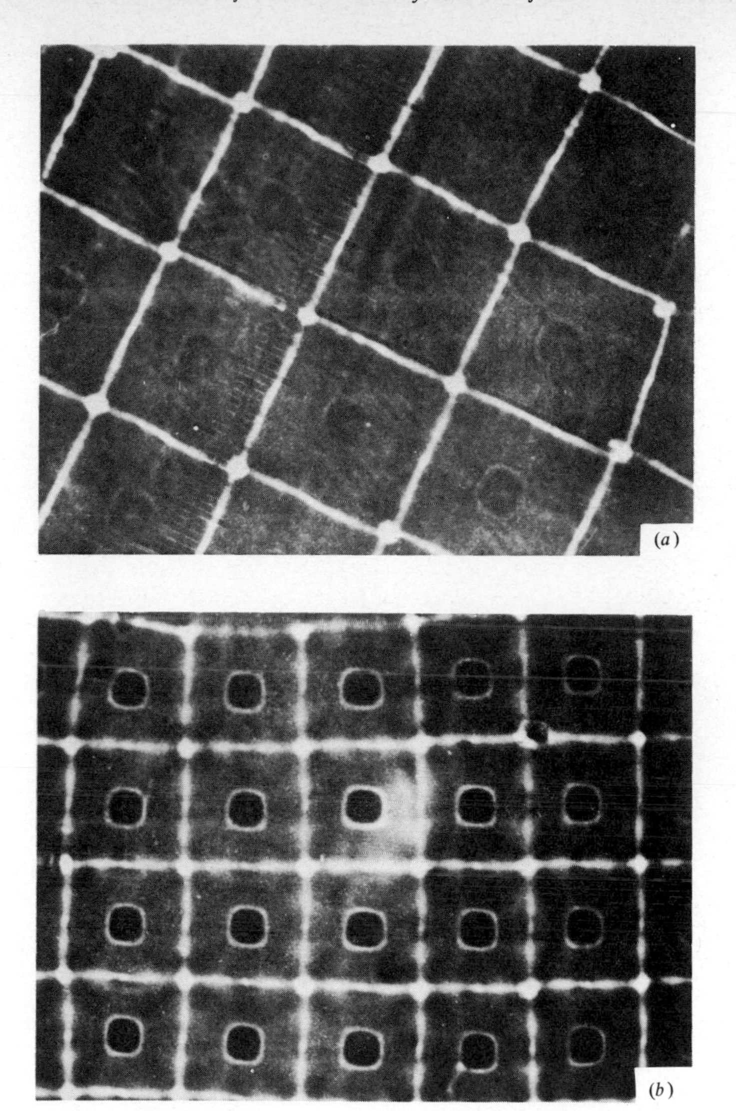

Plate 15. The square grid pattern in a cholesteric liquid crystal induced by (a) a magnetic field and (b) an electric field. (Rondelez.[73])

where we take z normal to the film. For small perturbations we may put

$$\left.\begin{aligned} n_x &= \cos(q_0 z + \varphi) \simeq \cos q_0 z - \varphi \sin q_0 z \\ n_y &= \sin(q_0 z + \varphi) \simeq \sin q_0 z + \varphi \cos q_0 z \\ n_z &= \theta \cos q_0 z. \end{aligned}\right\} \qquad (4.6.4)$$

Substituting in (4.2.1) the local energy density

$$F=\frac{1}{2}\left[(k_{11}\sin^2 q_0z+k_{33}\cos^2 q_0z)\left(q_0\theta+\frac{\partial\varphi}{\partial x}\right)^2\right.$$
$$+k_{11}\cos^2 q_0z\left(\frac{\partial\theta}{\partial x}\right)^2-k_{11}\sin 2q_0z\frac{\partial\theta}{\partial z}\left(q_0\theta+\frac{\partial\varphi}{\partial x}\right)$$
$$\left.+k_{22}\left(\frac{\partial\varphi}{\partial z}+\sin q_0z\cos q_0z\frac{\partial\theta}{\partial x}\right)^2+k_{33}\cos^4 q_0z\left(\frac{\partial\theta}{\partial x}\right)^2\right], \tag{4.6.5}$$

and the total energy

$$\mathscr{F}=\int F\,\mathrm{d}V.$$

Now

$$\frac{\partial\mathscr{F}}{\partial\varphi}=\int A\,\mathrm{d}V\quad\text{and}\quad\frac{\partial\mathscr{F}}{\partial\theta}=\int B\,\mathrm{d}V, \tag{4.6.6}$$

where

$$A=-(k_{11}\sin^2 q_0z+k_{33}\cos^2 q_0z)\left(\frac{\partial^2\varphi}{\partial x^2}+q_0\frac{\partial\theta}{\partial x}\right)$$
$$-k_{22}\left(\frac{\partial^2\varphi}{\partial z^2}+q_0\cos 2q_0z\frac{\partial\theta}{\partial x}\right)$$
$$+\tfrac{1}{2}(k_{11}-k_{22})\sin 2q_0z\frac{\partial^2\theta}{\partial x\,\partial z} \tag{4.6.7}$$

and

$$B=q_0\left(q_0\theta+\frac{\partial\varphi}{\partial x}\right)(k_{11}+k_{33})\cos^2 q_0z$$
$$-k_{33}\cos^4 q_0z\frac{\partial^2\theta}{\partial x^2}-\tfrac{1}{4}k_{22}\sin^2 2q_0z\frac{\partial^2\theta}{\partial x^2}$$
$$-\tfrac{1}{2}(k_{22}-k_{11})\sin 2q_0z\frac{\partial^2\varphi}{\partial x\,\partial z}$$
$$+k_{11}q_0\sin 2q_0z\frac{\partial\theta}{\partial z}-k_{11}\cos^2 q_0z\frac{\partial^2\theta}{\partial z^2}. \tag{4.6.8}$$

For a given θ, $\mathscr{F}$ is a minimum when $\partial\mathscr{F}/\partial\varphi=0$ or $A=0$. We consider the perturbations θ and φ to be dependent on z and x (where x is any arbitrary direction in the plane of the layers) and write them in the form

$$\left.\begin{aligned}\theta &= \theta_0 \sin k_x x \cos k_z z\\ \varphi &= (\varphi_0 + \varphi_1 \cos 2q_0 z) \cos k_x x \sin k_z z\end{aligned}\right\} \tag{4.6.9}$$

where $k_z = m\pi/d$, d being the film thickness and m an integer. We shall confine our discussion to $m = 1$. In practice,

$$k_z \ll k_x \ll q_0 \tag{4.6.10}$$

and we shall assume this to be true in what follows. The condition $\partial\mathscr{F}/\partial\varphi = 0$, yields

$$\begin{aligned}&\tfrac{1}{2}(k_{11}+k_{33})(k_x^2\varphi_0 - q_0 k_x \theta_0)\\ &\quad + k_{22}k_z^2\varphi_0 + \tfrac{1}{4}(k_{33}-k_{11})k_x^2\varphi_1 = 0,\end{aligned} \tag{4.6.11}$$

$$\begin{aligned}&k_{22}(4q_0^2\varphi_1 - q_0 k_x\theta_0) + \tfrac{1}{2}(k_{33}-k_{11})(k_x^2\varphi_0 - q_0 k_x\theta_0)\\ &\quad + \tfrac{1}{2}(k_{33}+k_{11})k_x^2\varphi_1 = 0,\end{aligned} \tag{4.6.12}$$

where we make a 'coarse-grained' approximation, i.e., take into consideration only the slowly varying parts of A. The minimum energy density

$$\begin{aligned}B = [&\tfrac{1}{2}q_0(q_0\theta_0 - k_x\varphi_0)(k_{11}+k_{22}) + \tfrac{1}{8}(3k_{33}+k_{22})k_x^2\theta_0\\ &-\tfrac{1}{4}q_0k_x\varphi_1(2k_{22}+k_{33}-k_{11})] \cos k_z z \sin k_x x.\end{aligned} \tag{4.6.13}$$

From (4.6.11) and (4.6.12) we get

$$\varphi_1 \simeq \frac{k_x}{4q_0}\theta_0 \tag{4.6.14}$$

and

$$\begin{aligned}\frac{q_0\theta_0}{k_x\varphi_0} \simeq &\left[\frac{1}{2}(k_{11}+k_{33}) + k_{22}\frac{k_z^2}{k_x^2}\right]\\ &\times\left[\frac{1}{2}(k_{11}+k_{33}) - \frac{1}{16}(k_{33}-k_{11})\frac{k_x^2}{q_0^2}\right]^{-1} \simeq 1\end{aligned}$$

because of (4.6.10). Therefore

$$\varphi_0 \simeq q_0\theta_0/k_x. \tag{4.6.15}$$

Thus, in the coarse-grained approximation the energy density becomes

$$F_{\text{cg}} = \frac{1}{2}\left(k_{22}\frac{q_0^2k_z^2}{k_x^2} + \frac{3}{8}k_{33}k_x^2\right)\theta^2. \tag{4.6.16}$$

An alternative derivation of (4.6.16) has been given by de Gennes. We consider the cholesteric to be a quasi-layered structure and write the energy density in terms of the displacement $u(r)$ of each plane in the following form:

$$F=\frac{1}{2}B\left(\frac{\partial u}{\partial z}\right)^2+\frac{1}{2}K\left(\frac{\partial^2 u}{\partial x^2}+\frac{\partial^2 u}{\partial y^2}\right)^2+\ldots \tag{4.6.17}$$

where B is an elastic coefficient associated with the compression of the layers. Terms involving $(\partial u/\partial x)^2$ and $(\partial u/\partial y)^2$ are not included as they correspond to a uniform rotation of the layers and do not contribute to the free energy. Introducing a unit vector $\mathbf{h}$ along the helical axis, (4.6.17) may be expressed as

$$F=\frac{1}{2}B(P/P_0-1)^2+\frac{1}{2}K(\nabla\cdot\mathbf{h})^2 \tag{4.6.18}$$

where P is the local value of the pitch. We know that the twist energy of deformation of the cholesteric structure can be written in terms of the pitch as $\frac{1}{2}k_{22}(q-q_0)^2$ where $q=2\pi/P$ and $q_0=2\pi/P_0$. Comparing this with the first term on the right-hand side of (4.6.18), it is clear that

$$B\simeq k_{22}q_0^2. \tag{4.6.19}$$

In order to evaluate K, one may use the following argument. Suppose that the film is rolled up into a cylinder, i.e., the screw axes are oriented radially about the cylinder axis and the layers form coaxial cylinders. In cylindrical coordinates, the components of the director are now

$$n_r=0,\qquad n_\psi=\cos\theta(r),\quad \text{and}\quad n_z=\sin\theta(r)$$

and the local free energy is given by

$$F=\frac{1}{2}k_{22}\left(\frac{\mathrm{d}\theta}{\mathrm{d}r}-q_0-\frac{\sin\theta\cos\theta}{r}\right)^2+\frac{1}{2}k_{33}\frac{\cos^4\theta}{r^2}. \tag{4.6.20}$$

The optimum value of $\theta(r)$ compatible with the periodicity $\theta(r)=\theta(r+P_0)$ corresponds to

$$\frac{\mathrm{d}\theta}{\mathrm{d}r}=q_0+\frac{1}{r}\sin\theta\cos\theta.$$

Averaging over $\cos^4\theta$,

$$K=\tfrac{3}{8}k_{33}, \tag{4.6.21}$$

so that

$$F_{\mathrm{cg}}=\frac{1}{2}k_{22}q_0^2\left(\frac{\partial u}{\partial z}\right)^2+\frac{3}{16}k_{33}\left(\frac{\partial^2 u}{\partial x^2}+\frac{\partial^2 u}{\partial y^2}\right)^2. \tag{4.6.22}$$

We may take u as our variable and write

$$u = u_0 \cos k_x x \cos k_z z, \tag{4.6.23}$$

where $k_z = \pi/d$. It is seen at once that (4.6.22) becomes equivalent to (4.6.16) if we replace θ by $k_x u$.

In the presence of a magnetic field applied normal to layers the total free energy

$$F = F_{cg} + F_m,$$

where

$$F_m = -\frac{1}{4}\chi_a\left(\frac{\partial u}{\partial x}\right)^2 H^2, \tag{4.6.24}$$

$\chi_a = \chi_\parallel - \chi_\perp$ being the anisotropy of diamagnetic susceptibility. Applying the condition $\partial F/\partial u = 0$ and using (4.6.22), (4.6.23) and (4.6.24)

$$\left(k_{22}q_0^2\frac{k_z^2}{k_x^2} + \frac{3}{8}k_{33}k_x^2 - \frac{1}{2}\chi_a H^2\right)u = 0. \tag{4.6.25}$$

Thus $H \to \infty$ when $k_x \to \infty$ as also when $k_x \to 0$. This is because the perturbation u is made up of two components, bend and twist: $k_x \to \infty$ excites the bend mode while $k_x \to 0$ excites the twist mode whose amplitude diverges in the limit. The optimum wave-vector corresponds to an admixture of both modes and is given by

$$k_x^4 = \frac{8k_{22}}{3k_{33}}q_0^2 k_z^2, \tag{4.6.26}$$

or

$$k_x \propto (P_0 d)^{-1/2}, \tag{4.6.27}$$

and the threshold field

$$H_H^2 = \frac{1}{\chi_a}(6k_{22}k_{33})^{1/2} q_0 k_z, \tag{4.6.28}$$

or

$$H_H \propto (P_0 d)^{-1/2}. \tag{4.6.29}$$

It is interesting to note that this threshold field is lower than that for a conical distortion or that for cholesteric–nematic (unwinding) transition. For a conical distortion, the theory is closely similar to that discussed in § 3.4.2 and has been treated by Leslie[68]; the critical field is given by

$$H_F^2 = \frac{1}{\chi_a}(k_{11}k_z^2 + k_{33}q_0^2).$$

For the cholesteric–nematic transition, we have from (4.6.3)

$$H_G^2 = \frac{\pi^2}{4\chi_a} k_{22} q_0^2,$$

and in view of (4.6.10), H_H is much less than H_F or H_G.

The experimentally observed square grid pattern corresponds to two such distortions, which are orthogonal.

4.6.3. Electric field along the helical axis

Electric field effects are more complicated because of conduction.[74] The Carr–Helfrich instability (see § 3.10.2) may be expected to take place in this case too, only the bend and twist distortions are now coupled. Moreover, the fluid motion along z can occur only by the process of permeation (§ 4.5.1).

We shall consider the DC case first (neglecting, of course, any charge injection at the boundaries). If $\sigma_{\parallel h}$ and $\sigma_{\perp h}$ are the conductivities along and perpendicular to the helical axis (which, as before, is taken to be parallel to z), the electric current

$$J_x = \sigma_{\perp h} E_x - (\sigma_{\parallel h} - \sigma_{\perp h}) E_0 \frac{\partial u}{\partial x}, \tag{4.6.30}$$

which should be zero ($\nabla \cdot \mathbf{J} = 0$). Here E_0 is the applied field and E_x that caused by the Carr–Helfrich mechanism. Therefore

$$E_x = -E_0 \frac{\partial u}{\partial x} \frac{\sigma_{\perp h} - \sigma_{\parallel h}}{\sigma_{\perp h}}. \tag{4.6.31}$$

The charge density is given by

$$\rho_c = \frac{1}{4\pi} \nabla \cdot (\boldsymbol{\varepsilon} \cdot \mathbf{E}) = \frac{\varepsilon_{\perp h} E_0}{4\pi} \frac{\partial^2 u}{\partial x^2} \left(\frac{\sigma_{\parallel h}}{\sigma_{\perp h}} - \frac{\varepsilon_{\parallel h}}{\varepsilon_{\perp h}} \right) \tag{4.6.32}$$

and the electric force

$$f_c = \rho_c E_0. \tag{4.6.33}$$

The dielectric torque

$$\Gamma_{\text{diel}} = \frac{\varepsilon_{\parallel h} - \varepsilon_{\perp h}}{4\pi} E_0^2 \left(\frac{E_x}{E_0} + \frac{\partial u}{\partial x} \right)$$

from which it follows that the contribution to the vertical force is

$$f_{\text{diel}} = \frac{\partial \Gamma_{\text{diel}}}{\partial x} = \frac{\varepsilon_{\parallel h} - \varepsilon_{\perp h}}{4\pi} E_0^2 \frac{\sigma_{\parallel h}}{\sigma_{\perp h}} \frac{\partial^2 u}{\partial x^2}.$$

The net electric force

$$f_{\text{elec}} = f_c + f_{\text{diel}}.$$

From (4.6.22) the elastic restoring force

$$f_{\text{elas},z} = -(k_{22}q_0^2k_z^2 + \tfrac{3}{8}k_{33}k_x^4)u.$$

The threshold field is obtained by setting $f_{\text{elec}} + f_{\text{elas}} = 0$. The optimum wave-vector k_x is given by (4.6.26) and the threshold field by

$$E_{\text{th}}^2 = \frac{8\pi^3}{\varepsilon_{\parallel\text{h}}} \frac{\sigma_{\perp\text{h}}}{\sigma_{\perp\text{h}} - \sigma_{\parallel\text{h}}} (\tfrac{3}{2}k_{22}k_{33})^{1/2}(P_0 d)^{-1}$$

$$= \frac{8\pi^3}{\varepsilon_\perp} \frac{\sigma_\parallel + \sigma_\perp}{\sigma_\parallel - \sigma_\perp} (\tfrac{3}{2}k_{22}k_{33})^{1/2}(P_0 d)^{-1} \tag{4.6.34}$$

where $\sigma_\parallel$ and $\sigma_\perp$ are the conductivities parallel and perpendicular to the preferred molecular direction and $\varepsilon_\parallel$ and $\varepsilon_\perp$ are similarly defined. It may be noted here that we have neglected the contribution of the viscous torque altogether. This is because, as remarked earlier, fluid motion takes place only by permeation, and moreover the distortions are infinitesimal and of such long wavelength ($k_x \ll q_0$) that the effect of shear flow will be very small indeed. The dependence of the spatial periodicity of the pattern and the threshold field on d and P_0 is similar to the magnetic field case, and is borne out by experiments[71,72,75] (figs. 4.6.4 and 4.6.5).

Hurault[74] has extended the treatment to AC fields. The method is somewhat analogous to the one discussed in § 3.10 for nematic instabilities and leads, in the conduction regime, to the following threshold for distortion:

$$\overline{E_{\text{th}}^2} = \frac{8\pi^3}{\varepsilon_\perp} \frac{\varepsilon_\parallel + \varepsilon_\perp}{\varepsilon_\parallel - \varepsilon_\perp} \frac{1 + \omega^2\tau^2}{1 - \zeta + \omega^2\tau^2} (\tfrac{3}{2}k_{22}k_{33})^{1/2}(P_0 d)^{-1}, \tag{4.6.35}$$

where

$$1 - \zeta = \frac{\sigma_\parallel - \sigma_\perp}{\sigma_\parallel + \sigma_\perp} \frac{\varepsilon_\parallel + \varepsilon_\perp}{\varepsilon_\parallel - \varepsilon_\perp}$$

and τ is the dielectric relaxation time given by

$$\frac{1}{\tau} = 4\pi \frac{\sigma_\parallel + \sigma_\perp}{\varepsilon_\parallel + \varepsilon_\perp}.$$

For negative dielectric anisotropy ($\varepsilon_\parallel - \varepsilon_\perp < 0$) the conduction regime occurs when $\omega < \omega_c$, where

$$\omega_c = 4\pi(\zeta - 1)^{1/2} \frac{\sigma_\parallel + \sigma_\perp}{\varepsilon_\parallel + \varepsilon_\perp}.$$

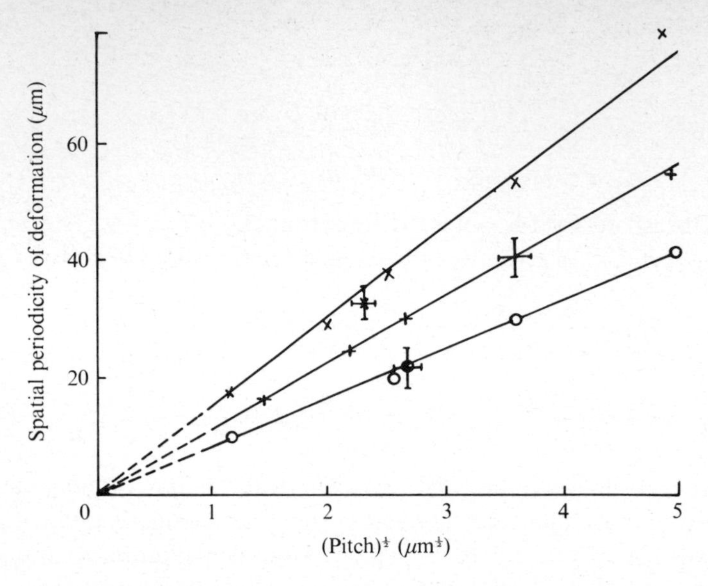

Fig. 4.6.4. The spatial periodicity of the Helfrich deformation plotted against the square root of the pitch in mixtures of nematic MBBA and cholesteryl nonanoate. AC field frequency 10 Hz. Sample thicknesses 30 μm (○), 50 μm (+) and 105 μm (×). A few typical error bars are shown. Lines represent the variation predicted by theory. (After Arnould-Netillard and Rondelez.[75])

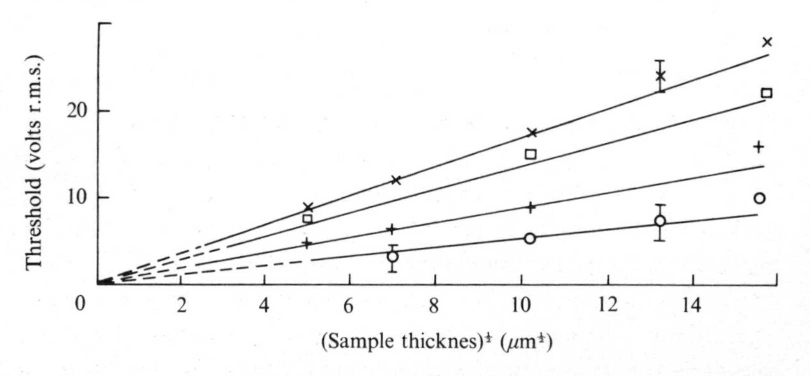

Fig. 4.6.5. Threshold voltage versus the square root of the sample thickness for the Helfrich deformation (see fig. 4.6.4). Pitch = 3.5 μm (×), 6.5 μm (□), 13 μm (+) and 23 μm (○). A few typical error bars are shown. AC field frequency 3 Hz. Lines represent the variation predicted by theory. (After Arnould-Netillard and Rondelez.[75])

These results are in quantitative agreement with observations[75] (fig. 4.6.6). However, the behaviour at higher frequencies does not appear to be fully understood.

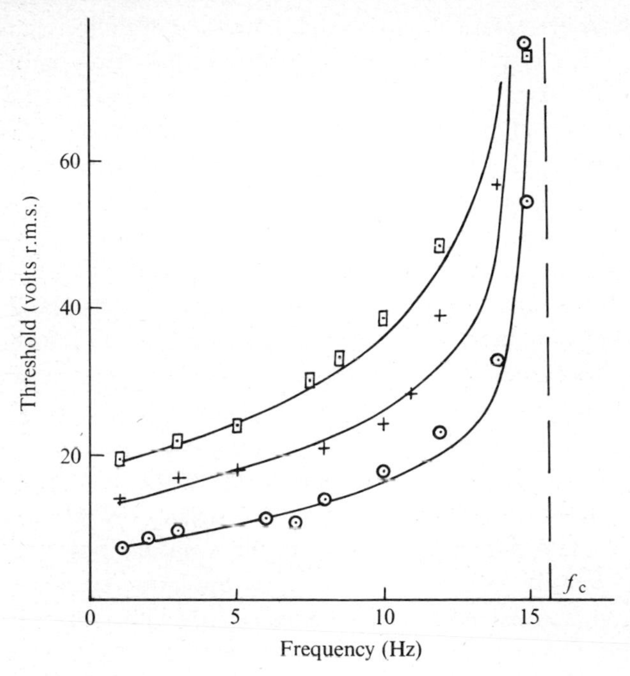

Fig. 4.6.6. Threshold voltage versus frequency for the Helfrich deformation (see fig. 4.6.4). Sample thickness 250 μm. Pitch $= 6.5$ μm (⊡), 13 μm (+) and 23 μm (⊙). Solid curves represent theoretical variation for $\zeta = 1.5$. The cut-off frequency f_c is 15.5 Hz for all samples. (After Arnould-Netillard and Rondelez.[75])

4.6.4. The storage mode (or the memory effect)

In the case of dielectrically negative molecules, the square grid pattern changes as the voltage is raised above the threshold and the helical axis rotates by 90°. (The tilted structure is metastable and relaxes to the planar texture if the field is switched off but only after a long time.) At even higher voltages turbulence sets in and the system goes over to the dynamic scattering mode. On switching off, the liquid crystal relaxes to the focal conic texture and the scattering persists. This has been described as the *storage mode* or the *memory effect* in cholesteric liquid crystals.[76] An audio frequency (~10 kHz) pulse then restores it to the planar texture. In combination with a photoconductor, this effect has been made use of to construct image storage panels.[77]

4.7. Anomalous optical rotation in the isotropic phase

We have seen in § 2.4 that the orientational correlations between the molecules give rise to certain remarkable pretransitional effects in the isotropic phase of nematic liquid crystals. Similar correlations exist in the cholesteric phase as well, except that by virtue of the chiral nature of the interactions the local order lacks a centre of inversion. Cheng and Meyer[78] discovered that these correlations give rise to an enhancement of the optical activity in the isotropic phase. The correlation length increases as the temperature approaches the isotropic–cholesteric transition point, the local helical ordering builds up and the optical rotation increases accordingly. The magnitude of this effect is just barely observable in most cholesteric materials as the anisotropy of molecular polarizability of these compounds is usually rather small. For this reason Cheng and Meyer used a specially synthesized nematogen with an optically active end group: *p*-ethoxybenzal-*p*′-(β-methylbutyl)aniline. This molecule forms a cholesteric phase and has the advantage of having a high anisotropy. Cheng and Meyer's experimental results are shown in fig. 4.7.1. The natural optical activity of the molecule (in the absence of correlations), determined by measuring the rotatory power of dilute solutions of varying concentration and extrapolating to 100 per cent concentration, was just about $1°\ \mathrm{cm}^{-1}$ as compared with a total rotation of nearly $40°\ \mathrm{cm}^{-1}$ close to the transition, proving that correlations play the predominant role.

The theory of Cheng and Meyer is rather elaborate and will not be discussed in detail here. We shall merely indicate the major steps in the calculations. We consider a system of identical molecules and for simplicity neglect the contribution of the natural optical activity to the total rotation. If $\mathbf{E}_0$ is the externally applied field, the net field $\mathbf{F}$, acting on a molecule at $\mathbf{x}$, is

$$\mathbf{F}(\mathbf{x}) = \mathbf{E}_0 + \int_v^V \mathbf{P}(\mathbf{x}') \cdot \mathbf{G}(\mathbf{x}-\mathbf{x}', \mathbf{k}_0)\, \mathrm{d}^3x',$$

where $\mathbf{P}(\mathbf{x}) = N\boldsymbol{\alpha}(\mathbf{x}) \cdot \mathbf{F}(\mathbf{x})$ is the polarization (or the dipole moment per unit volume) at $\mathbf{x}$, $\boldsymbol{\alpha}$ is the polarizability of the molecule at $\mathbf{x}$, $\mathbf{G}(\mathbf{x}-\mathbf{x}', \mathbf{k}_0)$ a tensor representing the field at $\mathbf{x}$ due to a dipole at $\mathbf{x}'$, $\mathbf{k}_0$ the wave-vector of the incident radiation, v the volume of the Lorentz cavity which is not supposed to contribute to the effective field and V the total volume.

$$\mathbf{P}(\mathbf{x}) = \mathbf{P}_0(\mathbf{x}) + \delta\mathbf{P}(\mathbf{x}),$$

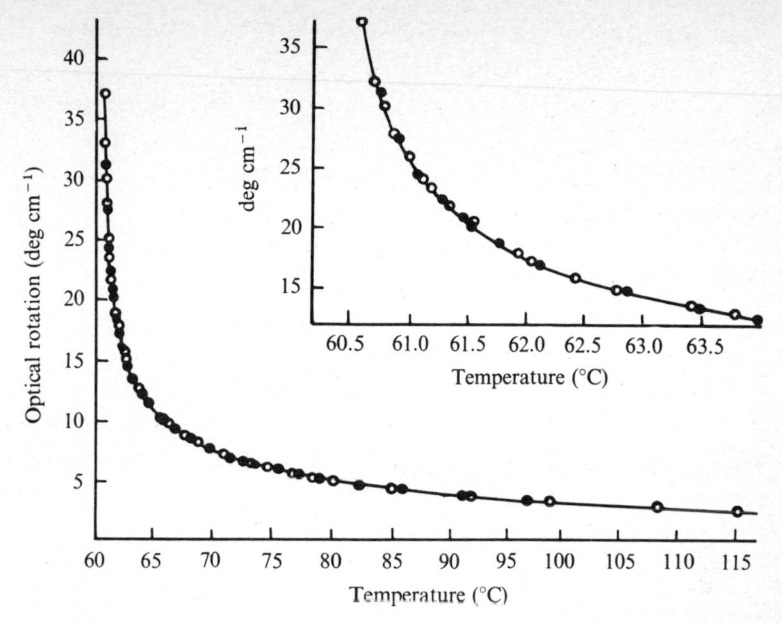

Fig. 4.7.1. Anomalous optical rotation in the isotropic phase of a cholesteric liquid crystal. Open and closed circles are measurements on two different samples appropriately normalized. Cholesteric–isotropic transition temperature 60.57 °C. $\lambda = 0.6328\ \mu$m. (After Cheng and Meyer.[78])

Writing where $\mathbf{P}_0$ is the polarization in the absence of correlations and $\delta\mathbf{P}$ a small correction term,

$$\langle\delta\mathbf{P}\rangle = N\langle\delta\boldsymbol{\alpha}(\mathbf{x})\cdot\mathbf{F}_0(\mathbf{x})\rangle + N^2\int_v^V \mathrm{d}^3x'\langle\delta\boldsymbol{\alpha}(\mathbf{x})\cdot\mathbf{G}\cdot\delta\boldsymbol{\alpha}(\mathbf{x}')\rangle\cdot\mathbf{F}(\mathbf{x}'),$$

neglecting higher powers of $\delta\boldsymbol{\alpha}$. Assuming a Lorentz–Lorenz type of relationship for the polarization field in the medium, expressing $\delta\boldsymbol{\alpha}$ in terms of the tensor order parameters $\mathbf{s}$ (see, for example, § 2.3.1) the susceptibility

$$\chi = \frac{n^2+2}{3}\Big[N\alpha + \Big(\frac{2}{3}N\alpha_{\mathrm{a}}\Big)^2\int \mathrm{d}^3R\,\mathbf{G}(\mathbf{R},\mathbf{k}_0)\cdot\exp(\mathrm{i}n\mathbf{k}_0\cdot\mathbf{R})$$
$$\times\langle\mathbf{s}(0)\cdot\mathbf{s}(\mathbf{R})\rangle\Big],$$

where n is the refractive index of the isotropic phase, α the mean molecular polarizability and $\alpha_{\mathrm{a}} = \alpha_\parallel - \alpha_\perp$ the polarizability anisotropy.

Expressing the integral in terms of the Fourier components,

$$\chi = \frac{n^2+2}{3}\left[N\alpha + \left(\frac{2}{3}N\alpha_a\right)^2 \int \frac{d^3q}{(2\pi)^3}\mathbf{G}(n\mathbf{k}_0+\mathbf{q}) \cdot \langle \mathbf{s}^*(\mathbf{q}) \cdot \mathbf{s}(\mathbf{q})\rangle\right].$$

Hence the dielectric tensor may be expressed as

$$\boldsymbol{\varepsilon} = \bar{\boldsymbol{\varepsilon}} + \boldsymbol{\Lambda},$$

where $\boldsymbol{\Lambda}$ is a tensor of the form

$$\Lambda_{ij} = \Lambda'_{ij} + i\Lambda''_{ij}.$$

In order that the medium be non-absorbing we must have

$$\varepsilon^*_{ij} = \varepsilon_{ji},$$

so that for an isotropic medium

$$\boldsymbol{\varepsilon} = \begin{bmatrix} \bar{\varepsilon} + \Lambda'_{xx} & i\Lambda''_{xy} & i\Lambda''_{xz} \\ -i\Lambda''_{xy} & \bar{\varepsilon} + \Lambda'_{yy} & i\Lambda''_{yz} \\ -i\Lambda''_{xz} & -i\Lambda''_{yz} & \bar{\varepsilon} + \Lambda'_{zz} \end{bmatrix}.$$

Such a system will exhibit circular birefringence,

$$n^2_\pm = [\bar{\varepsilon} + \Lambda'_{xx} \pm \Lambda''_{xy}]$$

or an optical rotation

$$\rho = \frac{\pi}{\lambda}(n_+ - n_-) = \frac{\pi}{n\lambda}\Lambda''_{xy}$$

$$= \frac{4\pi^2}{n\lambda}\left(\frac{n^2+2}{3}\right)\left(\frac{2}{3}N\alpha_a\right)^2 \int \frac{d^3q}{(2\pi)^3} G_{\alpha\beta}\, \mathrm{Im}\langle s^*_{x\beta}\, s_{y\beta}\rangle.$$

The rotation exists because of the correlations $\langle s^*_{x\alpha}\, s_{y\beta}\rangle$ which are non-vanishing in the case of a cholesteric. The averages may be evaluated on the basis of de Gennes's model[79] (see § 2.4). To allow for the non-centrosymmetric ordering in the cholesteric, we include an additional term of the form $\mathbf{s} \cdot \nabla \times \mathbf{s}$ in the free energy of the isotropic phase

$$F_{IC} = F_{IN} + 2q_0 L' e_{\alpha\beta\gamma} s_{\alpha\mu} \frac{\partial s_{\beta\mu}}{\partial x_\gamma}$$

where F_{IN} is the free energy per unit volume in the isotropic phase of a nematic given by (2.4.15) and (2.4.22), q_0 is a pseudo-scalar and L' is a constant. The average can then be worked out. For example

$$\langle s'_{xz}\, s''_{yz}\rangle = -\langle s''_{xz}\, s'_{yz}\rangle = \frac{k_B T}{4A} \frac{\gamma_2 q\xi_2}{[1+(q\xi_2)^2]^2 - (\gamma_2 q\xi_2)^2},$$

where

$$A = a(T - T^*),$$
$$\xi_1^2 = L_1/A,$$
$$\xi_2^2 = (L_1 + \tfrac{1}{2}L_2)/A,$$
$$\gamma_2 = 2L'q_0^2\xi_1^2/L_1\xi_2,$$

and L_1, L_2 are the constants occurring in (2.4.22). Therefore the optical rotation increases rapidly as the temperature approaches T^*.

4.8. Some factors influencing the pitch

We conclude this chapter with a brief survey of the experimental facts concerning the dependence of the pitch on temperature, composition, etc.

4.8.1. Dependence of pitch on temperature: applications to thermography

In most pure cholesteric materials, the pitch is a decreasing function of the temperature. An elementary picture of the temperature dependence of the pitch can be given in analogy with the theory of thermal expansion in crystals.[80] Assuming anharmonic angular oscillations of the molecules about the helical axis, the mean angle between successive layers

$$\langle\theta\rangle = Ak_BT/2I\omega_0^4$$

where A is the coefficient of the cubic anharmonicity term,[81] ω_0 the angular frequency and I the moment of inertia of the molecule. Thus the pitch $P(\propto 1/\langle\theta\rangle)$ may be expected to decrease slightly with temperature. However, in many substances, the rate of variation is extremely high. Fig. 4.8.1 presents the experimental data for a pure compound, cholesteryl nonanoate, and its mixtures with some cholesteryl esters.[82] It is now established that if the cholesteric phase is preceded by a smectic phase at a lower temperature, as is the case with cholesteryl nonanoate, the pitch increases very rapidly as the sample is cooled to the smectic–cholesteric transition point. This is due to the growth of smectic-like clusters in the cholesteric phase (see § 5.4.2).

The strong temperature dependence of the pitch has practical applications in thermography, as was first demonstrated by Fergason.[83] The material has to be so chosen that the pitch is of the order of the wavelength of visible light in the temperature range of interest. This is achieved by preparing suitable mixtures. Small variations of temperature

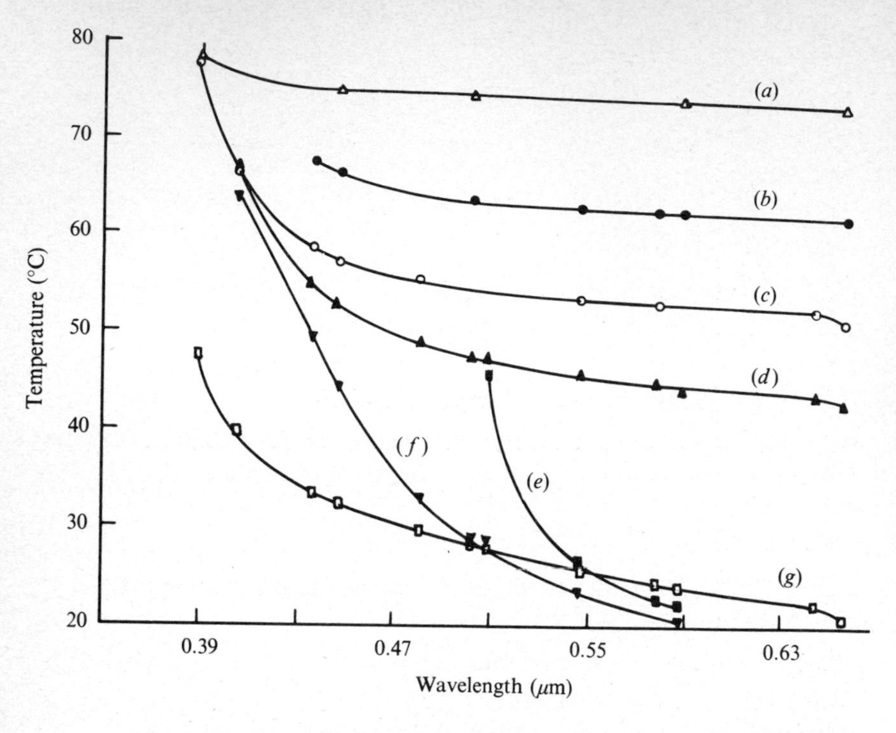

Fig. 4.8.1. Wavelength of maximum scattering as a function of temperature for a number of mixtures of cholesteryl esters and cholesteryl nonanoate: (*a*) pure cholesteryl nonanoate (CN); (*b*) 20 per cent cholesteryl hydrocinnamate, 80 per cent CN; (*c*) 20 per cent cholesteryl butyrate, 80 per cent CN; (*d*) 20 per cent cholesteryl propionate, 80 per cent CN; (*e*) 20 per cent cholesteryl chloride, 80 per cent CN; (*f*) 20 per cent cholesteryl acetate, 80 per cent CN; (*g*) 20 per cent cholesteryl methyl carbonate, 80 per cent CN. (After Fergason, Goldberg and Nadalin.[82])

are shown up as changes in the colour of the scattered light and can be used for visual display of surface temperatures,[84] imaging of infrared[85] and microwave[86] patterns, etc.

In thermography and other applications one invariably works with polydomain samples. Fergason has given the following simple expression for the wavelength λ of Bragg scattering from a polydomain film:

$$\lambda = nP \cos[\tfrac{1}{2} \sin^{-1}(n^{-1} \sin \varphi_i) + \tfrac{1}{2} \sin^{-1}(n^{-1} \sin \varphi_s)]$$

where φ_i and φ_s are the angles of incidence and scattering as shown in fig. 4.8.2, and n the mean refractive index.

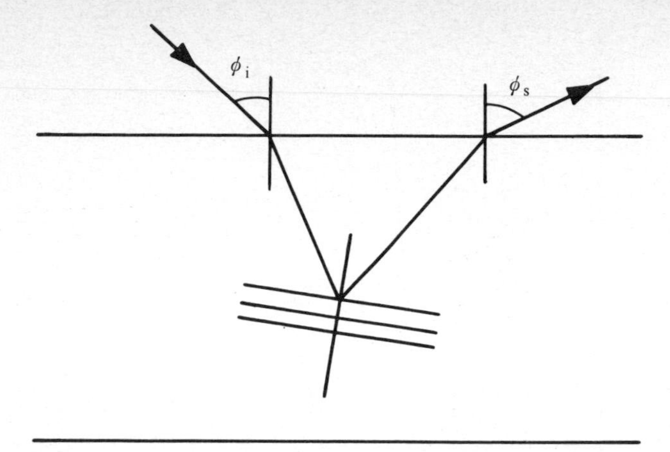

Fig. 4.8.2. Bragg scattering from a polydomain sample of cholesteric liquid crystal.

4.8.2. Dependence of pitch on pressure

Pollmann and Stegemeyer[87] have investigated the effect of pressure on the pitch of cholesteryl oleyl carbonate (COC) mixed with cholesteryl chloride. The pitch increases very rapidly with pressure, the effect being more pronounced the greater the concentration of COC (fig. 4.8.3). This appears at first quite surprising, but, in fact, the explanation is straightforward.[101] We have already remarked that the pitch increases anomalously as the temperature approaches the cholesteric–smectic transition point. Now pure COC exhibits a smectic *A* phase below 14 °C at atmospheric pressure. This temperature may of course be somewhat lower in the case of the mixture. However, as the pressure is raised the temperature of transition goes up[88,89] (see fig. 5.4.3), so that the pitch at room temperature may be expected to rise accordingly.

4.8.3. Mixtures: dependence of pitch on composition

A mixture of right- and left-handed cholesterics adopts a helical structure whose pitch is sensitive to temperature and composition.[90] For a given composition, there is an inversion of the rotatory power as the temperature is varied, indicating a change of handedness of the helix.[15] The inverse pitch exhibits a linear dependence on temperature, passing through zero at the nematic point where there is an exact compensation of the right- and left-handed forms (fig. 4.1.20). The theory of the optical behaviour of such a system as a function of the pitch has already been discussed at length in § 4.1.1.

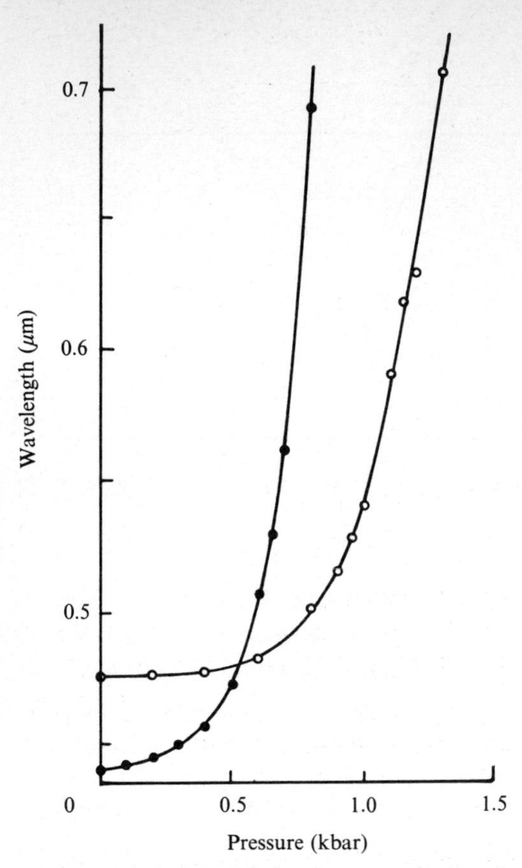

Fig. 4.8.3. Pressure dependence of wavelength of maximum reflexion in mixtures of cholesteryl oleyl carbonate (COC) and cholesteryl chloride (CC) at room temperature. Open circles, COC/CC = 74.8/25.2 (mole per cent). Full circles, COC/CC = 80.1/19.9 (mole per cent). (After Pollmann and Stegemeyer.[87])

A similar reversal of handedness takes place as the composition is varied[91] (fig. 4.8.4). The inverse pitch shows a nearly linear relationship with composition around the nematic point, but there are significant departures when one of the components has a smectic phase at a lower temperature.[92] The anomaly may again be attributed to smectic-like short range order.

A nematic liquid crystal readily adopts a helical configuration if a small quantity of a cholesteric is added to it. For low concentrations of the cholesteric, the inverse pitch is a linear function of the concentration, but at higher concentrations the linear law is not obeyed. Fig. 4.8.5 shows the twist per unit length (or inverse pitch) versus concentration in nematic

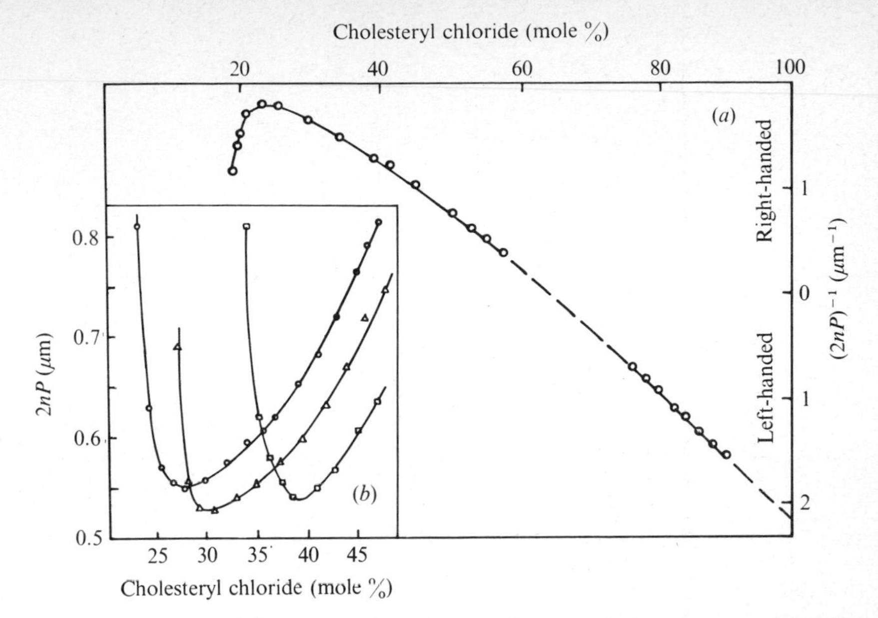

Fig. 4.8.4. (*a*) Relationship between inverse pitch and composition in cholesteryl chloride–cholesteryl nonanoate mixtures at room temperature. (After Adams and Haas.[91]) (*b*) Pitch versus composition in mixtures of cholesteryl chloride with cholesteryl laurate (□), cholesteryl decanoate (△) and cholesteryl nonanoate (○). (After Adams, Haas and Wysocki.[92])

MBBA + cholesteryl propionate.[93] The curve attains a maximum at a certain concentration, beyond which it decreases. This shows that the twist per unit length of the pure propionate is *less* than that of a mixture with a small amount of MBBA. Saeva and Wysocki[94] have observed that a compensation occurs even in an MBBA + cholesteryl chloride mixture when the MBBA concentration is about 30 per cent by weight. They suggest that certain optically inactive molecules may become chiral in the helical environment of a cholesteric liquid crystal. In this case, the MBBA in the mixture would appear to have a chirality opposite to that of the cholesteryl chloride.

A small quantity of a non-mesomorphic optically active compound may also transform a nematic into a cholesteric.[95] However, the handedness of the helix does not appear to be directly related to the absolute configuration of the solute molecule, as has been shown by Saeva.[96] For example, (*S*)-s-amyl-*p*-aminocinnamate and (*S*)-2-(octyl)-*p*-aminocinnamate, both of which have the same absolute configuration, result in helices of opposite senses when dissolved in MBBA. Impurities, indeed

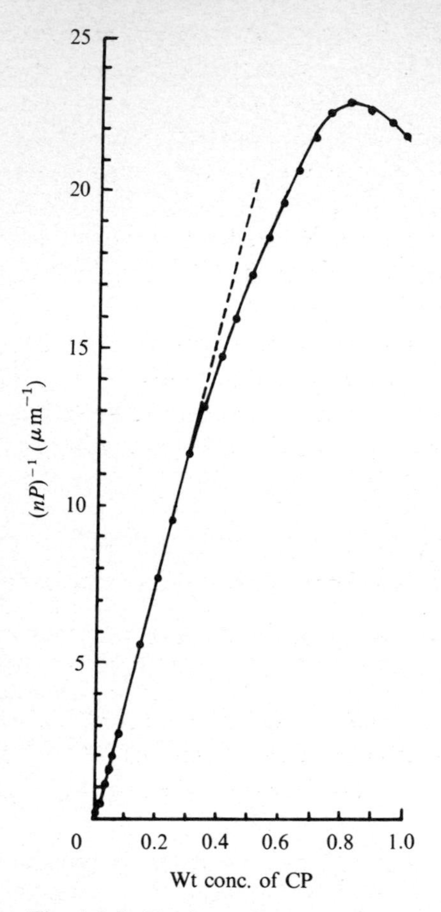

Fig. 4.8.5. Reciprocal of wavelength of maximum reflexion $(nP)^{-1}$, plotted against weight concentration of cholesteryl propionate (CP) in mixtures of nematic MBBA and CP. All measurements are at 8 °C below the cholesteric–isotropic transition temperature. (After Nakagiri, Kodama and Kobayashi.[93])

even the vapour of an organic liquid coming into contact with a cholesteric,[83] have a profound influence on the pitch.

4.8.4. Molecular models

It would be fair to say that the factors influencing the pitch are not at all well explained at the molecular level. A beginning has been made recently. Goossens[97] has extended the Maier–Saupe theory to take into account the chiral nature of the intermolecular coupling in the cholesteric phase, and has shown that cross terms involving the dipole–dipole and the dipole–quadrupole contributions to the dispersion energy must be included to explain the helicity. However, the theory in its present form leads to an increase of pitch with temperature whereas in most cholesterics the reverse is the case. Wulf[98] has made a model calculation to bring out the role of the asymmetry of the molecular shape in producing the twist. He has also correctly pointed out that the assumption that the local order in the cholesteric phase is uniaxial, while it may be a good enough approximation for most purposes (see § 4.1.5), cannot be strictly true as the information regarding the sense of the helix has to be transmitted from layer to layer.

Important evidence in regard to the factors responsible for the helical molecular arrangement has been reported recently by Coates and Gray.[99] They have demonstrated that a hydrogen–deuterium asymmetry in a molecule is sufficient to produce a cholesterogen. An example is given below:

$$N{\equiv}C-\langle\bigcirc\rangle-CH{=}N-\langle\bigcirc\rangle-CH{=}CH-\overset{\displaystyle O}{\overset{\|}{C}}-O-\underset{\underset{\displaystyle X}{|}}{\overset{\overset{\displaystyle H}{|}}{C}}-CH_2CH_2CH_3$$

X = H, nematic; X = D, cholesteric.

The C–H and C–D bond lengths are the same, 1.085 Å,[100] when one takes into account the anharmonicity of the vibrations. Thus it would seem that steric effects are not essential for the helical arrangement.

5

Smectic liquid crystals

5.1. Classification of the smectic phases

Present classification of smectic liquid crystals is based largely on the optical and miscibility studies of Sackmann and Demus.[1] The miscibility criterion relies on the postulate that two liquid crystalline modifications which are continuously miscible (without crossing a transition line) in the isobaric temperature–concentration diagram have the same symmetry and therefore can be designated by the same symbol. It is not clear whether this criterion is valid regardless of the differences in the molecular shapes and dimensions of the two components, but empirically Sackmann and Demus have found that in no case does a phase of a given symbol mix continuously with a phase of another symbol. The method is simple and has been used for the identification of a number of new phases, but it would be fair to say that except possibly for the classical smectics described by Friedel, the precise nature of the molecular order in the other modifications is not at all well understood at the microscopic level. Systematic X-ray investigations are currently in progress at several laboratories,[1–4] but the analyses have not yet reached the stage of yielding definitive information on the structures.

The notation of Sackmann and Demus is according to the order of the discovery of the different phases and bears no relation to the molecular packing:

Smectic A. The molecules are upright in each layer with their centres irregularly spaced. This is the least ordered of all the smectic mesophases.

Examples: Ethyl-*p*-ethoxybenzal-*p*-aminobenzoate (70–80.4 °C),[5]
Diethyl azoxybenzoate (114–120 °C).[1]

Smectic B. The molecules are upright as in smectic *A* but there is hexagonal order within each layer.

Examples: 4,4′-di-n-octadecyloxyazoxybenzene (94.8–100.1 °C),[6]
4,4′-diethylethoxybenzylideneaminocinnamate (81.4–119.3 °C).[6]

Sackmann and Demus regard the tilted B as belonging to the same class as the normal B and label it as B_c.[1] However, from X-ray evidence de Vries[3] is of the view that it is a separate phase and gives it the symbol H. The H phase has a very high degree of molecular order almost like that of a three-dimensional lattice.

Examples: 4-butyloxybenzal-4′-ethylaniline (40.5–51.0 °C),[7]
Terephthal-bis-4-n-butylaniline (113.0–144.1 °C).[8]

Smectic C. This is the tilted form of smectic A. The molecules are disordered within the layers. Three types of this modification are known:

(i) A structure with a temperature independent tilt angle.[9]

Examples: 4-4′-di-n-hexadecyloxyazoxybenzene (91.1–115.5 °C),[10]
n-octyloxybenzoic acid (100.7–107.5 °C).[11]

(ii) A structure with a temperature dependent tilt angle[9,12]; in this case a smectic A phase occurs at a higher temperature, the tilt angle with respect to the layer normal decreasing to zero at the C–A transition point.[8,9]

Example: Terephthal-bis-4-n-butylaniline (144.1–172.5 °C).[8]

(iii) A structure with a twist axis normal to the layers.[13]

Examples: *p,p′*-terephthal-bis-aminocinnamate (149–180 °C),[14]
bis-(*p*-6-methyloctyloxybenzylidene)-2-chloro-1,4-phenylenediamine (29–94.4 °C).[13]

Smectic D. This phase has been reported to be cubic,[15] and would appear to be an exception to the rule that all smectic structures have well defined stratifications. In at least one compound,[16] viz, the second example given below, this phase occurs between the A and C phases. The manner in which such a structural rearrangement takes place is still quite unclear.

Examples: 4′-n-hexadecyloxy-3-nitrodiphenyl-4-carboxylic acid (171.0–197.2 °C),[16]
4′-n-octadecyloxy-3′-nitrodiphenyl-4-carboxylic acid (158.9–195.0 °C).[16]

Smectic E. There appear to be two types of smectic E, the normal[1] and the tilted.[1,3,17] The X-ray diffraction patterns reveal a high degree of molecular order, but little else is known about the structure.

Examples: Dipropyl-*p*-terphenyl-4,4″-carboxylate (122.0–137.1 °C),[16]
2-(4-n-decyloxydiphenyl)-quinoxalline (97.0–120 °C).[16]

Smectic F. This phase occurs at a temperature below smectic *C* and has a structure closely related to it.

Example: 2-(4-n-pentylphenyl)-5-(4-n-pentyloxyphenyl)-pyrimidine (102.7–113.8 °C).[18]

Smectic G. X-ray diffraction studies indicate a highly ordered structure similar to that of smectic *E*.

Example: 2-(4-n-pentylphenyl)-5-(4-n-pentyloxyphenyl)-pyrimidine (79.0–102.7 °C).[18]

Microdensitometer traces of X-ray diffraction patterns obtained from some of these phases using randomly oriented specimens are presented in figs. 5.1.1 and 5.1.2.

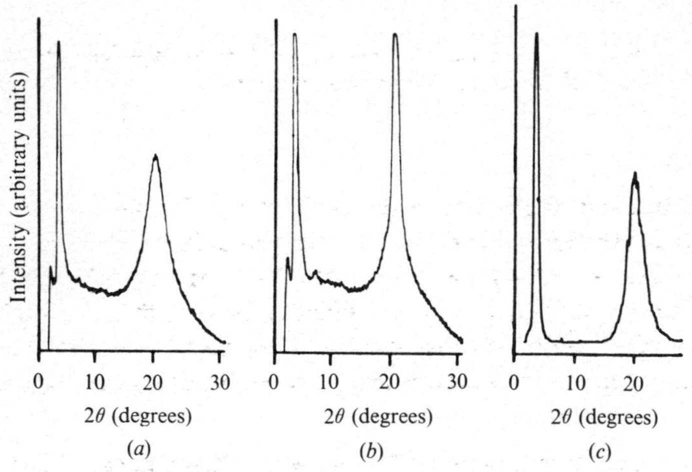

Fig. 5.1.1. Microdensitometer traces of X-ray 'powder' diffraction patterns: (*a*) smectic *A*, (*b*) smectic *B*, (*c*) smectic *C*. (After de Vries.[3])

It is evident that our knowledge of the structure of the smectics, especially of the newly discovered modifications, is quite limited. Furthermore, there is some doubt as to whether the more highly ordered of these phases (*B*, *E* and *H*, for example) should appropriately be described as soft solids. The suggestion has been made[19,20] that the shear modulus should vanish in smectic *B*, but Brillouin scattering experiments[21] show that this does not happen, at least at these high frequencies. The X-ray work of Levelut and Lambert[4] indicates that the *B* phase of terephthal-bis-4-n-butylaniline (TBBA) has no long range translational order perpendicular to the layers, suggesting that this phase is

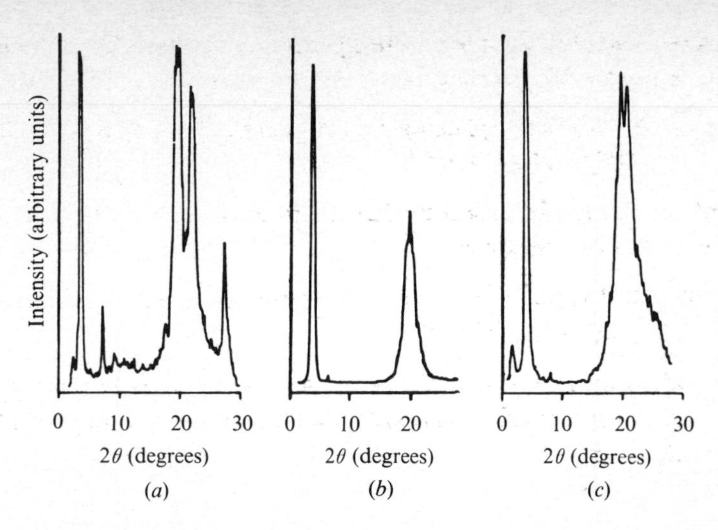

Fig. 5.1.2. Microdensitometer traces of X-ray 'powder' diffraction patterns: (*a*) smectic *E*, (*b*) smectic *F*, (*c*) smectic *G*. (After de Vries.[3])

probably a true mesophase, while the *E* phase of *p*-phenylbenzylidene-*p*-amino-n-pentylcinnamate does have three-dimensional periodicity. Mössbauer investigations[22] reveal that the *H* phase of 4-n-hexyloxybenzylidene-4′-n-propylaniline is less rigid than the solid. However, further studies on this and the other highly ordered phases are necessary before some of these questions can be answered unequivocally.

5.2. Extension of the Maier–Saupe theory to smectic *A*: McMillan's model

McMillan[23] has proposed a simple and elegant description of the smectic *A* phase by extending the Maier–Saupe theory to include an additional order parameter for characterizing the one-dimensional translational periodicity of a layered structure. A similar but somewhat more general treatment, based on the Kirkwood–Monroe theory of melting,[24] was developed independently by Kobayashi,[25] but McMillan's approach lends itself more easily to numerical calculations and comparison with experiment.

The anisotropic part of the pair potential is conveniently taken in the form

$$V_{12}(r_{12}, \cos\theta_{12}) = -(V_0/Nr_0^3\pi^{3/2})\exp[-(r_{12}/r_0)^2](3\cos^2\theta_{12}-1)/2, \qquad (5.2.1)$$

where the exponential term reflects the short range character of the interaction, r_{12} is the distance between the molecular centres and r_0 is of the order of the length of the rigid part of the molecule.

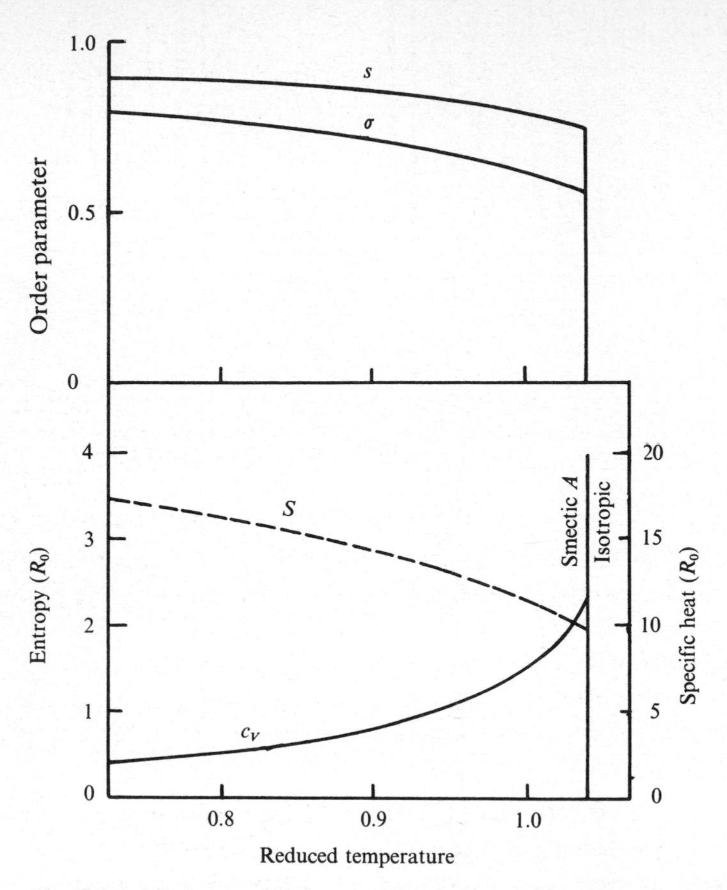

Fig. 5.2.1. Order parameters s and σ, entropy S and specific heat c_V, versus reduced temperature $k_B T/0.2202\, V_0$ predicted by the model for $\alpha = 1.1$ showing a first order smectic A–isotropic transition. S and c_V are expressed in terms of R_0, the gas constant. (After McMillan.[23])

If the layer thickness is a, we may write the self-consistent single particle potential, retaining only the leading term in the Fourier expansion, as follows:

$$V_1(z, \cos\theta) = -V_0[s + \sigma\alpha \cos(2\pi z/a)](3\cos^2\theta - 1)/2 \tag{5.2.2}$$

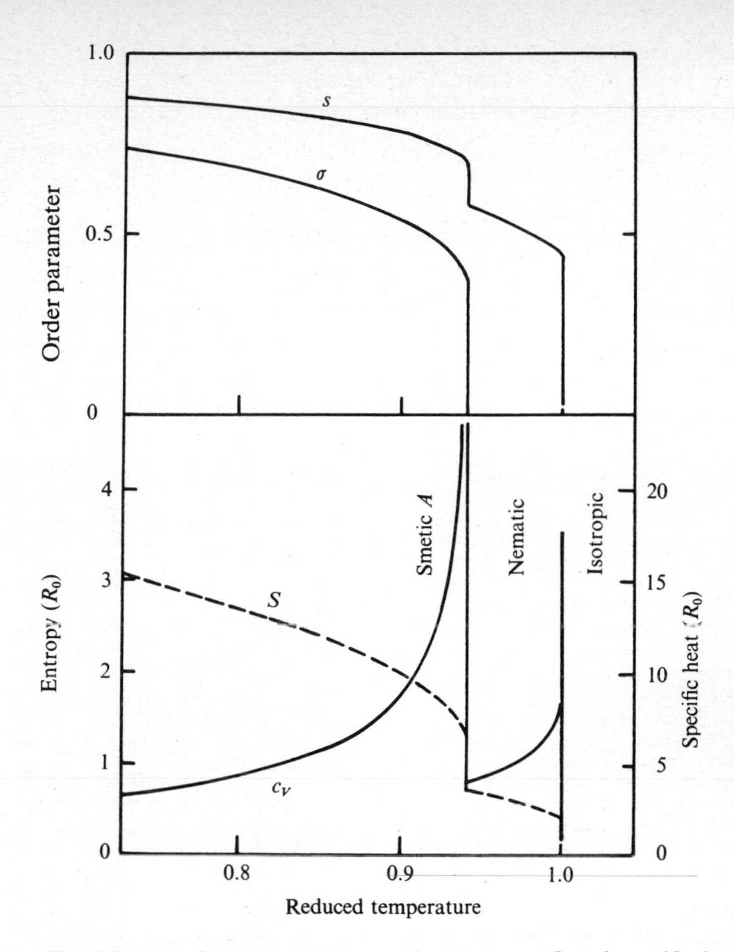

Fig. 5.2.2. Order parameters s and σ, entropy S and specific heat c_V versus reduced temperature for $\alpha = 0.85$ showing first order smectic A-nematic and nematic–isotropic transitions. (After McMillan.[23])

where

$$\alpha = 2 \exp[-(\pi r_0/a)^2]. \tag{5.2.3}$$

This form of the potential ensures that the energy is a minimum when the molecule is in the smectic layer with its axis along z. The order parameters s and σ will be defined presently.

The single particle distribution function is then

$$f_1(z, \cos\theta) = \exp[-V_1(z, \cos\theta)/k_B T] \tag{5.2.4}$$

and self consistency demands that

$$s = \langle(3\cos^2\theta - 1)/2\rangle \tag{5.2.5}$$

$$\sigma = \langle\cos(2\pi z/a)(3\cos^2\theta - 1)/2\rangle \tag{5.2.6}$$

where the angular brackets denote statistical averages over the distribution f_1. The parameter s defines the orientational order exactly as in the Maier–Saupe theory, while σ is a new order parameter which is a measure of the amplitude of the density wave describing the layered structure. The last two equations can be solved numerically to obtain the following types

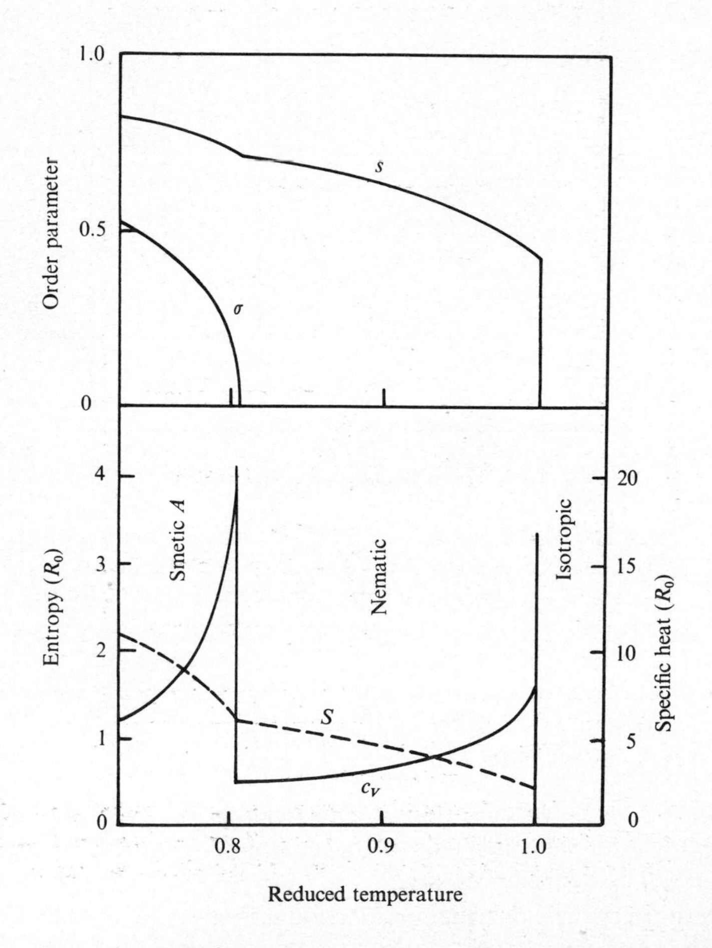

Fig. 5.2.3. Order parameters s and σ, entropy S and specific heat c_V versus reduced temperature for $\alpha = 0.6$ showing a second order smectic A–nematic transition and a first order nematic–isotropic transition. (After McMillan.[23])

of solutions:

(i) $\sigma = s = 0$ (isotropic phase)

(ii) $\sigma = 0, s \neq 0$ (nematic phase)

(iii) $\sigma \neq 0, s \neq 0$ (smectic phase).

The free energy of the system can be calculated in the usual manner:

$$F = U - TS$$

where

$$U = -\tfrac{1}{2}NV_0(s^2 + \alpha\sigma^2) \tag{5.2.7}$$

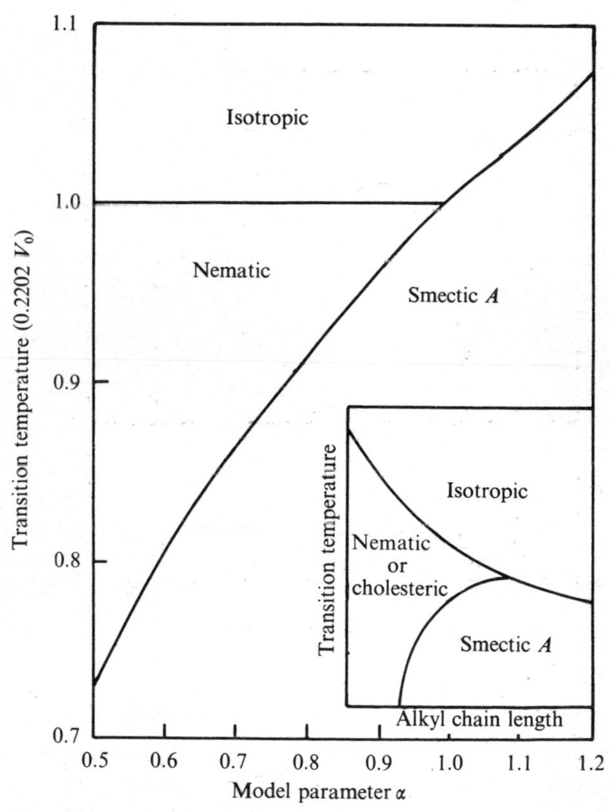

Fig. 5.2.4. Phase diagram for theoretical model parameter α. Inset: typical phase diagram for homologous series of compounds showing transition temperatures versus length of the alkyl end chains. (After McMillan.[23])

and

$$-TS = NV_0(s^2 + \alpha\sigma^2)$$
$$-Nk_BT \ln\left[a^{-1}\int_0^a dz \int_0^1 d(\cos\theta) f_1(z, \cos\theta)\right]. \tag{5.2.8}$$

The two parameters characterizing the material are V_0, which determines the nematic–isotropic transition temperature, and α, a dimensionless interaction strength, which can vary between 0 and 2. Experimentally the layer thickness a is of the order of the molecular length. Neglecting the odd–even effect (see § 2.3.4) the energy associated with the smectic ordering tends to increase if α (and hence a) increases. Thus α increases with increasing chain length of the alkyl tails.

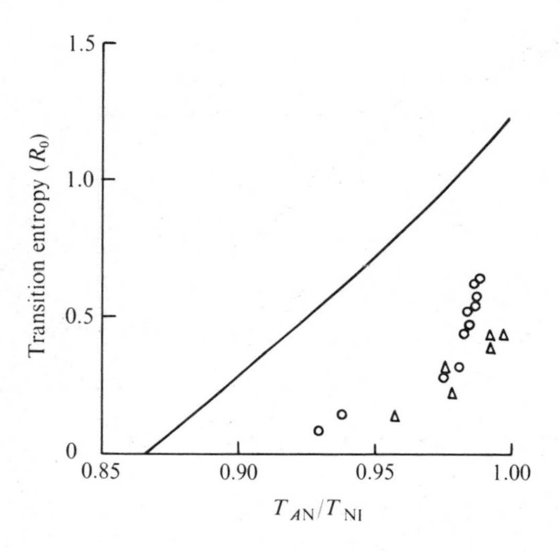

Fig. 5.2.5. Smectic A-nematic (or cholesteric) transition entropy versus ratio of transition temperatures T_{AN}/T_{NI}. Solid line is theoretical curve taken from fig. 5.2.4; open circles are experimental values of Davis and Porter (*Mol. Cryst. Liquid Cryst.* **10**, 1 (1970)); open triangles are data of Arnold (*Z. Physik. Chem.* (*Leipzig*), **239**, 283 (1968); *ibid.*, **240**, 185 (1969)). (After McMillan.[23])

Curves of the order parameters, entropy and specific heat for three representative values of α are presented in figs. 5.2.1, 5.2.2 and 5.2.3. For $\alpha > 0.98$, the smectic A transforms directly into the isotropic phase, while for $\alpha < 0.98$ there is a smectic A–nematic (A–N) transition followed by a nematic–isotropic transition at higher temperature. For $\alpha < 0.70$ the model predicts a second order A–N transition.

The phase diagram of transition temperature versus α or alkyl chain length is shown in fig. 5.2.4. There is broad agreement with the trends in thermodynamic data, though the theoretical A–N transition entropy versus T_{AN}/T_{NI} is somewhat higher than the observed values (fig. 5.2.5).

To improve the agreement, McMillan used, in a later paper,[26] the modified pair potential

$$V_{12}(r_{12}, \cos\theta_{12})$$
$$= -(V_0 N r_0^3 \pi^{3/2}) \exp[-(r_{12}/r_0)^2](\tfrac{3}{2}\cos^2\theta_{12} - \tfrac{1}{2} + \delta). \qquad (5.2.9)$$

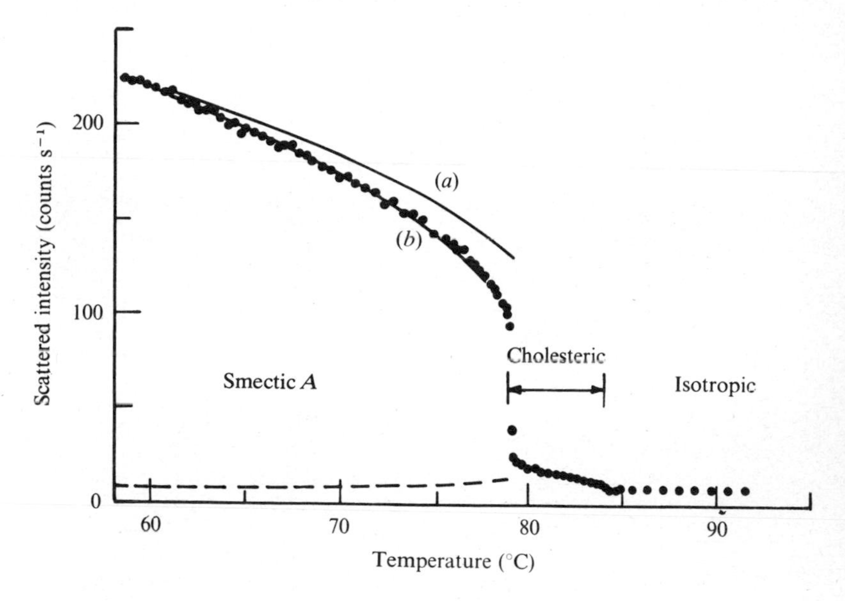

Fig. 5.2.6. Measured intensity of X-ray scattering at the Bragg angle versus temperature for cholesteryl myristate. The dashed line is the calculated diffuse scattering and fluctuation scattering contribution. The full lines represent the theoretical curves for the total intensity due to Bragg, diffuse and fluctuation scattering derived from (*a*) the simpler model potential and (*b*) the refined one. The theoretical intensity has been adjusted to be equal to the experimental value at the lowest temperature. (After McMillan.[26])

There are now three model potential parameters which are fixed by requiring the theory to fit T_{AN}, T_{NI} and S_{AN}. The results are essentially the same as those obtained with the simpler model but there are some quantitative improvements.

A direct method of studying the translational order (or the amplitude of the density wave) is by measuring the intensity of the Bragg scattering from the smectic planes. McMillan's experimental results on cholesteryl myristate[26] are shown in fig. 5.2.6 and as can be seen there is excellent

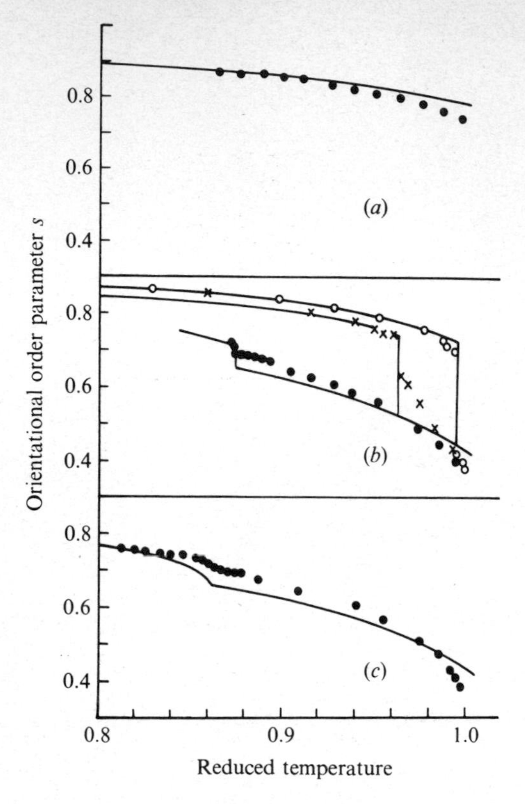

Fig. 5.2.7. Experimental values of the orientational order parameter s obtained from NMR measurements for the 4-n-alkoxybenzylidene-4′-phenylazoaniline series. (*a*) C_{14}; (*b*) C_{10} (○), C_7 (×), C_3 (●); (*c*) C_2. The solid curves give the values predicted by McMillan's model. (After Doane *et al.*[27])

agreement with the refined model. The X-ray intensities reveal an appreciable pretransitional smectic-like behaviour in the cholesteric (nematic) phase. This aspect of the problem will be dealt with in a later section.

The orientational order parameter in the smectic and nematic phases, studied by magnetic resonance and other techniques, also follow the predicted type of behaviour as the length of the alkyl end-chain is increased. In particular, a continuous change of s at T_{AN}, as expected of a second order transition, has been found (within experimental limits) in n-*p*-ethoxybenzylidene-*p*-phenylazoaniline[27] (fig. 5.2.7) and in *p*-cyanobenzylidene-*p*′-octyloxyaniline (CBOOA)[28] (fig. 5.2.8). However, the most recent calorimetric and volumetric[29] data indicate

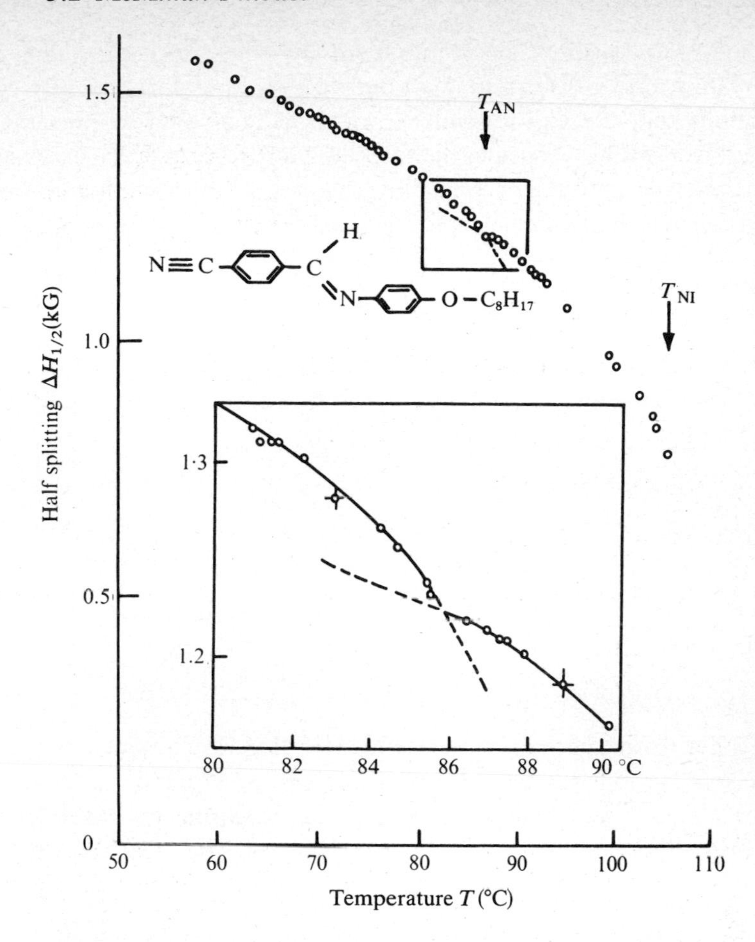

Fig. 5.2.8. Temperature dependence of the orientational order parameter s determined by ^{14}N quadrupolar splitting measurements for p-cyanobenzylidene-p'-octyloxyaniline (CBOOA). The inset shows the discontinuity in slope at the smectic A–nematic transition. (After Cabane and Clark.[28])

that the A–N transition in CBOOA may be only 'quasi second order'. The view has been put forward recently[31] that the A–N transition cannot be truly second order when the coupling with the director fluctuations is also taken into account.

5.3. Static and dynamic distortions in smectic *A*

5.3.1 Focal conic textures

The stratified structure of a smectic liquid crystal imposes certain restrictions on the types of deformation that can take place in it. A compression of the layers requires considerable energy – very much more than for a curvature elastic distortion in a nematic – and therefore the only deformations that are easily possible are those that tend to preserve the interlayer spacing. Consider the smectic A structure in which each layer is effectively a two-dimensional fluid with the director $\mathbf{n}$ normal to its surface. Assuming the layers to be incompressible, the integral

$$\frac{1}{a}\int_P^Q \mathbf{n}\cdot d\mathbf{r} \tag{5.3.1}$$

represents the number of layers crossed on going from P to Q, where a is the layer thickness.[32] In a dislocation-free sample, this number should be independent of the path chosen so that

$$\nabla\times\mathbf{n}=0$$

and hence

$$\mathbf{n}\cdot\nabla\times\mathbf{n}=0$$

and

$$\mathbf{n}\times(\nabla\times\mathbf{n})=0. \tag{5.3.2}$$

In other words, both twist and bend distortions are absent, leaving only the splay term in the Oseen–Frank free energy expression (3.3.7).

It is easily seen that by bending or corrugating the layers a splay deformation can be achieved without affecting the layer thickness (fig. 5.3.1). Moreover, as the layers can slide over one another, the smectic A structure adjusts itself readily to surface conditions. For example, when there is a centre of attachment at the glass surface the molecules adopt a radiating or fan-like arrangement and the smectic layers form a family of equi-spaced surfaces normal to the molecular directions. Under a

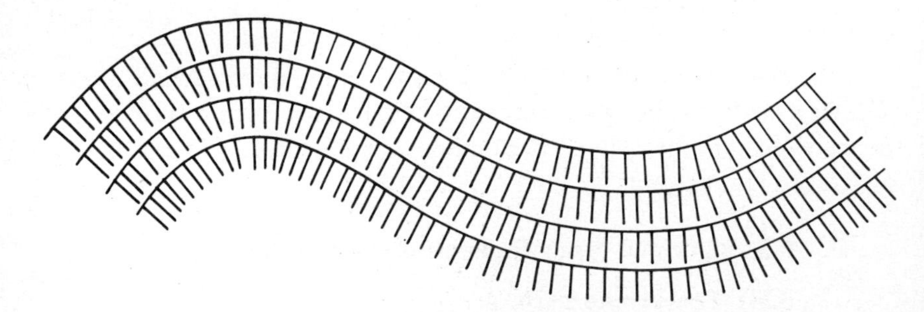

Fig. 5.3.1. Flexibility of smectic A layers: only such deformations as preserve the inter-layer spacing take place readily.

polarizing microscope such distortions give rise to beautiful optical patterns known as *focal conic textures*. These were studied in considerable detail by Friedel[33] to whom is due the explanation of their origin.

A property of a pair of focal conics is that the hyperbola is the locus of vertices of cones of revolution of which the ellipse is a common section (fig. 5.3.2). Similarly the ellipse is the locus of vertices of cones of revolution passing through the hyperbola. The molecules lie along the generators to a cone and the smectic layers are throughout perpendicular to these lines and constitute a series of parallel curved surfaces called *Dupin cyclides* (fig. 5.3.3). A mathematical proof of this result has been given by Geurst.[34]

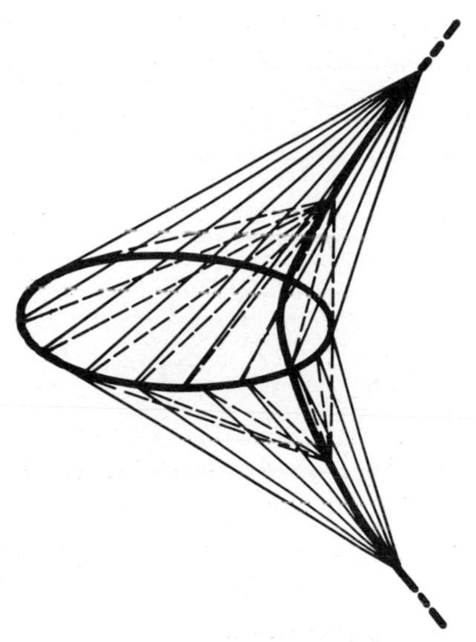

Fig. 5.3.2. Geometry of a pair of focal conics. (After Friedel.[33])

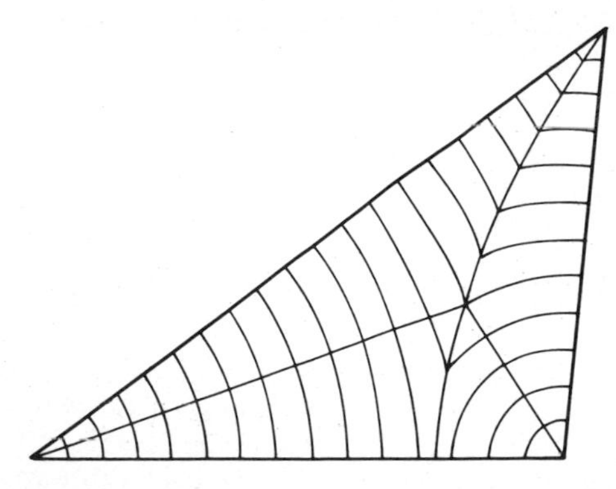

Fig. 5.3.3. Top half of the section of fig. 5.3.2 in the plane of the hyperbola, showing parts of Dupin cyclides. (After Bragg.[35])

We shall now see qualitatively how the entire liquid crystal film can be built up of such curved strata.[35] The rectangular space between the glass plates can be divided into sets of pyramids and tetrahedra as illustrated in fig. 5.3.4. The base of the pyramids may have any number of sides, but for simplicity we have shown only four-sided polygons. Each pyramid can itself be further subdivided into cones, the larger cones in the central region, smaller and smaller ones filling up the corners (fig. 5.3.5). As adjacent cones have a common generator, the cyclides in one cone pass continuously into those in the next one. As regards the tetrahedra, which

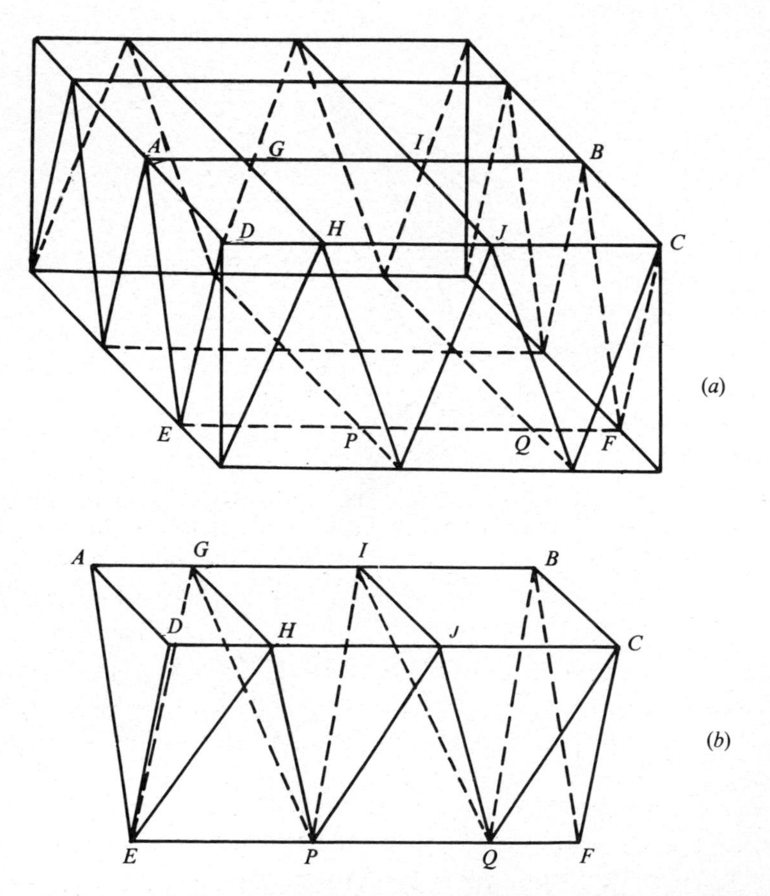

Fig. 5.3.4. (*a*) The rectangular space occupied by a liquid crystal film can be divided into wedges, the division being made in two ways, parallel to the two edges of the block. (*b*) One of the wedges obtained by cutting parallel to *AB*. The set of cuts parallel to *AD* divides this wedge into pyramids such as *PGHJI* and smaller wedges such as *IJPQ*. (After Bragg.[35])

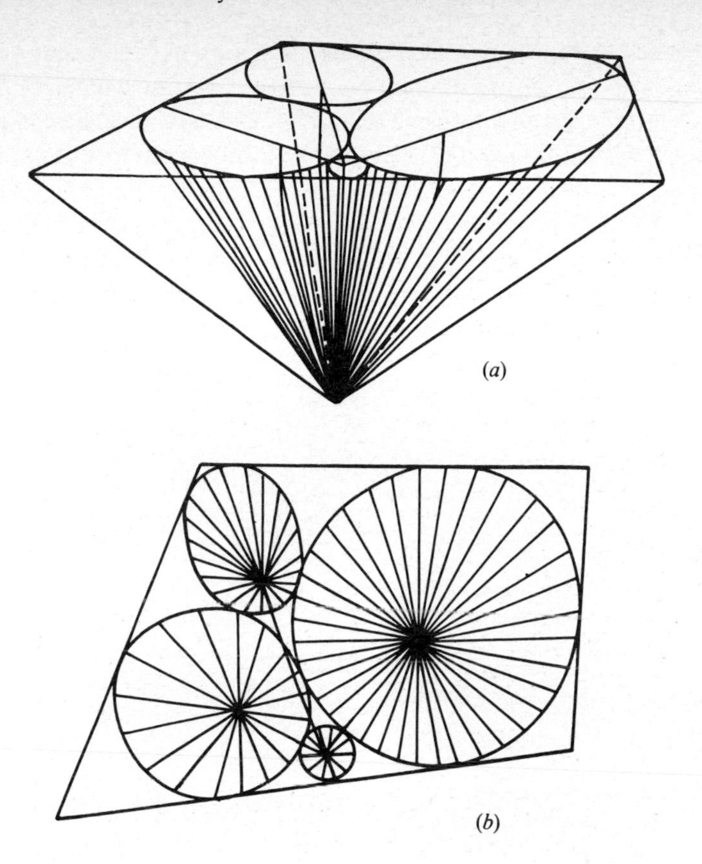

Fig. 5.3.5. (*a*) Arrangement of a set of cones within a pyramid. The hyperbolae belonging to the ellipses meet at the vertex of the pyramid. (*b*) The base of the pyramid and the axial directions radiating from the foci at the base of the hyperbolae. (After Bragg.[35])

are required besides the pyramids to fill up space, their top and bottom edges may be regarded as portions of a pair of focal conics so that the cyclides in this region intersect the four edges at right angles and therefore join up with the cyclides in the adjoining pyramids. In this manner, the entire space can be filled up with curved layers of uniform thickness, belonging to different sets but fitting on to each other exactly.

The observed optical effects confirm such an arrangement. The ellipses and the hyperbolae can actually be identified under a polarizing microscope and are found to bear a focal conic relationship to one another. The polygonal and the fan-shaped textures (plate 4) are just different views of the same arrangement: in the former the hyperbolae are lines in projection, in the latter it is the ellipses that are seen as lines.

A remarkable property of the smectic liquid crystal is that an open droplet, homeotropically aligned on a perfectly clean glass plate or on a fresh cleaved mica sheet, forms terraces (plate 16[36]). This is known as the stepped drop, *goutte à gradins*, and was discovered by Grandjean.[37]

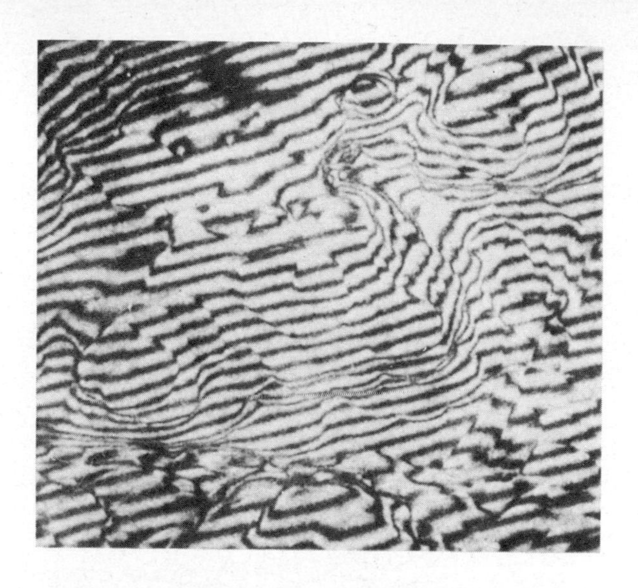

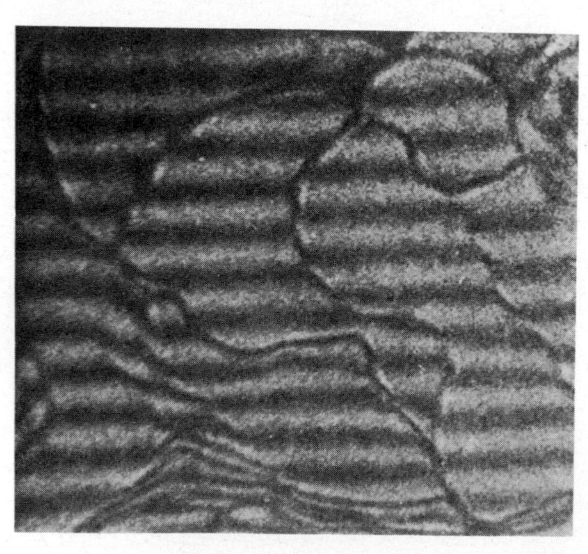

Plate 16. Optical interferometric patterns showing terraces on the surface of a homeotropically aligned smectic liquid crystal formed by a solution of potassium oleate in aqueous methyl alcohol. $\lambda = 0.5893\ \mu$m. (Reference 36.)

From surface energy considerations it can be shown that the occurrence of steps is a consequence of the layered structure.[38] Under sufficient magnification the terraces are seen to be fringed by a chain of focal conic patterns indicating that the layers are crumpled at the edges.

When the isotropic melt is cooled, the smectic phase often makes its appearance in the form of needle-shaped particles showing evidence of focal conic structure. Examples of such needles, termed *bâtonnets*, can be seen in plate 5(*b*).

5.3.2. Continuum theory of smectic *A*: de Gennes's model

A more complete description of smectic *A* needs to take into account the compressibility of the layers, though, of course, the elastic constant for compression may be expected to be quite large, almost comparable to that for a solid. The basic ideas of this model were put forward by de Gennes.[32] We consider an idealized structure which has no long range translational order within each smectic layer and which is optically uniaxial and non-ferroelectric. For small displacements u of the layers normal to their planes, the free energy in the presence of a magnetic field along z takes the form

$$F=\frac{1}{2}B\left(\frac{\partial u}{\partial z}\right)^2+\frac{1}{2}\chi_a H^2\left[\left(\frac{\partial u}{\partial x}\right)^2+\left(\frac{\partial u}{\partial y}\right)^2\right]+\frac{1}{2}k_{11}\left(\frac{\partial^2 u}{\partial x^2}+\frac{\partial^2 u}{\partial y^2}\right)^2$$

$$+\frac{1}{2}K'\left(\frac{\partial^2 u}{\partial z^2}\right)^2+\frac{1}{2}K''\frac{\partial^2 u}{\partial z^2}\left(\frac{\partial^2 u}{\partial x^2}+\frac{\partial^2 u}{\partial y^2}\right), \qquad (5.3.3)$$

where the first term is the elastic energy for the compression of the layers and χ_a the anisotropy of diamagnetic susceptibility. When $H=0$, there will be no terms in $(\partial u/\partial x)^2$ or $(\partial u/\partial y)^2$ as a uniform rotation about y or x does not affect the free energy. The last two terms are usually negligible and may be omitted. Also, the physically reasonable assumption is made that twist and bend distortions given by (5.3.2) are not allowed despite the fact that $\nabla\times\mathbf{n}$ does not strictly vanish when the layers are compressible.

Equation (5.3.3) is analogous to the Peierls–Landau[39] free energy expression for an unbounded 'one-dimensional' solid and leads to a logarithmic divergence of the mean square fluctuation $\langle u^2\rangle$ as $H\to 0$. Writing the free energy in terms of the Fourier components of u,

$$u_q=\int u(\mathbf{r})\exp(\mathrm{i}\mathbf{q}\cdot\mathbf{r})\,\mathrm{d}\mathbf{r}, \qquad (5.3.4)$$

$$F=\tfrac{1}{2}\sum_q |u_q|^2[Bq_z^2+k_{11}(q_\perp^2+\xi^{-2})q_\perp^2], \qquad (5.3.5)$$

where $q_\perp^2 = q_x^2 + q_y^2$ and $\xi = (k_{11}/\chi_a)^{1/2} H^{-1}$ is the magnetic coherence length. Applying the equipartition theorem

$$\langle |u_q|^2 \rangle = \frac{k_B T}{B q_z^2 + k_{11}(q_\perp^2 + \xi^{-2}) q_\perp^2}, \tag{5.3.6}$$

from which the mean square fluctuation

$$\langle u^2(\mathbf{r}) \rangle = \frac{k_B T}{4\pi (B k_{11})^{1/2}} \log \frac{\xi}{a}, \tag{5.3.7}$$

where a is the layer spacing. As $H \to 0$, $\langle u^2 \rangle \to \infty$, implying that such a structure cannot be stable. This is a fundamental difficulty regarding the description of long range order in smectic A and as yet there appears to be no completely satisfactory answer. However, we shall not discuss this basic question but consider only the application of this model to long-wavelength distortions in bounded samples.

5.3.3. The Helfrich deformation

If a magnetic field is applied parallel to the smectic planes in a homeotropically aligned sample and $\chi_a > 0$, one may expect a Helfrich type of deformation to set in above a critical field, as in the case of cholesterics (fig. 4.6.3). Assuming a distortion of the form

$$u(x, z) = u_0 \sin k_z z \cos k_x x, \tag{5.3.8}$$

where $k_z = \pi/d$ and d is the sample thickness, and evaluating the average free energy, one obtains

$$F = (k_x^2 u_0^2/8)\{B[(k_z/k_x)^2 + k_x^2 \lambda^2] - \chi_a H^2\}, \tag{5.3.9}$$

where

$$\lambda = (k_{11}/B)^{1/2} \tag{5.3.10}$$

is a characteristic length of the material of the order of the layer thickness. Proceeding as in § 4.6.2, the optimum value of the distortion wave-vector is

$$k_x^2 = k_z/\lambda = \pi/\lambda d. \tag{5.3.11}$$

In other words, the spatial periodicity of the deformation is proportional to the geometric mean of the layer thickness and the sample thickness. The critical field given by

$$[2\pi k_{11}/(\chi_a \lambda d)]^{1/2} \tag{5.3.12}$$

is however rather large compared with that for cholesterics. For example, taking $\lambda = 20$ Å and $d = 1$ mm, $H_c \sim 50$ kG, and thus far no experimental studies of this effect appear to have been carried out.

The same type of distortion can however be achieved more easily by mechanical means, i.e., by increasing the separation between the glass plates[40,41] (fig. 5.3.6). In this case we note that a bending of the layers

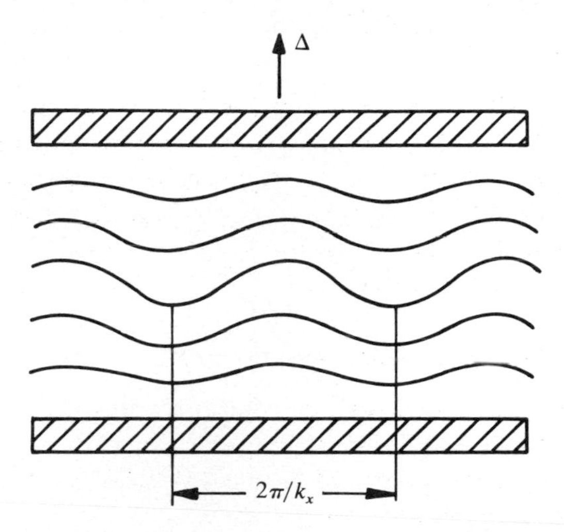

Fig. 5.3.6. The Helfrich deformation in a smectic *A* film subjected to a mechanical dilatation Δ.

alters the effective layer spacing along z and therefore makes a second order contribution to the layer dilatation in that direction. Hence the free energy expression needs a correction term:

$$F = \frac{1}{2}B\left[\frac{\partial u}{\partial z} - \frac{1}{2}\left(\frac{\partial u}{\partial x}\right)^2\right]^2 + \frac{1}{2}k_{11}\left(\frac{\partial^2 u}{\partial x^2}\right)^2. \tag{5.3.13}$$

The displacement u is now given by

$$u = \alpha z + u_0 \sin k_z z \cos k_x x, \tag{5.3.14}$$

where $\alpha = \Delta/d$, Δ being the plate displacement. The problem is analogous to the magnetic case except that $B\alpha$ replaces $\chi_a H^2$. The threshold value of the strain

$$\alpha_c = 2\pi\lambda/d \tag{5.3.15}$$

and the plate displacement $\Delta = 2\pi\lambda \sim 150$ Å which is easily realized in practice. Experiments to verify these conclusions have been done in the smectic A phase of CBOOA.[40,42] The plate separation was increased in a controlled manner by piezoelectric ceramics. When the dilatation reached a certain value, there appeared two transient bright spots in the diffracted laser beam confirming the onset of a spatially periodic distortion above a threshold strain. A transient periodic pattern was also visible under a microscope. It was verified that $k_x^2 \propto d^{-1}$ in accordance with (5.3.11) (fig. 5.3.7). The measurements yielded $\lambda = (22 \pm 3)$ Å for CBOOA at 78 °C. Also assuming $k_{11} \sim 10^{-6}$, B was estimated to be 2×10^7 cgs.

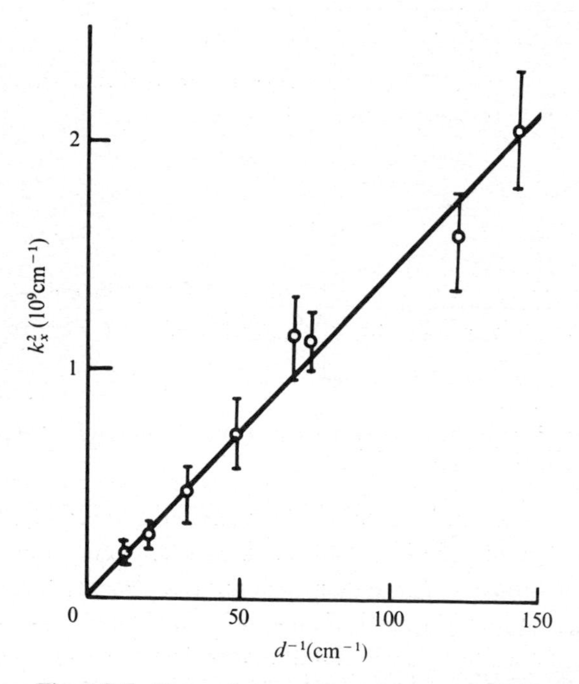

Fig. 5.3.7. Dependence of the wave-vector of the Helfrich deformation in the smectic A phase of CBOOA on the sample thickness d. The slope yields a value of $\lambda = (2.2 \pm 0.3)$ nm. (After Durand.[42])

5.3.4. Fluctuations and Rayleigh scattering

If a smectic A sample is homeotropically aligned between two glass plates having a separation d, the boundary conditions require that $q_z = m\pi/d$, where m is an integer. We shall confine the discussion to $m = 1$. When $H = 0$ we have from (5.3.6)

$$\langle |u_q|^2 \rangle = \frac{k_B T}{B q_z^2 + k_{11} q_\perp^4}. \tag{5.3.16}$$

We know that the intensity of light scattering is proportional to the mean square fluctuation of the director (see § 3.9):

$$\langle|\delta n|^2\rangle = \langle|u_q q|^2\rangle = \frac{k_B T}{B q_z^2 q_\perp^{-2} + k_{11} q_\perp^2}$$

$$= \frac{k_B T}{k_{11}[q_\perp^2 + (q_c^4/q_\perp^2)]}, \tag{5.3.17}$$

where

$$q_c = (q_z/\lambda)^{1/2} = (\pi/\lambda d)^{1/2}.$$

The elastic energy is minimized when $q_\perp = q_c$, which represents the optimum wave-vector. When $q_z = 0$,

$$\langle|\delta n|^2\rangle = \frac{k_B T}{k_{11} q_\perp^2}, \tag{5.3.18}$$

which is a large quantity as in a nematic liquid crystal, since it involves only the splay coefficient. On the other hand, when q_z and $q_\perp$ are comparable,

$$\langle|\delta n|^2\rangle \sim \frac{k_B T}{B(q_z/q_\perp)^2} \tag{5.3.19}$$

is quite small; consequently, in certain geometries – for example, when a homeotropically aligned sample is held against an extended source of light and viewed normally – the medium will not appear very turbid.

Thus in a *perfect* sample of smectic A, the most important contribution to light scattering comes from the undulation mode with the wave-vector parallel to the layers ($q_z = 0$). However unless very special precautions are taken, it is difficult to observe this scattering, for even small irregularities on the surfaces of the glass plates cause static distortions of the layers which extend to appreciable depths inside the specimen[43,41] (fig. 5.3.8). The reason for the high penetration length of a surface distortion is easily understood. A corrugation of the layer which involves a nematic-like splay elastic constant and requires very little energy can be relaxed only by compression, which needs considerable energy and therefore occurs over a large distance. If a static undulation of this type is written as

$$u = u_0 \exp(-z/L) \cos qx,$$

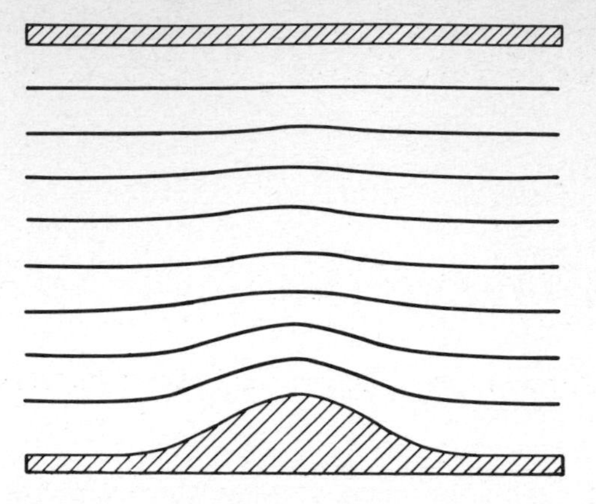

Fig. 5.3.8. Distortion caused by an irregularity on the glass surface. The bending of the layers can be relaxed only by compression, which requires considerable energy, so that the surface distortion extends to appreciable depths inside the specimen.

where L is the 'attenuation' length, substitution in the free energy expression gives

$$Lq^2 = (B/k_{11})^{1/2} = 1/\lambda,$$

or

$$L = (\lambda q^2)^{-1}$$

(whereas in nematics, $L \sim q^{-1}$). These long range static undulations give rise to an intense scattering of light which completely swamps that due to the thermal fluctuations. Indeed in the first experiments[44,45] it was the static effects that were observed, as was confirmed by a temporal analysis of the scattered light using a laser beat spectrometer and a correlator. However, more recently[46] the scattering due to thermal fluctuations has been detected by choosing optical glass plates of very high quality. The temporal analysis showed an exponentially decaying correlation function characteristic of a dynamic undulation.

These modes are highly damped, the relaxation time τ being of the order of 10^{-3} s. To discuss the damping quantitatively we have to consider the hydrodynamics of smectic A. The most general formulation of the theory is due to Martin, Parodi and Pershan,[19] but we shall present the relevant equations in a simplified form using de Gennes's notation.

5.3.5. Damping rate of the undulation mode

We write the energy density in a more general form[32] to include volume dilatation θ:

$$F = \frac{1}{2}A_0\theta^2 + \frac{1}{2}B_0\left(\frac{\partial u}{\partial z}\right)^2 + C_0\theta\frac{\partial u}{\partial z}$$

$$+\frac{1}{2}\chi_a H^2\left[\left(\frac{\partial u}{\partial x}\right)^2 + \left(\frac{\partial u}{\partial y}\right)^2\right] + \frac{1}{2}k_{11}\left(\frac{\partial^2 u}{\partial x^2} + \frac{\partial^2 u}{\partial y^2}\right)^2. \tag{5.3.20}$$

For static isothermal deformations, F may be minimized with respect to θ to give

$$\theta = -\frac{C_0}{A_0}\frac{\partial u}{\partial z},$$

so that

$$B = B_0 - \frac{2C_0^2}{A_0} \tag{5.3.21}$$

and (5.3.20) reduces to the simpler form (5.3.3) assumed previously.

If v_i is the velocity of the particle, the equation of motion is

$$\rho\dot{v}_i = -p_{,i} + g_i + t'_{ji,j} \tag{5.3.22}$$

where g_i is the force on the layers and t'_{ji} the viscous stress tensor which, in contrast to the nematic case, is assumed to be symmetric and dependent on the velocity gradients only. In the problem under consideration we have

$$p = -\frac{\partial F}{\partial \theta} \tag{5.3.23}$$

and

$$g = -\frac{\delta F}{\delta u}. \tag{5.3.24}$$

The force g normal to the layers will be associated with *permeation* effects. The idea of permeation was put forward originally by Helfrich[47] to explain the very high viscosity coefficients of cholesteric and smectic liquid crystals at low shear rates (see figs. 4.5.1 and 5.3.9). In cholesterics, permeation falls conceptually within the framework of the Ericksen–Leslie theory[49] (see § 4.5.1), but in the case of smectics, it invokes an entirely new mechanism reminiscent of the drift of charge carriers in the hopping model for electrical conduction (fig. 5.3.10).

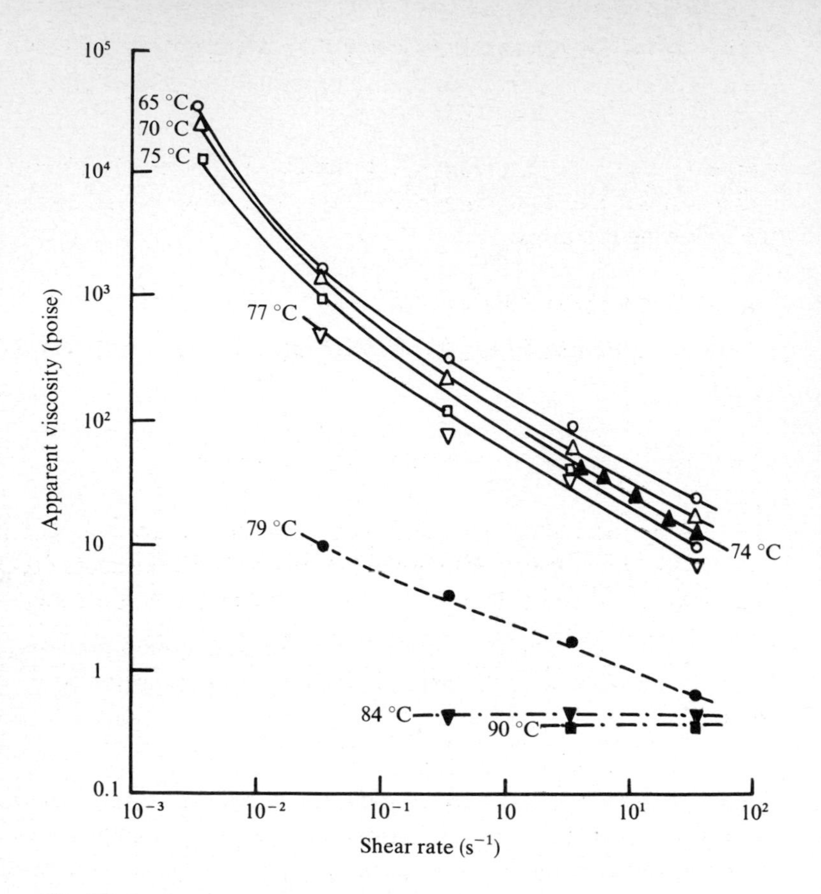

Fig. 5.3.9. Apparent viscosity versus shear rate for cholesteryl myristate at different temperatures: —— smectic *A*, – – – cholesteric, –·–·– isotropic. (After Sakamoto, Porter and Johnson.[48])

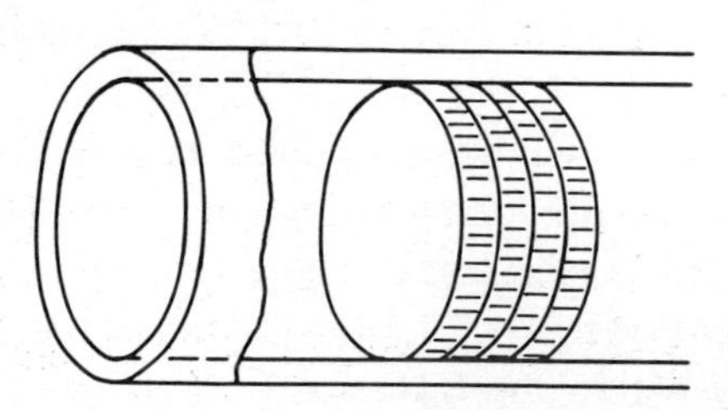

Fig. 5.3.10. Helfrich's model of permeation in smectic liquid crystals. The flow takes place normal to the layers in a manner similar to the drift of charge carriers in the hopping model for electrical conduction.

The rate of entropy production may be written as (see § 3.1.4)

$$T\dot{S} = t'_{ji}d_{ij} + g(\dot{u} - v_z), \tag{5.3.25}$$

where $d_{ij} = \frac{1}{2}(v_{i,j} + v_{j,i})$, and $(\dot{u} - v_z)$ describes the permeation. Treating t'_{ji} and g as fluxes and d_{ij} and $\dot{u} - v_z$ as forces, and expanding t'_{ji}, one obtains the following relations:

$$t'_{ji} = \mu_1 \delta_{ji} d_{kk} + \mu_2 \delta_{jz} \delta_{iz} d_{zz} + \mu_3 d_{ji} + \mu_4(\delta_{jz} d_{zi} + \delta_{iz} \delta_{zj}) + \mu_5 \delta_{jz} \delta_{iz} d_{kk}, \tag{5.3.26}$$

$$\dot{u} - v_z = \nu_{\mathrm{p}} g, \tag{5.3.27}$$

where $\mu_1 \ldots \mu_5$ are five viscosity coefficients and ν_{p} the permeation coefficient.

As far as the highly damped undulation modes are concerned, the volume dilatation can justifiably be neglected and the isothermal approximation is probably satisfactory. The equation of motion then reduces to

$$\rho \dot{v}_z = -\eta q^2 v_z + g, \tag{5.3.28}$$

where

$$g = -k_{11} q^4 u = (\dot{u} - v_z)/\nu_{\mathrm{p}} \tag{5.3.29}$$

and

$$\eta = \tfrac{1}{2}(\mu_3 + \mu_4). \tag{5.3.30}$$

Neglecting the inertial term in (5.3.28) and eliminating v_z,

$$\dot{u} + \frac{k_{11} q^2 u}{\eta} + k_{11} \nu_{\mathrm{p}} q^4 u = 0.$$

For small q, the last term may be neglected (except very near the boundary, but we shall ignore the boundary layer). This results in a purely damped mode whose relaxation rate is

$$\frac{1}{\tau} = \frac{k_{11} q^2}{\eta}. \tag{5.3.31}$$

More generally, taking into account the thickness d of the sample one may write

$$\frac{1}{\tau} = \frac{k_{11}}{\eta}\left(q_\perp^2 + \frac{q_{\mathrm{c}}^4}{q_\perp^2}\right), \tag{5.3.32}$$

where q_c is defined as in (5.3.17). The relaxation rate should have a minimum value

$$\frac{1}{\tau_c}=\frac{2\pi k_{11}}{\eta\lambda d}=\frac{2k_{11}}{\eta}q_c^2, \tag{5.3.33}$$

where

$$q_\perp = q_c = \left(\frac{\pi}{\lambda d}\right)^{1/2}. \tag{5.3.34}$$

Experiments with specimens of different thicknesses have confirmed this prediction[46] (fig. 5.3.11). At very low $q_\perp$, the boundary effects quench the fluctuations and τ^{-1} increases sharply. At high $q_\perp$, τ^{-1} becomes a linear function of $q_\perp^2$ from the slope of which k_{11}/η may be determined. The value is comparable to that for a nematic ($\sim 2\times 10^{-6}$ cgs) as is to be expected. The relaxation time τ_c decreases linearly with decreasing sample thickness as predicted by (5.3.33), and the wave-vector q_c also shows the expected dependence on d. The measurements yield a value of λ which is in reasonable agreement with that obtained from (5.3.15).

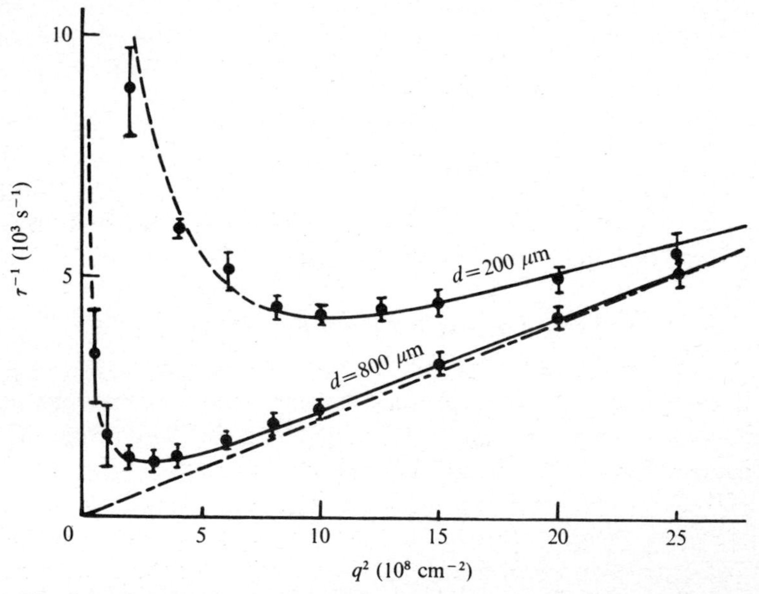

Fig. 5.3.11. Damping rate of the undulation mode in the smectic *A* phase of CBOOA determined by laser beat spectroscopy for two different sample thicknesses, 200 and 800 μm. Solid lines represent the theoretical curves calculated from (5.3.32) and (5.3.33). (After Ribotta, Salin and Durand.[46])

5.3.6. Ultrasonic propagation and Brillouin scattering

For an arbitrary direction of the wave-vector there are two acoustic modes.[32] Neglecting viscous effects and permeation, putting $\gamma = \partial u/\partial z$, and using the conservation law

$$\operatorname{div} v + \dot{\theta} = 0, \tag{5.3.35}$$

we obtain the following equations:

$$\left.\begin{aligned} \rho\frac{\mathrm{d}^2 u}{\mathrm{d}t^2} &= (B_0 + C_0)\frac{\partial\gamma}{\partial z} + (A_0 + C_0)\frac{\partial\theta}{\partial z}, \\ \rho\frac{\mathrm{d}v_x}{\mathrm{d}t} &= C_0\frac{\partial\gamma}{\partial x} + A_0\frac{\partial\theta}{\partial x}, \\ \rho\frac{\mathrm{d}v_y}{\mathrm{d}t} &= C_0\frac{\partial\gamma}{\partial y} + A_0\frac{\partial\theta}{\partial y}, \end{aligned}\right\} \tag{5.3.36}$$

γ and θ being the independent variables. This leads to the secular equation

$$\begin{aligned} \omega\{[\rho\omega^2 - (B_0 + C_0)q_z^2][\rho\omega^2 - A_0 q_\perp^2] \\ - (A_0 + C_0)q_z^2(\rho\omega^2 + C_0 q_\perp^2)\} = 0. \end{aligned} \tag{5.3.37}$$

We are interested here in propagating modes and ignore the case $\omega = 0$. Solving for the velocities $c = \omega/q$, we get

$$c_1^2 + c_2^2 = \rho^{-1}[A_0 \cos^2\varphi + (A_0 + B_0 + 2C_0)\sin^2\varphi], \tag{5.3.38}$$

$$c_1^2 c_2^2 = \rho^{-2}(B_0 A_0 - C_0^2)\sin^2\varphi\cos^2\varphi, \tag{5.3.39}$$

where φ is the angle between q and its projection on the layer.

Ultrasonic velocity measurements have been reported for oriented smectic A samples of diethyl 4,4′-azoxydibenzoate. The first studies by Lord[50] for two values of φ (0° and 90°) established the anisotropy of the velocity of propagation. More recently, Miyano and Ketterson[51] have investigated the angular dependence of the sound velocity and from a least squares fit with (5.3.37) have been able to determine the elastic coefficients A_0, B_0 and C_0. By analogy with the elasticity theory of solids,[52] we may write

$$C_{11} = A_0$$

$$C_{11} - C_{13} = -C_0$$

$$C_{33} - C_{11} - 2C_{13} = B_0.$$

The estimates of Miyano and Ketterson show that A_0 is much larger than B_0 or C_0 (fig. 5.3.12). This enables us to give a simple physical interpretation of the two branches described by (5.3.38) and (5.3.39): the density

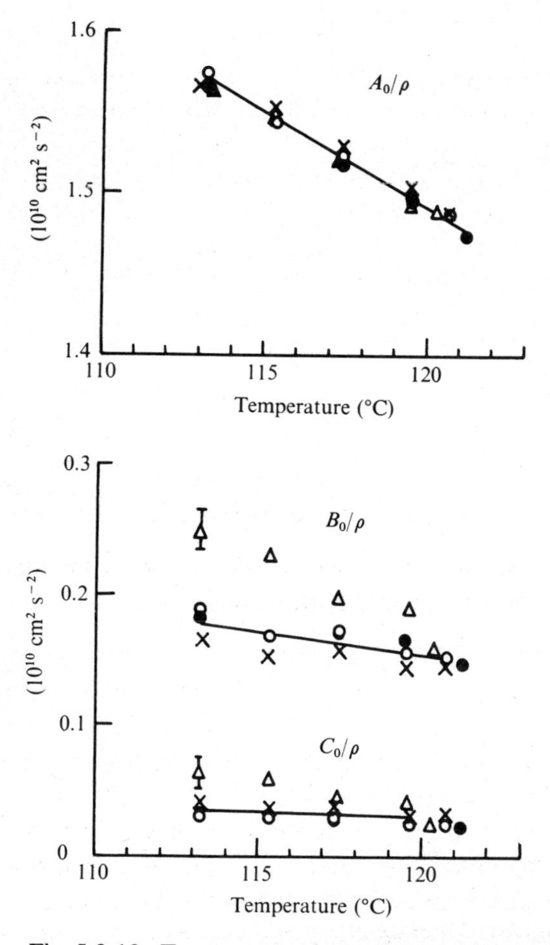

Fig. 5.3.12. Temperature dependence of the elastic constants of the smectic *A* phase of diethyl 4,4′-azoxydibenzoate determined by ultrasonic velocity measurements; ○ 2 MHz; ● 5 MHz; △ 12 MHz; × 20 MHz. (After Miyano and Ketterson.[51])

and layer oscillations are effectively uncoupled. Hence one of the branches corresponds to the normal longitudinal wave whose velocity can be seen from (5.3.38) to be

$$c_1 \simeq (A_0/\rho)^{1/2} \tag{5.3.40}$$

which is practically independent of the direction of propagation. The other branch corresponds to changes in the layer spacing, without appreciable density changes, and may be compared with the phonon branch in superfluids known as *second sound.*[53] The velocity of this mode

$$c_2 \simeq (B_0/\rho)^{1/2} \sin\varphi \cos\varphi \tag{5.3.41}$$

is strongly orientation dependent. It becomes zero for propagation along the layers as well as perpendicular to them. When the wave-vector is parallel to the layers ($q_z = 0$), it becomes the highly damped undulation mode which we have already examined in detail. When q is normal to the layers permeation effects set in and the wave is again strongly damped.[19] Neglecting density changes for this mode, it is clear from (5.3.35) that for q along z, $v_z = 0$. Thus from (5.3.27)

$$\dot{u} = \nu_p g, \tag{5.3.42}$$

or, using (5.3.24),

$$\dot{u} = \nu_p B \frac{\partial^2 u}{\partial z^2}. \tag{5.3.43}$$

Therefore, when the wave-vector of second sound is normal to the layers, it becomes a purely dissipative mode with a relaxation rate

$$\frac{1}{\tau} = \nu_p B q^2. \tag{5.3.44}$$

A direct confirmation of the existence of these two branches has been found by Liao, Clark and Pershan[21] from their Brillouin scattering experiments on monodomain samples of β-methyl butyl p[(p-methoxybenzylidene)amino] cinnamate. This compound shows nematic, smectic A and smectic B phases. Choosing both the incident and the scattered light to be polarized either as ordinary or extraordinary waves, they observed two peaks corresponding to the two modes, the angular dependence of which is in excellent agreement with the theory (fig. 5.3.13).

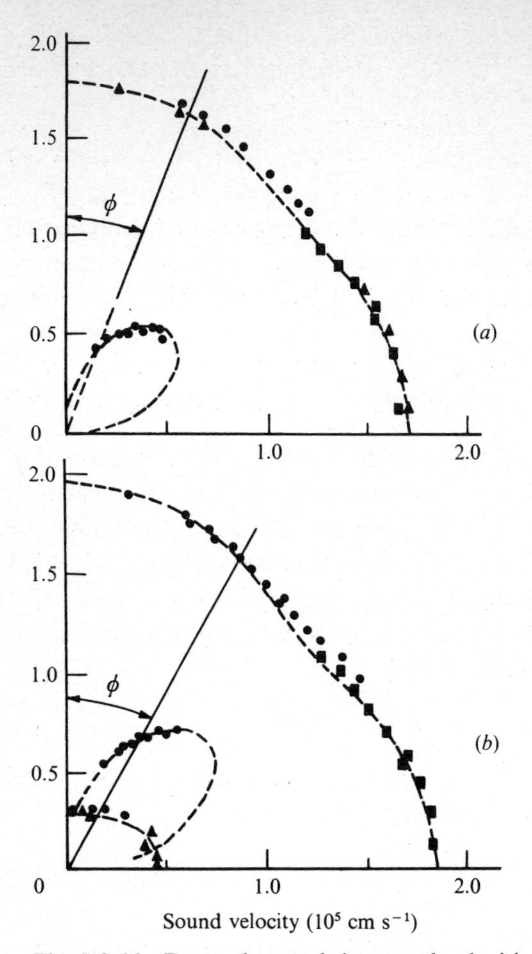

Fig. 5.3.13. Dependence of the sound velocities on the polar angle φ from Brillouin scattering experiments on β-methyl butyl p[(p-methoxybenzylidene)amino] cinnamate. (a) Smectic A ($T = 60.7\,°C$), (b) smectic B($T = 48.1\,°C$). The dashed lines are calculated from theory. The presence of a third component in (b) indicates that the shear modulus does not vanish in smectic B at these very high frequencies. ●, ▲ and ■ represent measurements at different scattering angles. (After Liao, Clark and Pershan.[21])

5.4. The smectic *A*–nematic transition

5.4.1. Pretransition effects in the nematic phase

In § 5.2, we have referred briefly to McMillan's X-ray evidence[26] for the growth of smectic-like short range order in the nematic (cholesteric) phase near the A–N transition (see fig. 5.2.6). This pretransition effect manifests itself more strikingly in the temperature dependence of the elastic properties of the nematic phase, as was first demonstrated by Gruler.[54] The twist and bend distortions, which are normally disallowed in the smectic phase, become increasingly difficult as the smectic clusters build up and these two elastic constants rise much more rapidly than expected from the simple s^2 law discussed in § 2.3.6. We shall now outline a Landau type of phenomenological description of these fluctuations and the associated pretransition effects. This approach to the problem is due to de Gennes[55] and to McMillan[26] both of whom recognized the analogy with similar phenomena in superfluids. We shall follow de Gennes's treatment.

We shall suppose for the present that the A–N transition is of second order, which as we have seen in § 5.2 is a possibility predicted by McMillan's simple microscopic theory. We start with the density wave in the smectic phase

$$\rho(z)=\rho_0(1+2^{-1/2}|\psi|\cos(q_s z-\varphi)) \tag{5.4.1}$$

where ρ_0 is the mean density, $|\psi|$ the amplitude and $q_s=2\pi/a$ the wave-vector of the density wave, a the interlayer spacing and φ a phase factor which gives the position of the layers. Thus the smectic order can be fully specified by the complex parameter

$$\psi=|\psi|\exp(i\varphi). \tag{5.4.2}$$

Near the transition, the free energy may be expanded in powers of ψ and its gradients. For a fixed orientation of the director,

$$F=\alpha|\psi|^2+\beta|\psi|^4+\frac{1}{2M_V}\left(\frac{\partial\psi}{\partial z}\right)^2 + \frac{1}{2M_T}\left[\left(\frac{\partial\psi}{\partial x}\right)^2+\left(\frac{\partial\psi}{\partial y}\right)^2\right]. \tag{5.4.3}$$

From symmetry considerations, it is clear that only even powers of ψ may be included. The coefficient β is always positive; at a certain temperature T^*, which is the second order transition point, $\alpha=0$. Accordingly, as explained in § 2.4.1, we may set, in the mean field approximation,

$$\alpha=a(T-T^*). \tag{5.4.4}$$

In the smectic phase ($T<T^*$), the amplitude of the density wave may be taken to be constant, so that only φ varies. The gradient terms of F therefore become

$$F_g=\frac{\psi_0^2}{2M_V}\left(\frac{\partial\varphi}{\partial z}\right)^2+\frac{\psi_0^2}{2M_T}\left[\left(\frac{\partial\varphi}{\partial x}\right)^2+\left(\frac{\partial\varphi}{\partial y}\right)^2\right]. \tag{5.4.5}$$

Comparing this with (5.3.3) it is at once clear that φ is related to the layer displacement u: $|\varphi|^2 = q_s^2|u|^2$ and $B = |\psi|^2 q_s^2/M_V$. The terms $\partial\varphi/\partial x$ and $\partial\varphi/\partial y$ represent the tilt of the layers with respect to the director. If the director orientation is not fixed, it is the relative tilt between the layers and the director that should be considered and therefore (5.4.5) takes the generalized form

$$F = \alpha|\psi|^2 + \beta|\psi|^4 + \frac{1}{2M_V}\left(\frac{\partial\psi}{\partial z}\right)^2$$

$$+\frac{1}{2M_T}|\nabla_T\psi + \mathrm{i}q_s\delta n\psi|^2 \qquad (5.4.6)$$

where ∇_T is the gradient operator in the plane of the layers. This equation is reminiscent of the Landau–Ginsburg expression[(56)] for the free energy of superconductors; $\mathbf{n}$ corresponds to the vector potential $\mathbf{A}$, $\nabla\times\mathbf{A}$ being the local magnetic field.

The analogy may be extended further. By including the Frank elastic free energy terms in (5.4.5), we may define (as already shown in § 5.3.3) a characteristic length $\lambda = (k/B)^{1/2}$. Making use of the condition $\partial F/\partial\psi_0 = 0$ and ignoring the difference between M_T and M_V,

$$\lambda^2 = \frac{2Mk\beta}{q_s^2|\alpha|}, \qquad (5.4.7)$$

where k is an appropriate elastic constant. For a twist or bend, both of which involve $\nabla\times\mathbf{n}$, λ may be interpreted as the depth to which the distortion penetrates into the smectic material. Thus λ is equivalent to the *penetration depth* of the magnetic field in superconductors.

Above T^*, we may ignore the term involving the fourth power in ψ and from the equipartition theorem obtain in the usual manner

$$\langle|\psi(q)|^2\rangle = k_B T/[\alpha + (1/2M_V)q_z^2 + (1/2M_T)(q_x^2 + q_y^2)], \qquad (5.4.8)$$

where the half-widths or the associated coherence lengths may be related to M_V, M_T and α as follows:

$$\xi_V^2 = \frac{1}{2M_V\alpha}, \qquad (5.4.9)$$

$$\xi_T^2 = \frac{1}{2M_T\alpha}. \qquad (5.4.10)$$

Since $\alpha \to 0$ as $T \to T^*$, $\langle|\psi(q)|^2\rangle$, ξ_V and ξ_T diverge near the transition. The variation of $\langle|\psi(q)|^2\rangle$ reveals itself in the intensity of the Bragg scattering (see fig. 5.2.6).

Critical divergence of the elastic constants To discuss the critical behaviour of the twist and bend elastic constants in the nematic phase, we observe that the Frank free energy expression should include the contribution due to smectic short range order:

$$F = \tfrac{1}{2}k^0(\nabla\mathbf{n})^2 + F_s(\psi), \qquad (5.4.11)$$

where k^0 is the usual nematic elastic constant in the absence of smectic-like order, and $F_s(\psi)$ when averaged over all ψ is of the form

$$F_s \sim \frac{q_s^2}{M}\langle|\psi|^2\rangle\xi^2(\nabla\mathbf{n})^2, \qquad (5.4.12)$$

where we have replaced δn by $\xi\nabla\mathbf{n}$ and ignored the difference between M_V and M_T. For a correlated region of volume ξ^3, it can be shown, using (5.4.8), that in the mean field approximation

$$\langle|\psi|^2\rangle \propto \frac{k_B T}{\alpha\xi^3}. \qquad (5.4.13)$$

Thus from (5.4.11), (5.4.12) and (5.4.13), the effective elastic constant for twist or bend will be $k^0 + \delta k$, where

$$\delta k \propto \frac{k_B T}{a^2}\xi. \qquad (5.4.14)$$

Since the coherence length diverges rapidly near the A–N transition, the elastic constants for twist and bend should also show critical behaviour. In the mean field approximation

$$\xi \propto (T - T^*)^{-1/2}. \qquad (5.4.15)$$

However, invoking the analogy with superfluids, de Gennes predicts

$$\xi \propto (T - T^*)^{-2/3}. \qquad (5.4.16)$$

The theory continues to be valid even if the A–N transition is weakly first order, except that T^* now represents a hypothetical second order transition temperature slightly below T_{AN}. Halperin, Lubensky and Ma[31] argue that when the director fluctuations are taken into account the A–N transition should be at least weakly first order and that in such a case only a classical critical behaviour is likely to be observed.

Experimentally, the first determination of the critical exponent was by McMillan[30] from X-ray scattering studies. His measurements on p-n-octyloxybenzylidene-p-toluidine, which exhibits a first order A–N transition, agreed with the mean field theory. On the other hand, CBOOA gave $\alpha \propto (T - T^*)^{1.5}$ and also showed appreciable anisotropy in the temperature variation of the longitudinal and transverse coherence lengths. CBOOA was originally believed to have a second order A–N transition, but recent heat capacity and volumetric data[29] indicate that the transition may be very weakly first order.

The increases in k_{22} and k_{33} in the nematic phase near an A–N transition, over and above that given by the usual s^2 law, have been observed by a number of investigators.[57–61] Fig. 5.4.1 presents the data of Cheung, Meyer and Gruler[57] for CBOOA; k_{33} diverges rapidly while k_{11} exhibits normal behaviour. Unfortunately, the values of the critical exponent reported by the different authors do not agree even for the same compound and it is not yet possible to say unequivocally whether the mean field approximation is valid or not. For CBOOA, δk_{33} (determined by the Freedericksz method[57]) and δk_{22} (from Rayleigh scattering studies[58]) have been reported to vary as $(T-T^*)^{-0.66}$. However, a later

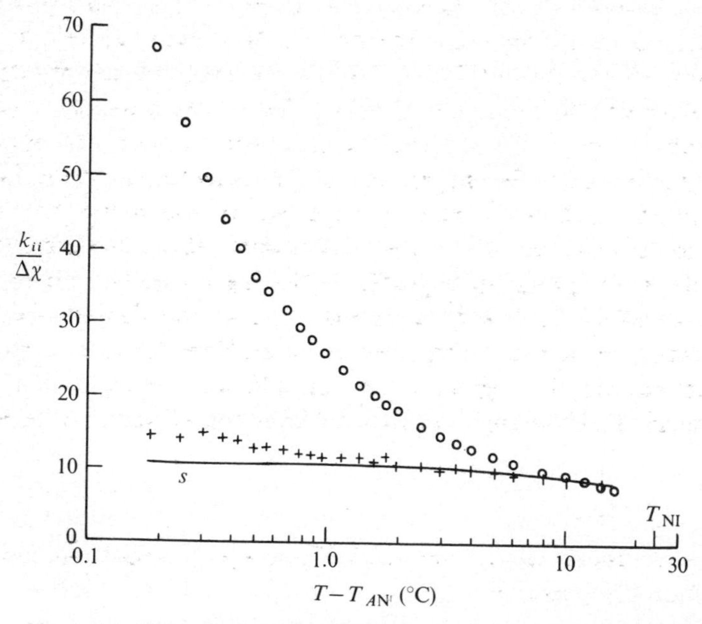

Fig. 5.4.1. The temperature dependence of the splay and bend elastic constants, k_{11} (+) and k_{33} (○) respectively, in the nematic phase of CBOOA prior to the smectic A–nematic transition. The values are plotted as $k_{ii}/\Delta\chi$ where $\Delta\chi$ is the anisotropy of the diamagnetic susceptibility of the nematic. The full line shows the order parameter s, normalized to fit $k_{11}/\Delta\chi$ at high temperature. The splay constant k_{11} deviates only very slightly from ordinary nematic behaviour while the bend constant k_{33} exhibits a critical increase near T_{AN} due to pretransition fluctuations. (After Cheung, Meyer and Gruler.[57])

determination of δk_{33} on a pure sample of the same compound (also by the Freedericksz method, but employing a least squares analysis of the data[59]) has yielded $(T-T^*)^{-1/2}$, i.e., the mean field value. This is in agreement with a more recent measurement, again on the same compound, of δk_{22} and δk_{33} (both by the Freedericksz method and using a least squares analysis[60]). Cladis[59] has found that the behaviour of CBOOA depends very much on the purity of the compound, the temperature variation of δk_{33} ranging from $(T-T^*)^{-1/2}$ for the pure material to $(T-T^*)^{-1}$ for impure samples. Another determination[61] of k_{33} using the Freedericksz method has given an exponent of -0.65 for *p*-butoxybenzylidene-*p*′-octylaniline, a compound with a weakly first order *A*–N transition. A subsequent study on the same material has revealed a most unexpected feature: the splay elastic constant k_{11} shows an appreciable pretransition increase.[62]

Critical divergence of the viscosity coefficients Certain nematic viscosity coefficients also exhibit critical behaviour. The origin of this effect may be explained physically as follows. Take, for example, the frictional torque associated with the twist viscosity coefficient λ_1 defined by (3.3.14). The formation of smectic clusters results in an extra torque due to the flow of the liquid normal to the smectic planes. This extra torque increases as the clusters become larger and longer lived, and in consequence there is a net enhancement of the effective λ_1. To estimate this enhancement,[63,64] consider a slowly rotating magnetic field having an angular velocity Ω. We have seen in § 3.6.2 that the director follows the field at a constant inclination, the torque being given by $-\lambda_1\Omega$. Now, the layered structure of the smectic-like regions makes an additional contribution to the torque, say Γ_s. If the angle between the layer normal and the director is θ, then

$$\theta = \Omega\tau_\psi,$$

where τ_ψ is the relaxation time of $|\psi|$. The torque Γ_s may be derived from (5.4.12) with $\theta = \xi\nabla\mathbf{n}$:

$$\Gamma_s = \frac{\partial F_s}{\partial\theta} = \frac{q_s^2}{M}\langle|\psi|^2\rangle\tau_\psi\Omega.$$

Using the value of $\langle|\psi|^2\rangle$ from (5.4.13), the excess viscosity

$$\delta\lambda_1 \propto \tau_\psi/\xi.$$

The temperature dependence of $\delta\lambda_1$ depends on both τ_ψ and ξ. The theory has been worked out in detail by McMillan[63] using the mean field approximation and by Brochard[64] assuming dynamical scaling laws. The critical exponents for the divergence of the viscosity as predicted by the two theories are different:

$$\delta\lambda_1 \sim (T-T^*)^{-0.50} \quad \text{(mean field)} \tag{5.4.17}$$

$$\delta\lambda_1 \sim (T-T^*)^{-0.33} \quad \text{(dynamical scaling).} \tag{5.4.18}$$

Another difference between these two approaches lies in the behaviour of the *director* relaxation time τ. For example, for a twist deformation we know that (§ 3.8.1)

$$\tau^{-1} = k_{22}q^2/\lambda_1.$$

It is seen that in the mean field approach τ is independent of temperature since near T^* both k_{22} and λ_1 diverge similarly. On the other hand, with helium-like exponents

$$\tau^{-1} \propto (T-T^*)^{-0.33}.$$

The experimental situation is as confused as in the case of the elastic constants. An anomalous increase of λ_1 has been observed[62,65–69] but the value of the exponent is uncertain. The principal experimental difficulty is that the normal nematic viscosity (in the absence of correlations) is itself strongly temperature dependent and the anomalous part forms only a relatively small contribution. Huang *et al.*[67] have determined λ_1 of CBOOA by studying the dynamics of the Freedericksz deformation and have reported an exponent of 0.37 ± 0.05. Using the same technique, d'Humières and Léger[62] find that in butyloxybenzylidene octylaniline the divergence is too weak to fix any specific value for the exponent; however, they do find an increase in the relaxation rate τ^{-1} close to the transition, favouring the predictions of the dynamical scaling laws. Wise, Olah and Doane[68] claim that the divergence of λ_1 of 4-n-butyloxybenzylidene-4′-heptylaniline, derived by the pulsed NMR technique, yields a critical exponent of 0.4 ± 0.1. Salin, Smith and Durand[69] and Chu and McMillan[69] find that the relaxation rate of a pure twist mode in CBOOA determined by laser beat spectroscopy obeys the classical mean field theory.

Thus the precise form of the critical divergence of the elastic and viscous constants in the vicinity of an A–N transition still remains an open question.

5.4.2. Pretransition effects in the cholesteric phase

In the case of a cholesteric near a smectic A transition, we may write

$$F=\tfrac{1}{2}k_{22}^{0}q^{2}-k_{22}^{0}q_{0}q+F_{s}(\psi) \tag{5.4.19}$$

omitting the 'background' term, where q_0 is the equilibrium value of the twist per unit length in the absence of smectic-like short range order and q the actual value. Therefore

$$F=\tfrac{1}{2}(k_{22}^{0}+\delta k_{22})q^{2}-k_{22}^{0}q_{0}q. \tag{5.4.20}$$

Minimizing with respect to q, we get

$$q=\frac{q_{0}k_{22}^{0}}{k_{22}^{0}+\delta k_{22}} \tag{5.4.21}$$

which decreases rapidly as the temperature drops to the smectic A–cholesteric transition point, or, in other words, the pitch $P=2\pi/q$ increases as the temperature is lowered. Fig. 5.4.2 presents the temperature dependence of the pitch of cholesteryl nonanoate,[70] a compound

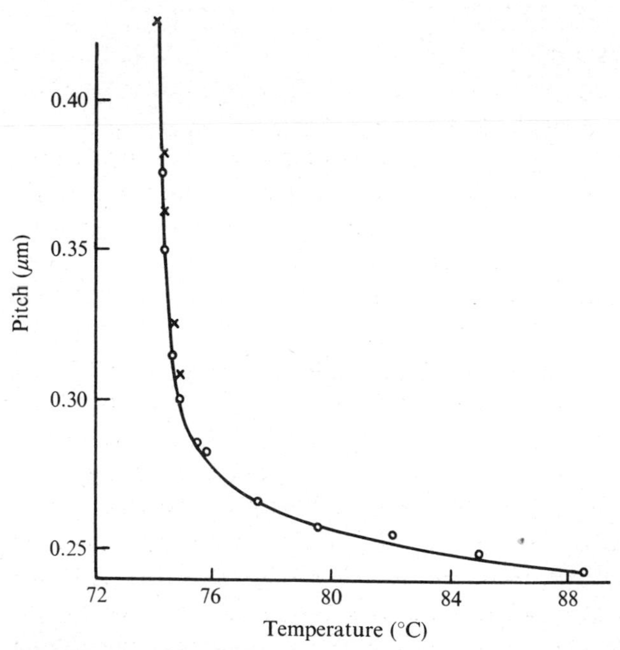

Fig. 5.4.2. Pitch versus temperature in cholesteryl nonanoate prior to the smectic A-cholesteric transition (74 °C); the crosses are values obtained from observations of the Grandjean–Cano walls and the circles from the wavelengths of maximum reflexion. (After Kassubek and Meier.[70])

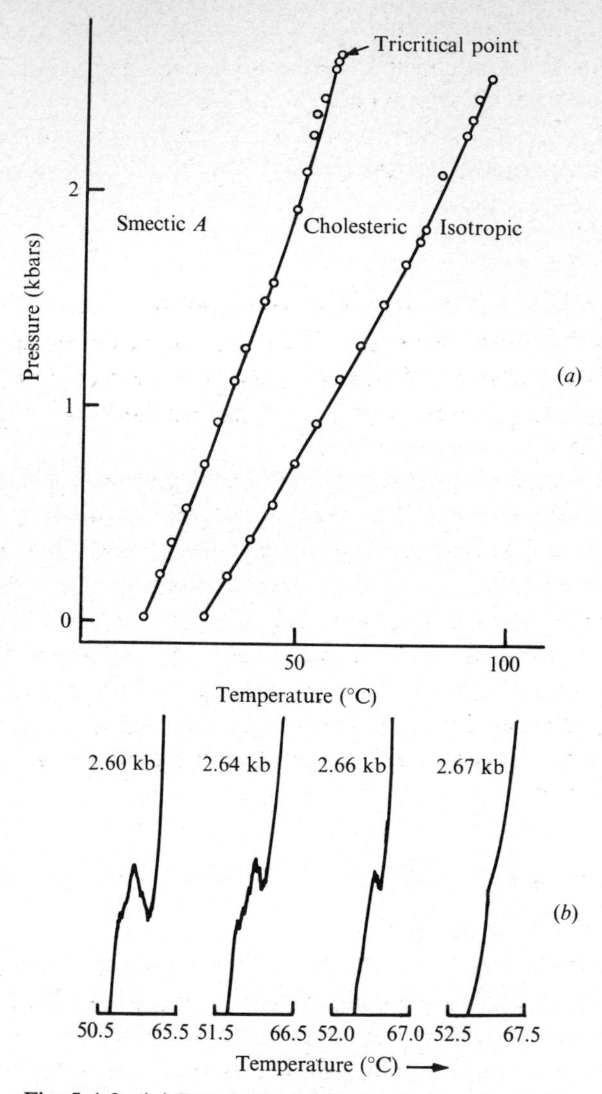

Fig. 5.4.3. (*a*) Experimental phase diagram for cholesteryl oleyl carbonate. (*b*) The raw differential thermal analysis traces of the smectic *A*–cholesteric transition at different pressures in the vicinity of the tricritical point. (After reference 73.)

which shows the smectic *A* phase below 74 °C. The high temperature-sensitivity of the pitch in this region has applications in thermography, as we have already seen in § 4.8.1.

Pindak, Huang and Ho[71] have investigated in some detail the critical divergence of the pitch in cholesteryl nonanoate. They have taken into account the intrinsic variation of the pitch with temperature by studying mixtures of cholesteryl nonanoate with cholesteryl chloride (in which the smectic–cholesteric transition points were low enough for pretransition effects to be neglected) and extrapolating to zero concentration of the chloride. Their experiment gives

$$\delta k_{22} \propto (T-T^*)^{-0.67}.$$

5.4.3. Effect of pressure: the tricritical point

Recent studies[72,73] have established that the character of an A–N transition can be changed from first to second order by the application of pressure. This phenomenon is known to take place in He^3–He^4 mixtures and is referred to as *tricritical* behaviour.[74]

The first observation of this effect in liquid crystal systems was made by Keyes, Weston and Daniels[72] in cholesteryl oleyl carbonate. From optical transmission measurements as a function of pressure and temperature, they concluded that the smectic A–cholesteric transition in this compound becomes second order (or very nearly so) at 2.66 kbar and 60.3 °C. This result has since been confirmed by Shashidhar and Chandrasekhar[73] who used a more sensitive technique to detect the phase transitions, viz, differential thermal analysis. The phase diagram is shown in fig. 5.4.3. The significance of this observation at the molecular level is not fully understood.

5.5. Smectic *C*

5.5.1. Optical properties

We now turn our attention to smectic C, which is the tilted form of smectic A. The molecules are disordered within the layers, but inclined with respect to the layer normal. At first sight the structure appears to have uniaxial symmetry about the preferred molecular direction, but in fact this is not the case for the following reason.[8] The tilt angle ω is coupled with the layer thickness whereas the aximuthal angle φ is not (fig. 5.5.1). Therefore, at any given temperature, the amplitude of the ω-oscillations of the director are small compared with those of the φ-oscillations, with the result that the uniaxial symmetry about the mean molecular direction disappears. There exists now a plane of symmetry as indicated in the figure. Because of this, and also because of possible anisotropic polarization field effects, smectic C is optically biaxial. The optic axial angle $2V$ is generally quite small, of the order of 10°, and is practically independent of temperature.

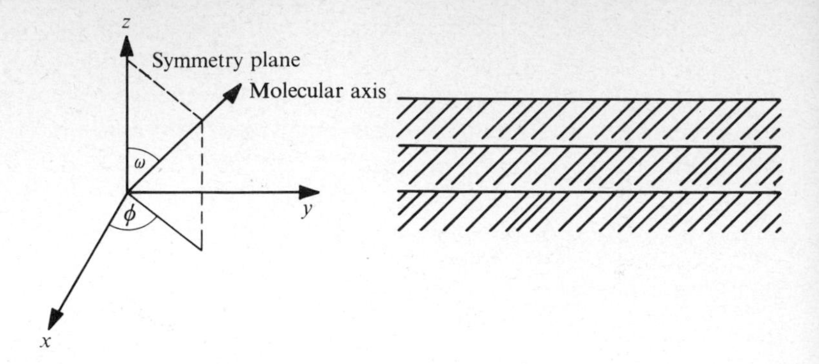

Fig. 5.5.1. The smectic *C* structure.

If the smectic *C* phase is followed by an *A* phase at a higher temperature, the tilt angle decreases gradually to zero at the *C–A* transition point T_{CA}.[8,9] Fig. 5.5.2 presents the temperature variation of the tilt angle in the smectic *C* phase of TBBA determined by optical observations. If the *C* phase is not followed by an *A*, the tilt angle is temperature independent and usually about 45°.[9]

Smectic *C* bears some interesting similarities to the nematic liquid crystal, as was first pointed out by Saupe.[75,76] Suppose one defines a unit vector **n** to represent the preferred orientation of the projection of the molecules on the basal (xy) plane, then it is clear that **n** can be compared to the nematic director in a homogeneously aligned sample. For example, the orientational fluctuations of **n** can be large and the smectic *C* appears quite turbid. Also, as in the case of a nematic, smectic *C* exhibits schlieren textures,[1] except that as **n** and −**n** correspond to tilts in opposite directions, they are not equivalent and only disclinations with integral values of s are possible (see § 3.5.1). Inversion walls of the first and second kind also occur.[1] It has also been suggested[77] that in principle a Freedericksz transition should be observable in smectic *C*, but no experiments have so far been carried out.

Another point of similarity with the nematic is that the smectic *C* can be easily twisted by the addition of optically active molecules.[75] Pure compounds showing a twisted smectic *C* phase have also been discovered.[13] The structure is illustrated in fig. 5.5.3. The optical properties of such a structure resemble those of a cholesteric, though there are some obvious differences. In the cholesteric, the local dielectric properties can be represented by an ellipsoid of revolution with $\varepsilon_a \neq \varepsilon_b = \varepsilon_c$, the principal axis Oc being parallel to the helical axis Oz. In the twisted smectic *C* the local dielectric ellipsoid is triaxial, $\varepsilon_a \neq \varepsilon_b \neq \varepsilon_c$, with Oa making an

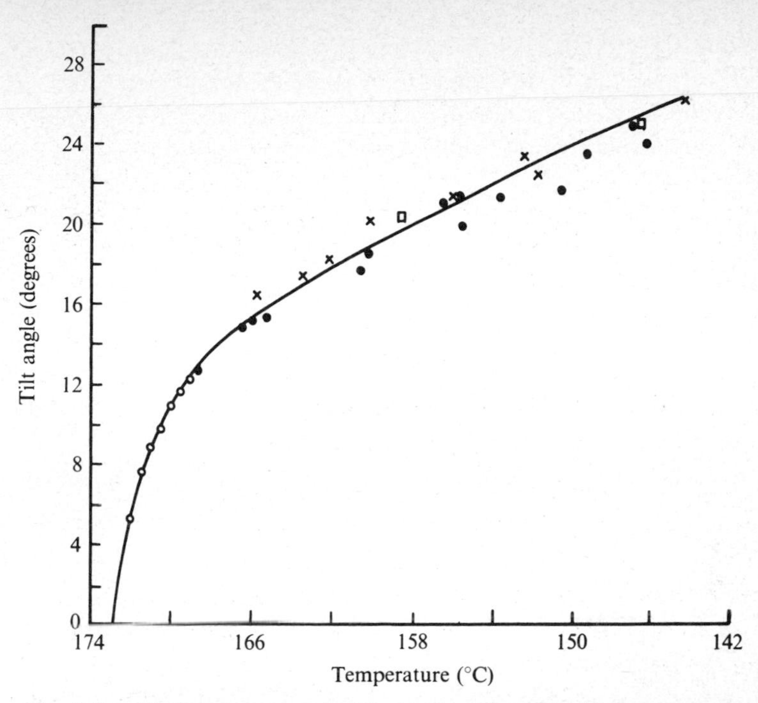

Fig. 5.5.2. Temperature variation of the tilt angle in the smectic *C* phase of terephthal-bis-4-n-butylaniline (TBBA) determined by optical methods from phase retardation measurements at 450 nm (●), 525 nm (×), 625 nm (□), and from the interference figure (○). (After Taylor, Arora and Fergason.[8])

Fig. 5.5.3. Schematic diagram of the structure of twisted smectic *C*.

angle ω with Oz. As far as light propagation along the twist axis is concerned, the twisted smectic C is identical to the cholesteric, but at oblique incidence some additional features may be expected.[78,79] No detailed experimental studies have yet been reported.

5.5.2. Continuum theory of smectic C

To construct a continuum theory of smectic C, we have to take into account firstly the orientational fluctuations of the director about the layer normal (z axis), and secondly, as in smectic A, the distortions of the layers themselves. Expressions for the former were given by Saupe,[75,76] but the complete theory including the latter contributions and the coupling between the two was derived by the Orsay group.[80] We choose a cartesian coordinate system such that the projection of the mean molecular direction on the basal (xy) plane is along x. If the layer displacement along z is represented by u, we observe that

$$\frac{\partial u}{\partial y} = \Omega_x,$$

$$\frac{\partial u}{\partial x} = -\Omega_y,$$

where Ω_x and Ω_y represent rotations about the x and y axes respectively. Therefore

$$\frac{\partial \Omega_x}{\partial x} + \frac{\partial \Omega_y}{\partial y} = 0. \tag{5.5.1}$$

Making use of (5.5.1), and assuming the layers to be incompressible, the free energy of elastic distortion may be written in the form

$$\begin{aligned} F = {} & \frac{1}{2}\left[A\left(\frac{\partial \Omega_x}{\partial x}\right)^2 + A_{12}\left(\frac{\partial \Omega_y}{\partial x}\right)^2 + A_{21}\left(\frac{\partial \Omega_x}{\partial y}\right)^2\right] \\ & + \frac{1}{2}\left[B_1\left(\frac{\partial \Omega_z}{\partial x}\right)^2 + B_2\left(\frac{\partial \Omega_z}{\partial y}\right)^2 + B_3\left(\frac{\partial \Omega_z}{\partial z}\right)^2\right. \\ & \left. + 2B_{13}\frac{\partial \Omega_z}{\partial x}\frac{\partial \Omega_z}{\partial z}\right] + C_1\frac{\partial \Omega_x}{\partial x}\frac{\partial \Omega_z}{\partial x} + C_2\frac{\partial \Omega_x}{\partial y}\frac{\partial \Omega_z}{\partial y}. \end{aligned} \tag{5.5.2}$$

Here the A terms describe curvature distortions of the smectic planes, the B terms the distortions of the director when the smectic planes are unperturbed, and the C terms the coupling between these two types of distortions. All the coefficients are approximately of the same order of magnitude as the nematic elastic constants. A term of the type $\frac{1}{2}B(\partial u/\partial z)^2$

may also be included to allow for the compression of the layers, but we shall neglect it in the present discussion.

The fluctuations of the director are evidently related to fluctuations in Ω. From (5.5.2) and the equipartition theorem, we obtain for a general wave-vector **q**

$$\langle|\Omega_z(q)|^2\rangle = \frac{k_B T}{B_1 q_x^2 + B_2 q_y^2 + B_3 q_z^2 + 2B_{13} q_x q_z}. \tag{5.5.3}$$

Light scattering studies using monodomain samples of smectic C have been reported by Galerne *et al.*[81] They have confirmed that the intensity of scattering arising from the director fluctuations in the vertical (scattering) plane (the k_e, k_e' configuration, e standing for extraordinary) is extremely weak, whereas it is quite large due to fluctuations normal to the scattering plane (the k_e, k_o' or k_o, k_e' configuration). Now, for the k_e, k_o' configuration, we may write

$$(B_1 \cos^2\theta + B_2 \sin^2\theta + 2B_{13} \sin\theta \cos\theta) q^2 I$$
$$= B(\theta) q^2 I = \text{constant}, \tag{5.5.4}$$

where $\theta = \tan^{-1}(q_z/q_x)$ and I the intensity. Thus a plot of $q_x I^{1/2}$ versus $q_z I^{1/2}$ should give an ellipse. This has been verified to be true (fig. 5.5.4). The damping time of these modes also shows a similar angular dependence. From an analysis of the data Galerne *et al.* have estimated that the viscosity coefficients are an order of magnitude larger than for a typical nematic. The viscous behaviour of smectic C has been discussed by Martin, Parodi and Pershan.[19]

Again, as in smectic A, the undulation mode ($q_z = 0$) should make an important contribution to the scattering, but for reasons explained in § 5.3.4 highly perfect samples are necessary to observe this.

5.5.3. The smectic C–smectic A transition

We have noted that when the smectic C phase is followed by an A phase at a higher temperature, the tilt angle decreases gradually to zero. Also, as there can be large fluctuations in φ the complete order parameter has two components and may be written as

$$\psi = \omega \exp(i\varphi). \tag{5.5.5}$$

This again brings out the analogy with superfluids.[56] In the ordered phase ($T < T_{CA}$), de Gennes[55] predicts

$$\omega = |\psi| \sim (\Delta T)^\beta, \tag{5.5.6}$$

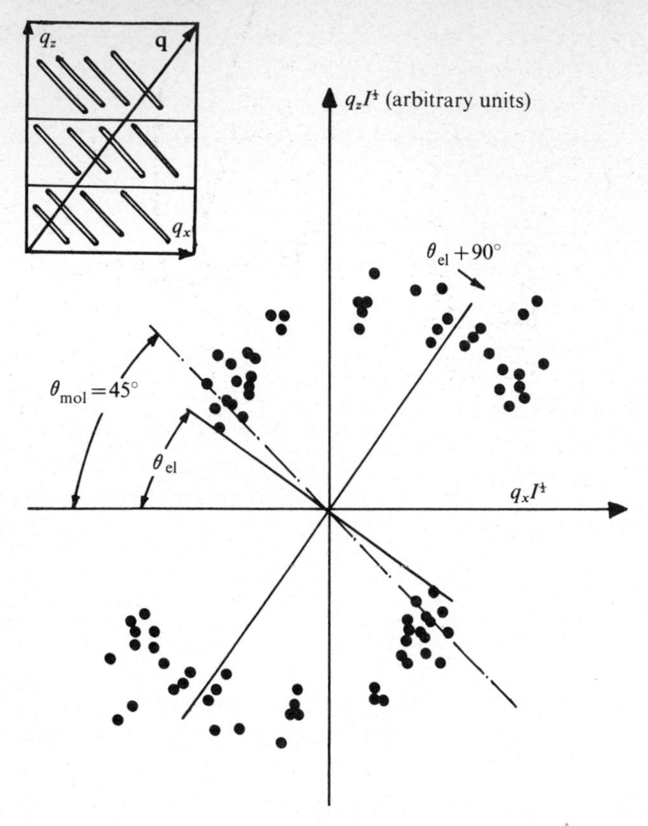

Fig. 5.5.4. Plot of $q_x I^{1/2}$ versus $q_z I^{1/2}$ from light scattering measurements in the smectic C phase of di(4-n-decyloxybenzal)-2-chloro-1-4-phenylene diamine. The wave-vector **q** relative to the layers is indicated in the inset at the top left hand corner. The experimental points lie on an ellipse as expected theoretically. However, the minor axis of the ellipse does not coincide exactly with the molecular axis, which is assumed to be inclined at 45° to the layer normal. (After Galerne *et al.*[81])

where $\Delta T = T_{CA} - T$, T_{CA} being the second order transition temperature, and $\beta \simeq 0.35$. In the mean field approximation, $\beta = 0.5$. There have been two estimates of β for TBBA. Wise, Smith and Doane,[9] who employed the NMR free induction decay technique, have reported a value of $\beta = 0.4 \pm 0.04$, which supports the helium analogy. On the other hand, from a study of the NMR splitting of the probe molecule CH_2Cl_2 in TBBA, Luz and Meiboom[82] have found that $\beta = 0.5 \pm 0.1$, which favours the classical theory. Above T_{CA}, one may expect to observe an anomalous light scattering due to the formation of smectic C clusters, and perhaps some anomalous magnetic effects also.

Ribotta, Meyer and Durand[83] have observed that a compression applied normal to the layers of a smectic A induces a transition to smectic C when the stress exceeds a threshold value. The effect is particularly easy to observe very near T_{CA}. If s is the strain normal to the layers, the energy density may be written as

$$F=\tfrac{1}{2}B(s^2+2\alpha s\omega^2)+\tfrac{1}{2}a\omega^2+\tfrac{1}{4}b\omega^4 \tag{5.5.7}$$

where $\omega \sim |\psi|$ is taken as the order parameter, and B is the elastic constant for compression. It is assumed here that the reduced layer thickness due to the tilting of the molecules is $l(1-\alpha\omega^2)$, where l is the normal layer thickness and α a constant which depends on the molecular axial ratio. For a second order C–A transition $b>0$, and in the high temperature phase $a>0$ and $\omega=0$. Clearly for a large enough compression (i.e., large negative s), there is a finite tilt angle

$$\omega=[-(a+2\alpha sB)/b]^{1/2} \tag{5.5.8}$$

when s is greater than a threshold value

$$s_{\text{th}}=-a/2\alpha B. \tag{5.5.9}$$

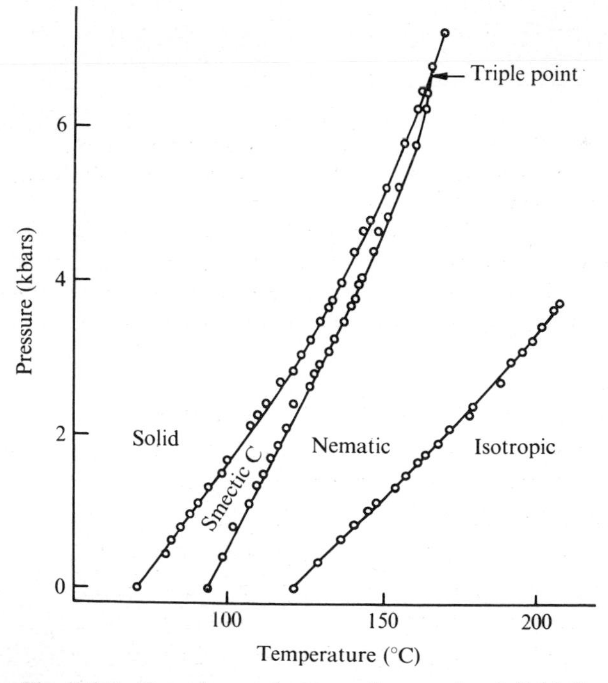

Fig. 5.5.5. Experimental phase diagram for 4,4′-bis(heptyloxy)azoxybenzene. (After reference 73.)

Since $a \sim (\Delta T)^{\gamma}$, it should be possible to use this method to determine the critical exponent γ.

5.5.4. Effect of pressure on the thermal stability of smectic *C*

We conclude our discussion of smectic *C* with a brief reference to a very recent experiment on the effect of pressure.[73] At atmospheric pressure, 4,4′-heptyloxyazoxybenzene shows two mesophases, smectic *C* and nematic. As the pressure is increased, the temperature range of smectic *C* decreases, and above a critical pressure it disappears altogether. The phase diagram is shown in fig. 5.5.5. This is the inverse of the phenomenon of pressure-induced mesomorphism discussed in § 2.1.3. From the structural point of view a simple explanation that suggests itself for the suppression of the smectic *C* phase is that the tilt angle keeps increasing with pressure till the smectic layering collapses and only the nematic order remains. This question can be decided by a direct determination of the variation of the layer thickness with pressure.

References

Chapter 1

1 F. Reinitzer, *Montash Chem.* **9**, 421 (1888).

2 O. Lehmann, *Z. Krist.* **18**, 464 (1890).
For a historical account of the early work, see H. Kelker, *Mol. Cryst. Liquid Cryst.* **21**, 1 (1973).

3 W. Kast, *Landolt-Bornstein Tables*, vol. 2, 6th edition, p. 266, Springer-Verlag (1969).

4 D. Demus and H. Demus, *Flüssige Kristalle in Tabellen*, VEB Deutscher Verlag für Grundstoffindustrie, Leipzig (1973).

5 G. Friedel, *Ann. Physique*, **18**, 273 (1922).

6 I. G. Chistyakov and W. M. Chaikowsky, *Mol. Cryst. Liquid Cryst.* **7**, 269 (1969).

7 A. de Vries, *Mol. Cryst. Liquid Cryst.* **10**, 219 (1970); in *Proceedings of the International Liquid Crystals Conference*, Bangalore, December 1973, *Pramana supplement I*, p. 93.

8 G. W. Stewart, *Trans. Faraday Soc.* **29**, p. 982 (1933).

9 A. Saupe, *Angew. Chem. International Edition*, **7**, 97 (1968).

10 A. D. Buckingham, G. P. Ceasar and M. B. Dunn, *Chem. Phys. Letters*, **3**, 540 (1969).

11 S. Diele, P. Brand and H. Sackmann, *Mol. Cryst. Liquid Cryst.* **17**, 163 (1972).

12 D. Demus, S. Diele, M. Klapperstück, V. Link and H. Zaschke, *Mol. Cryst. Liquid Cryst.* **15**, 161 (1971).

13 D. Demus, G. Kunicke, J. Neelsen and H. Sackmann, *Z. Naturforsch.* **23a**, 84 (1968).

14 See, for example, H. Sackmann and D. Demus, *Mol. Cryst. Liquid Cryst.* **21**, 239 (1973).

15 K. Herrmann, *Z. Krist.* **92**, 49 (1935).
The problem has been re-examined very recently, applying group theoretical methods, by N. Boccara, *Ann. Physique*, **76**, 72 (1973).

16 J. D. Bernal and I. Fankuchen, *J. Gen. Physiol.* **25**, 111 (1941).

17 A. J. Mabis, *Acta Cryst.* **15**, 1152 (1962).

18 V. Luzzati and A. Tardieu, *Annual Review of Physical Chemistry*, **25**, 79 (1974).
A. E. Skoulios and V. Luzzati, *Acta Cryst.* **14**, 278 (1961).
A. S. C. Lawrence, *Mol. Cryst. Liquid Cryst.* **7**, 1 (1960).
S. Friberg, *J. Am. Oil Chem. Soc.* **48**, 578 (1971).
P. Ekwall, L. Mandell and K. Fontell, *Mol. Cryst. Liquid Cryst.* **8**, 157 (1969).

F. B. Rosevear, *J. Soc. Cosmet. Chem.* **19**, 581 (1968).
P. A. Winsor, *Mol. Cryst. Liquid Cryst.* **12**, 141 (1971).
G. H. Brown, J. W. Doane and V. D. Neff, *A Review of the Structure and Physical Properties of Liquid Crystals*, Butterworths (1971).
19 V. Luzzati, A. Tardieu, T. Gulik-Krzywickyi, E. Rivas and F. Reiss-Husson, *Nature*, **220**, 485 (1968).
20 K. D. Lawson and T. J. Flautt, *J. Am. Chem. Soc.* **89**, 5489 (1967).
21 C. L. Khetrapal, A. C. Kunwar, A. S. Tracey and P. Diehl, in *NMR – Basic Principles and Progress*, vol. 9 (eds. P. Diehl, E. Fluck and R. Kosfeld), Springer-Verlag (1975).
22 P. A. Winsor, *Chem. Rev.* **68**, 1 (1968).
23 C. Robinson, *Trans. Faraday Soc.* **52**, 571 (1956).
24 D. Chapman, *Faraday Soc. Symp.* no. 5, 163 (1971); *Pure and Applied Chemistry*, **38**, 59 (1974).
E. J. Ambrose, *Faraday Soc. Symp.* no. 5, 175 (1971); in *Proceedings of the International Liquid Crystals Conference*, Bangalore, December 1973, *Pramana supplement I*, p. 480.
G. T. Stewart, *Mol. Cryst.* **1**, 563 (1966); *Mol. Cryst. Liquid Cryst.* **7**, 75 (1969).
D. M. Small and M. Bourgis, *Mol. Cryst.* **1**, 541 (1966).
V. Luzzati, in *Biological Membranes* (ed. D. Chapman), Academic (1968).
25 H. Arnold, *Z. Phys. Chem.* (Leipzig), **226**, 146 (1964).
26 G. W. Gray, in *Liquid Crystals and Ordered Fluids*, vol. 2 (eds. J. F. Johnson and R. S. Porter), p. 617, Plenum (1974).
27 D. Vorländer, *Z. Phys. Chem.* *A***126**, 449 (1927).
28 W. G. Merret, G. D. Cole and W. W. Walker, *Mol. Cryst. Liquid Cryst.* **15**, 105 (1971).
29 E. Friedel, *C. R. Acad. Sci.* **180**, 269 (1925).
30 W. L. McMillan, *Phys. Rev.* *A***6**, 936 (1972).
31 T. R. Taylor, S. L. Arora and J. L. Fergason, *Phys. Rev. Letters*, **25**, 722 (1970).
J. M. Schnur, J. P. Sheridan and M. Fontana, in *Proceedings of the International Liquid Crystals Conference*, Bangalore, December 1973, *Pramana supplement I*, p. 175.
32 G. W. Gray, *Mol. Cryst. Liquid Cryst.* **21**, 161 (1973); see also *Molecular Structure and the Properties of Liquid Crystals*, Academic (1962). Earlier work in this field has been reviewed by G. H. Brown and W. G. Shaw, *Chem. Rev.* **57**, 1049 (1957).

Chapter 2

1 J. D. Bernal and D. Crowfoot, *Trans. Faraday Soc.* **29**, 1032 (1933).
2 W. R. Krigbaum, Y. Chatani and P. G. Barber, *Acta Cryst.* *B***26**, 97 (1970).
3 J. L. Galigne and J. Falgueirettes, *Acta Cryst.* *B***24**, 1523 (1968).
4 J. A. Pople and F. E. Karasz, *J. Phys. Chem. Solids*, **18**, 28 (1961); F. E. Karasz and J. A. Pople, *J. Phys. Chem. Solids*, **20**, 294 (1961).

5 S. Chandrasekhar, R. Shashidhar and N. Tara, *Mol. Cryst. Liquid Cryst.* **10**, 337 (1970).

6 S. Chandrasekhar, R. Shashidhar and N. Tara, *Mol. Cryst. Liquid Cryst.* **12**, 245 (1972).

7 J. E. Lennard-Jones and A. F. Devonshire, *Proc. Roy. Soc.* *A***169**, 317 (1939); *ibid.*, *A***170**, 464 (1939).

8 See, for example, A. R. Ubbelohde, *Melting and Crystal Structure*, Clarendon, Oxford (1965).

9 See, for example, R. Fowler and E. A. Guggenheim, *Statistical Thermodynamics*, Cambridge University Press (1965).

10 R. H. Wentorf, R. J. Buehler, J. O. Hirschfelder and C. F. Curtiss, *J. Chem. Phys.* **18**, 1484 (1950).

11 L. M. Amzel and L. N. Becka, *J. Phys. Chem. Solids*, **30**, 521 (1969). See also G. W. Smith, *Advances in Liquid Cryst.* **1**, 189 (1975).

12 D. Demus and R. Rurainski, *Z. Phys. Chem.* (Leipzig), **253**, 53 (1973).

13 E. McLaughlin, A. Shakespeare and A. R. Ubbelohde, *Trans. Faraday Soc.* **60**, 25 (1964).

14 H. Arnold, *Z. Phys. Chem.* (Leipzig), **226**, 146 (1964).

15 W. Maier and A. Saupe, *Z. Naturforsch.* **15**(*a*), 287 (1960).

16 E. Bauer and J. Bernamont, *J. Phys. Radium*, **7**, 19 (1936).

17 J. Mayer, T. Waluga and J. A. Janik, *Phys. Letters*, **41***A*, 102 (1972).

18 M. J. Press and A. S. Arrott, *Phys. Rev.* *A***8**, 1459 (1973).

19 N. V. Madhusudana, R. Shashidhar and S. Chandrasekhar, *Mol. Cryst. Liquid Cryst.* **13**, 61 (1971).

20 A. P. Kapustin and N. T. Bykova, *Sov. Phys. – Crystallogr.* **11**, 297 (1966).

21 S. Chandrasekhar and R. Shashidhar, *Mol. Cryst. Liquid Cryst.* **16**, 21 (1972).

22 S. Chandrasekhar, S. Ramaseshan, A. S. Reshamwala, B. K. Sadashiva, R. Shashidhar and V. Surendranath, in *Proceedings of the International Liquid Crystals Conference*, Bangalore, December 1973, *Pramana supplement I*, p. 117.
R. Shashidhar and S. Chandrasekhar, *J. de Physique*, **36**, *C*1-49 (1975).

23 L. Onsager, *Ann. N.Y. Acad. Sci.* **51**, 627 (1949).

24 R. Zwanzig, *J. Chem. Phys.* **39**, 1714 (1963).

25 J. P. Straley, *J. Chem. Phys.* **57**, 3694 (1972).

26 J. E. Mayer, and M. G. Mayer, *Statistical Mechanics*, John Wiley (1940).

27 J. E. Mayer, *Handbuch der Physik*, vol. 12, p. 73, Springer-Verlag (1958).

28 L. K. Runnels and C. Colvin, *J. Chem. Phys.* **53**, 4219 (1970).

29 J. P. Straley, *Mol. Cryst. Liquid Cryst.* **22**, 333 (1973).

30 M. A. Cotter, *J. Chem. Phys.* (to be published).

31 P. J. Flory, *Proc. Roy. Soc.*, *A***234**, 73 (1956).

32 E. A. di Marzio, *J. Chem. Phys.* **35**, 658 (1961).

33 R. Alben, *Mol. Cryst. Liquid Cryst.* **13**, 193 (1971).

34 H. Reiss, H. L. Frisch and J. L. Lebowitz, *J. Chem. Phys.* **31**, 369 (1959).

35 M. A. Cotter and D. E. Martire, *J. Chem. Phys.* **52**, 1902 (1970); *ibid.*, **52**, 1909 (1970); *ibid.*, **53**, 4500 (1970).

36 G. Lasher, *J. Chem. Phys.* **53**, 4141 (1970).

37 A. Isihara, *J. Chem. Phys.* **19**, 114 (1951).

38 L. K. Runnels and C. Colvin, in *Liquid Crystals*, vol. 3 (eds. G. H. Brown and M. M. Labes), p. 299, Gordon & Breach (1972).

39 J. D. Brooks and G. H. Taylor, *Carbon*, **3**, 185 (1965).

40 A. Wulf and A. G. De Rocco, *J. Chem. Phys.* **55**, 12 (1971).

41 G. I. Agren and D. E. Martire, *J. Chem. Phys.* **61**, 3959 (1974).

42 R. Alben, *J. Chem. Phys.* **59**, 4299 (1973).

43 G. I. Agren and D. E. Martire, *J. de Physique*, **36**, *C*1-141 (1975).

44 C. A. Croxton, *Liquid State Physics: A Statistical Mechanical Introduction*, Cambridge University Press (1974).

45 M. Born, *Sitz d. Phys. Math.* **25**, 614 (1916).

46 W. Maier and A. Saupe, *Z. Naturforsch.* **13***a*, 564 (1958); *ibid.*, **14***a*, 882 (1959); *ibid.*, **15***a*, 287 (1960).

For an introduction to intermolecular forces, see J. O. Hirschfelder, C. F. Curtiss and R. B. Bird, *Molecular Theory of Gases and Liquids*, Part 3, Wiley (N.Y.), Chapman and Hall (London) (1954).

47 V. Tsvetkov, *Acta Physicochim.* (USSR), **16**, 132 (1942).

48 A. Saupe and W. Maier, *Z. Naturforsch.* **16***a*, 816 (1961).

49 P. Chatelain, *Bull. Soc. Franc. Miner. Crist.* **78**, 262 (1955).

50 S. Chandrasekhar and N. V. Madhusudana, *J. de Physique*, **30**, *C*4-24 (1969).

51 N. V. Madhusudana, 'A statistical theory of the nematic phase', thesis, Mysore University, India (1970).

52 H. S. Subramhanyam and D. Krishnamurti, *Mol. Cryst. Liquid Cryst.* **22**, 239 (1973).

I. Haller, H. A. Huggins, H. R. Lilienthal and T. R. Mcguire, *J. Phys. Chem.* **77**, 950 (1973).

Y. Poggi, G. Labrunie and J. Robert, *C. R. Acad. Sci. B* **277**, 561 (1973).

C. C. Huang, R. S. Pindak and J. T. Ho, *J. de Physique Letters*, **35**, *L*-185 (1974).

S. Jen, N. A. Clark, P. S. Pershan and E. B. Priestley, *Phys. Rev. Letters*, **31**, 1552 (1973).

J. Cheng and R. B. Meyer, *Phys. Rev. A* **9**, 2744 (1974).

S. A. Shaya and H. Yu, *J. de Physique*, **36**, C1-59 (1975).

53 R. D. Spence, H. A. Moses and P. L. Jain, *J. Chem. Phys.* **21**, 380 (1953).

R. D. Spence, H. S. Gutowsky and C. H. Holon, *J. Chem. Phys.* **21**, 1891 (1953).

54 H. Lippmann and K. H. Weber, *Ann. Physik*, **20**, 265 (1957).

55 B. Cabane and W. G. Clark, *Phys. Rev. Letters*, **25**, 91 (1970); *Sol. State Commun.* **13**, 129 (1973).

56 J. C. Rowell, W. D. Phillips, L. R. Melby and M. Panar, *J. Chem. Phys.* **43**, 3442 (1965).

W. D. Phillips, J. C. Rowell and L. R. Melby, *J. Chem. Phys.* **41**, 2551 (1964).

57 A Saupe and G. Englert, *Phys. Rev. Letters*, **11**, 462 (1963).

58 A. Carrington and G. R. Luckhurst, *Mol. Phys.* **8**, 401 (1964).

59 A. Saupe, *Angew. Chem. International Edition*, **7**, 97 (1968); *Angew. Chemie*, **80**, 99 (1968).
60 A. Saupe, in *Magnetic Resonance* (eds. C. K. Coogan, N. S. Ham, S. N. Stuart, J. R. Pitbrow and G. V. H. Wilson), Plenum (1970).
L. C. Snyder, *J. Chem. Phys.* **43**, 4041 (1965).
A. D. Buckingham and K. A. McLauchlan, in *Progress in NMR Spectroscopy*, vol. 2 (eds. J. M. Emsley, J. Feeney and L. H. Sutcliffe), Pergamon (1967).
61 G. R. Luckhurst, *Quart. Rev.* **22**, 179 (1968); *Mol. Cryst. Liquid Cryst.* **21**, 125 (1973).
62 G. H. Brown, J. W. Doane and V. D. Neff, *A review of the Structure and Physical Properties of Liquid Crystals*, Butterworths (1971).
63 P. Diehl and C. L. Khetrapal, in *NMR – Basic Principles and Progress*, vol. 1 (eds. P. Diehl, E. Fluck and R. Kosfeld), Springer-Verlag (1969).
64 S. Chandrasekhar and N. V. Madhusudana, *Applied Spectroscopy Reviews*, **6**, 189 (1972).
65 J. R. McColl and C. S. Shih, *Phys. Rev. Letters*, **29**, 85 (1972).
66 H. Gasparoux, B. Regaya and J. Prost, *C. R. Acad. Sci.* **272***B*, 1168 (1971).
67 R. Alben, J. R. McColl and C. S. Shih, *Sol. State Commun.* **11**, 1081 (1972).
68 A. Saupe, *Z. Naturforsch.* **19***a*, 161 (1964).
69 M. A. Cotter (to be published).
See also B. Widom, *J. Chem. Phys.* **39**, 2808 (1963).
70 S. Chandrasekhar and N. V. Madhusudana, *Mol. Cryst. Liquid Cryst.* **17**, 37 (1972).
71 S. Chandrasekhar and N. V. Madhusudana, *Acta Cryst.* *A***27**, 303 (1971).
S. Chandrasekhar, N. V. Madhusudana and K. Shubha, *Faraday Soc. Symp.* no. 5, 26 (1971).
72 R. L. Humphries, P. G. James and G. R. Luckhurst, *J. Chem. Soc., Faraday Trans. 2*, **68**, 1031 (1972).
R. L. Humphries and G. R. Luckhurst, *Mol. Cryst. Liquid Cryst.* **26**, 269 (1974).
See also S. Jen, N. A. Clark, P. S. Pershan and E. B. Priestley, *Phys. Rev. Letters*, **31**, 1552 (1973).
73 S. Marcelja, *J. Chem. Phys.* **60**, 3599 (1974).
74 B. Deloche, B. Cabane and D. Jerome, *Mol. Cryst. Liquid Cryst.* **15**, 197 (1971).
75 S. Chandrasekhar and N. V. Madhusudana, *Mol. Cryst. Liquid Cryst.* **24**, 179 (1973).
76 C. Weygand and R. Gabler, *Z. Phys. Chem.* **48***B*, 148 (1941).
See also G. W. Gray, *Molecular Structure and the Properties of Liquid Crystals*, Academic (1962).
77 A. Pines, D. J. Ruben and S. Allison, *Phys. Rev. Letters*, **33**, 1002 (1974).
78 H. Gruler, *Z. Naturforsch.* **28***a*, 474 (1973).
79 P. J. Flory, *Statistical Mechanics of Chain Molecules*, Interscience (1969).
80 J. Van der Veen, W. H. de Jeu, M. W. M. Wanninkhof and C. A. M. Tienhoven, *J. Phys. Chem.* **77**, 2153 (1973).
81 W. Maier and G. Meier, *Z. Naturforsch.* **16***a*, 262 (1961).
82 W. Maier and G. Meier, *Z. Naturforsch.* **16***a*, 470 (1961).

83 M. Schadt, *J. Chem. Phys.* **56**, 1494 (1972).
84 C. J. F. Bottcher, *Theory of Electric Polarization*, Elsevier (1973).
85 L. Onsager, *J. Am. Chem. Soc.* **58**, 1486 (1936).
86 W. Maier and G. Meier, *Z. Naturforsch.* **16***a*, 1200 (1961).
87 A. Axmann, *Z. Naturforsch.* **21***a*, 290 (1966).
88 G. Meier and A. Saupe, *Mol. Cryst.* **1**, 515 (1966).
A. J. Martin, G. Meier and A. Saupe, *Faraday Soc. Symp.* no. 5, 119 (1971).
89 A. Saupe, *Z. Naturforsch.* **15***a*, 810 (1960).
90 A. Saupe, *Z. Naturforsch.* **15***a*, 815 (1960).
91 I. Haller and J. D. Litster, in *Liquid Crystals*, vol. 3, (eds. G. H. Brown and M. M. Labes), p. 85, Gordon & Breach (1972).
92 A. Saupe and J. Nehring, *J. Chem. Phys.* **56**, 5527 (1972).
93 R. G. Priest, *Mol. Cryst. Liquid Cryst.* **17**, 129 (1972).
94 R. G. Priest, *Phys. Rev. A***7**, 720 (1973).
95 P. Chatelain, *Acta Cryst*, **1**, 315 (1948); *Bull. Soc. Franc. Miner. Crist.* **77**, 353 (1954).
96 J. Zadoc-Kahn, *Ann. Physique (series* 2), **6**, 455 (1936).
97 V. Tsvetkov, *Acta Physicochim.* (*USSR*), **19**, 86 (1944).
98 N. A. Tolstoi and L. N. Fedotov, *J. Exp. Theor. Phys.* (*USSR*), **17**, 564 (1947).
99 V. N. Tsvetkov and E. I. Ryumtsev, *Sov. Phys. – Crystallogr.* **13**, 225 (1968).
100 B. Cabane and W. G. Clark, *Phys. Rev. Letters*, **25**, 91 (1970).
R. Blinc, D. E. O'Reilly, E. M. Peterson, G. Lahajnar and I. Levstek, *Sol. State Commun.* **6**, 839 (1968).
S. K. Ghosh, E. Tettamanti, P. L. Indovina, *Phys. Rev. Letters*, **29**, 638 (1972).
See also G. H. Brown, J. W. Doane and V. D. Neff, *A review of the structure and physical properties of liquid crystals*, Butterworths (1971).
101 G. Foex, *Trans. Faraday Soc.* **29**, 958 (1933).
102 P. G. de Gennes, *Mol. Cryst. Liquid Cryst.* **12**, 193 (1971).
103 L. D. Landau and E. M. Lifshitz, *Statistical Physics*, 2nd edition, Pergamon (1969).
104 See, for example, J. W. Beams, *Rev. Mod. Phys.* **4**, 133 (1932).
105 T. W. Stinson and J. D. Litster, *Phys. Rev. Letters*, **25**, 503 (1970).
106 N. V. Madhusudana and S. Chandrasekhar, in *Liquid Crystals and Ordered Fluids*, vol. 2 (eds. J. F. Johnson and R. S. Porter), p. 657, Plenum (1974).
107 B. R. Ratna, M. S. Vijaya, R. Shashidhar and B. K. Sadashiva, in *Proceedings of the International Liquid Crystals Conference*, Bangalore, December 1973, *Pramana supplement I*, p. 69.
B. R. Ratna and R. Shashidhar, *Pramana*, **6**, 278 (1976).
108 T. W. Stinson, J. D. Litster, and N. A. Clark, *J. de Physique*, **33**, *C*1-69 (1972).
109 G. Chu, C. S. Bak, F. L. Lin, *Phys. Rev. Letters*, **28**, 1111 (1972).
110 T. W. Stinson and J. D. Litster, *Phys. Rev. Letters*, **30**, 688 (1973).
111 P. Martinoty, S. Candau and F. Debeauvais, *Phys. Rev. Letters*, **27**, 1123 (1971).

112 H. A. Bethe, *Proc. Roy. Soc.* **149**, 1 (1935).
113 T. S. Chang, *Proc. Camb. Phil. Soc.* **33**, 524 (1937).
114 T. J. Krieger and H. M. James, *J. Chem. Phys.* **22**, 796 (1954).
115 N. V. Madhusudana and S. Chandrasekhar, *Sol. State Commun.* **13**, 377 (1973); *Pramana*, **1**, 12 (1973).
116 N. V. Madhusudana and S. Chandrasekhar, in *Proceedings of the International Liquid Crystals Conference*, Bangalore, December 1973, *Pramana supplement I*, p. 57.
117 N. V. Madhusudana, K. L. Savithramma and S. Chandrasekhar, *Pramana* (in press).
118 J. G. J. Ypma and G. Vertogen, *Sol. State Commun.* **18**, 475 (1976).
The method is analogous to the Bethe–Peierls–Weiss approximation in ferromagnetism; see J. S. Smart, *Effective field theories of magnetism*, Saunders (1966); B. Strieb, H. B. Callen and G. Horwitz, *Phys. Rev.* **130**, 1798 (1963).
119 The method yields the same results as the so-called 'constant coupling' approximation in ferromagnetism; see J. S. Smart[(118)]; B. Strieb, H. B. Callen and G. Horwitz.[(118)]
120 W. Helfrich, *Phys. Rev. Letters*, **24**, 201 (1970).
121 M. Schadt and W. Helfrich, *Mol. Cryst. Liquid Cryst.* **17**, 355 (1972).
122 A. J. Leadbetter, R. M. Richardson and C. N. Colling, *J. de Physique*, **36**, *C*1-37 (1975).
123 J. E. Lydon and C. J. Coakley, *J. de Physique*, **36** *C*1-45 (1975).
124 C. A. Croxton and R. P. Ferrier, *J. Physics*, *C***4**, 1921 (1971); *ibid.*, **4**, 1909 (1971); *ibid.*, **4**, 2433 (1971); *ibid.*, **4**, 2447 (1971).
125 C. A. Croxton and S. Chandrasekhar, in *Proceedings of the International Liquid Crystals Conference*, Bangalore, December 1973, *Pramana supplement I*, p. 237.
126 A. Ferguson and S. J. Kennedy, *Phil. Mag.* **26**, 41 (1938).
127 S. Krishnaswamy and R. Shashidhar, in *Proceedings of the International Liquid Crystals Conference*, Bangalore, December 1973, *Pramana supplement I*, p. 247.
128 W. L. McMillan, *Phys. Rev. A***4**, 1238 (1971).
129 S. Chandrasekhar, *Mol. Cryst.* **2**, 71 (1966).
130 D. Langevin, *J. de Physique*, **33**, 249 (1972).
D. Langevin and M. A. Bouchiat, *J. de Physique*, **33**, 101 (1972); *ibid.*, **33**, *C*1-77 (1972).

Chapter 3

1 C. W. Oseen, *Trans. Faraday Soc.* **29**, 883 (1933).
2 H. Zöcher, *Trans. Faraday Soc.* **29**, 945 (1933).
3 F. C. Frank, *Disc. Faraday Soc.* **25**, 19 (1958).
4 A. Anzelius, *Uppsala Univ. Arsskr., Mat. Och Naturvet.* **1** (1931).
5 J. L. Ericksen, *Arch. Rational Mech. Anal.* **4**, 231 (1960).
6 J. L. Ericksen, *Trans. Soc. Rheol.* **5**, 23 (1961).
7 F. M. Leslie, *Quart. J. Mech. Appl. Math.* **19**, 357 (1966); *Arch. Rational Mech. Anal.* **28**, 265 (1968).
8 *a*. P. C. Martin, O. Parodi and P. S. Pershan, *Phys. Rev. A***6**, 2401 (1972).

b. J. D. Lee and A. C. Eringen, *J. Chem. Phys.* **54**, 5027 (1971); *ibid.*, **55**, 4504, 4509 (1971); *ibid.*, **58**, 4203 (1973).
c. H. Schmidt and J. Jähnig, *Annals of Physics*, **71**, 129 (1972).
For a full list of references to papers on continuum theories, see review by J. D. Lee and A. C. Eringen in *Liquid Crystals and Ordered Fluids*, vol. 2 (eds. J. F. Johnson and R. S. Porter), p. 315, Plenum (1974).
9 See, for example, C. Truesdell and W. Noll, *The Non-linear Field Theories of Mechanics, Handbuch der Physik*, vol. 3/3, Springer-Verlag (1965).
10 L. Onsager, *Phys. Rev.* **37**, 405 (1931); *ibid.*, **38**, 2265 (1932).
11 O. Parodi, *J. de Physique*, **31**, 581 (1970).
12 C. Truesdell, *Rational Thermodynamics*, McGraw-Hill (1969).
13 See, for example, P. K. Currie, *Sol. State Commun.* **12**, 31 (1973).
14 See, For example, J. F. Nye, *Physical Properties of Crystals*, Oxford University Press (1957).
15 J. L. Ericksen, *Phys. Fluids*, **9**, 1205 (1966).
16 See, for example, W. L. Ferrar, *Algebra, A Text Book of Determinants, Matrices and Algebraic Forms*, Oxford University Press (1941).
17 J. L. Ericksen, *Arch. Rational Mech. Anal.* **10**, 189 (1962).
18 J. Nehring and A. Saupe, *J. Chem. Phys.* **54**, 337 (1971).
19 V. Freedericksz and V. Tsvetkov, *Phy. Z. Soviet Union*, **6**, 490 (1934).
See also, V. Freedericksz and V. Zolina, *Trans. Faraday Soc.* **29**, 919 (1933).
20 A. Saupe, *Z. Naturforsch.* **15***a*, 815 (1960).
21 H. Gruler, T. J. Scheffer and G. Meier, *Z. Naturforsch.* **27***a*, 966 (1972).
See also I. Haller, *J. Chem. Phys.* **57**, 1400 (1972).
22 P. G. de Gennes, *Mol. Cryst. Liquid Cryst.* **12**, 193 (1971).
23 N. V. Madhusudana, P. P. Karat and S. Chandrasekhar, in *Proceedings of the International Liquid Crystals Conference*, Bangalore, December 1973, *Pramana supplement I*, p. 225.
24 P. E. Cladis, *Phys. Rev. Letters*, **31**, 1200 (1973).
25 P. Chatelain, *Bull. Soc. Franc. Miner. Crist.* **66**, 105 (1943).
J. F. Dreyer, in *Third International Liquid Crystals Conference*, Berlin (1970).
D. Berreman, *Phys. Rev. Letters*, **28**, 1683 (1972).
It has been discovered that oblique deposition of certain materials, e.g. gold or silicon monoxide, on glass has the same effect as rubbing: J. L. Janning, *Appl. Phys. Letters*, **21**, 173 (1972).
26 A. Rapini and M. Papoular, *J. de Physique*, **30**, *C*4-54 (1971).
27 C. J. Gerritsma, W. H. de Jeu and P. Van Zanten, *Phys. Letters*, **36***A*, 389 (1971).
28 F. M. Leslie, *Mol. Cryst. Liquid Cryst.* **12**, 57 (1970).
29 M. Schadt and W. Helfrich, *Appl. Phys. Letters*, **18**, 127 (1971).
The first practical field effect device was made by Fergason in 1971; see S. L. Arora and J. L. Fergason, in *Proceedings of the International Liquid Crystals Conference*, Bangalore, December 1973, *Pramana supplement I*, p. 520.
30 W. Snyder and A. Saupe, in *Fourth International Liquid Crystal Conference*, Kent, USA (1972).
31 O. Lehmann, *Ann. Physik*, **2**, 649 (1900).

32 G. Friedel, *Ann. Physique*, **18**, 273 (1922).
33 J. Nehring and A. Saupe, *J. Chem. Soc., Faraday Trans.* 2, **68**, 1 (1972).
34 J. Friedel and M. Kléman, in *Fundamental Aspects of Dislocation Theory* (eds. J. A. Simmons, R. de Wit and R. Bullough), p. 607, *Nat. Bur. Stand.* (*US.*) Spec. Publ. 317, **1** (1970).
35 A. Saupe, *Mol. Cryst. Liquid Cryst.* **21**, 211 (1973).
36 J. L. Ericksen, in *Liquid Crystals and Ordered Fluids* (eds. J. F. Johnson and R. S. Porter), p. 189, Plenum (1970).
37 R. B. Meyer, *Phil. Mag.* **27**, 405 (1973).
38 C. Williams, P. Pieranski and P. E. Cladis, *Phys. Rev. Letters*, **29**, 90 (1972).
39 For a review on the viscometric studies of mesophase types see R. S. Porter, E. M. Barrall and J. F. Johnson, *J. Chem. Phys.* **45**, 1452 (1966).
40 M. Miesowicz, *Nature*, **158**, 27 (1946).
41 F. M. Leslie, G. R. Luckhurst and H. J. Smith, *Chem. Phys. Letters*, **13**, 368 (1972).
42 V. Tsvetkov, *Acta Physicochim.* (*USSR*), **10**, 557 (1939).
43 H. Gasparoux and J. Prost, *J. de Physique*, **32**, 65 (1971).
44 P. G. de Gennes, *J. de Physique*, **32**, 789 (1971).
45 R. J. Atkin, *Arch. Rational Mech. Anal.* **38**, 224 (1970).
46 J. L. Ericksen, *Trans. Soc. Rheol.* **13**, 9 (1969).
47 J. Fisher and A. G. Frederickson, *Mol. Cryst. Liquid Cryst.* **6**, 255 (1969).
48 H. C. Tseng, D. L. Silver and B. A. Finlayson, *Phys. Fluids*, **15**, 1213 (1972).
49 U. D. Kini and G. S. Ranganath, *Pramana*, **4**, 19 (1975).
50 B. A. Finlayson, in *Liquid Crystals and Ordered Fluids*, vol. 2 (eds. J. F. Johnson and R. S. Porter), p. 211, Plenum (1974).
51 Ch. Gähwiller, *Mol. Cryst. Liquid Cryst.* **20**, 301 (1973).
52 P. Martinoty and S. Candau, *Mol. Cryst. Liquid Cryst.* **14**, 243 (1971).
53 P. Pieranski, F. Brochard and E. Guyon, *J. de Physique*, **34**, 35 (1973). See also F. Brochard, *Mol. Cryst. Liquid Cryst.* **23**, 51 (1973).
54 L. Léger, *Sol. State Comm.* **10**, 697 (1972); in *Fourth International Liquid Crystals Conference*, Kent, USA (1972).
55 P. Chatelain, *Acta Cryst.* **4**, 453 (1951).
56 P. G. de Gennes, *C. R. Acad. Sci.* **266**, 15 (1968).
57 Orsay Liquid Crystals Group, *J. Chem. Phys.* **51**, 816 (1969).
58 Orsay Liquid Crystals Group, *Phys. Rev. Letters*, **22**, 1361 (1969).
L. Léger-Quercy, 'Etude expérimentale des fluctuations thermiques d'orientation dans un cristal liquide nématique par diffusion inélastique de la lumière', thesis, University of Paris, France (1970).
59 L. D. Landau and E. M. Lifshitz, *Statistical Physics*, 2nd edition, Pergamon (1969).
See also I. L. Fabelinskii, *Molecular Scattering of Light*, Plenum (1968).
60 W. Kast, *Angew. Chemie.* **67**, 592 (1955).
61 E. F. Carr, *J. Chem. Phys.* **38**, 1536 (1963); *ibid.*, **39**, 1979 (1963); *ibid.*, **42**, 738 (1965); *ibid.*, **43**, 3905 (1965); *Advances in Chemistry Series*, **63**, 76 (1967); *Mol. Cryst. Liquid Cryst.* **7**, 253 (1969).
62 V. N. Tsvetkov and G. M. Mikhailov, *Acta Physicochim.* (*USSR*), **8**, 77 (1938).
63 Y. Bjornstahl, *Z. Phys. Chem.* *A***175**, 17 (1933).

64 V. Naggiar, *Ann. Physique*, **18**, 5 (1943).
65 R. Williams, *J. Chem. Phys.* **39**, 384 (1963).
66 G. H. Heilmeier, L. A. Zanoni and L. A. Barton, *Proc. IEEE*, **56**, 1162 (1968).
67 A. P. Kapustin and L. K. Vistin, *Kristallografiya*, **10**, 118 (1965).
G. Elliot and J. G. Gibson, *Nature*, **205**, 995 (1965).
68 P. A. Penz, *Mol. Cryst. Liquid Cryst.* **15**, 141 (1971).
69 Orsay Liquid Crystals Group, *Mol. Cryst. Liquid Cryst.* **12**, 251 (1971); *Phys. Rev. Letters*, **39***A*, 181 (1972).
70 H. Gruler and G. Meier, *Mol. Cryst. Liquid Cryst.* **16**, 299 (1972).
71 W. Helfrich, *J. Chem. Phys.* **51**, 4092 (1969).
For a concise review of electric field effects see W. Helfrich, *Mol. Cryst. Liquid Cryst.* **21**, 187 (1973).
72 E. Dubois-Violette, P. G. de Gennes and O. Parodi, *J. de Physique*, **32**, 305 (1971).
73 I. W. Smith, Y. Galerne, S. T. Lagerwall, E. Dubois-Violette and G. Durand, *J. de Physique*, **36**, *C*1-237 (1975).
74 See, for example, L. D. Landau and E. M. Lifshitz, *Fluid Mechanics*, Pergamon (1966).
75 P. A. Penz and G. W. Ford, *Phys. Rev. A***6**, 414 (1972).
76 W. Greubel and U. Wolff, *Appl. Phys. Letters*, **19**, 213 (1971).
77 L. K. Vistin, *Sov. Phys. – Crystallogr.* **15**, 514 (1970).
78 R. B. Meyer, *Phys. Rev. Letters*, **22**, 918 (1969).
79 D. Schmidt, M. Schadt and W. Helfrich, *Z. Naturforsch.* **27***A*, 277 (1972).
80 P. G. de Gennes, *Phys. Letters*, **41***A*, 479 (1972).
81 P. A. Penz, *Phys. Rev. Letters*, **24**, 1405 (1970).

Chapter 4

1 M. C. Mauguin, *Bull. Soc. Franc. Miner. Crist.* **34**, 71 (1911).
2 C. W. Oseen, *Trans. Faraday Soc.* **29**, 883 (1933).
3 H. de Vries, *Acta Cryst.* **4**, 219 (1951).
4 H. Poincaré, *Théorie Mathématique de la Lumière*, vol. 2, chap. 12 (1892).
5 R. C. Jones, *J. Opt. Soc. Am.* **31**, 488 (1941); *ibid.*, **31**, 493 (1941).
6 G. N. Ramachandran and S. Ramaseshan, *Handbuch der Physik*, vol. 25/1, 'Crystal Optics' (1961).
7 R. C. Jones, *J. Opt. Soc. Am.* **32**, 486 (1942).
8 F. Abeles, *Ann. Physique*, **5**, 777 (1950).
9 S. Chandrasekhar and K. N. Srinivasa Rao, *Acta Cryst. A***24**, 445 (1968).
10 S. Chandrasekhar, G. S. Ranganath, U. D. Kini and K. A. Suresh, *Mol. Cryst. Liquid Cryst.* **24**, 201 (1973).
11 Analogous expressions for a twisted pile of crystal plates were derived by: L. Sohncke, *Math. Ann.* **9**, 504 (1876); F. Pockels, *Lehrbuch der Kristalloptik*, Teubner, Leipzig, p. 289 (1906); R. C. Jones, *J. Opt. Soc. Am.* **31**, 500 (1941).
12 C. Robinson, *Tetrahedron*, **13**, 219 (1961).
13 R. Cano and P. Chatelain, *C. R. Acad. Sci.* **259**, 352 (1964).
14 H. Baessler, T. M. Laronge and M. M. Labes, *J. Chem. Phys.* **51**, 3213 (1969).

15 E. Sackmann, S. Meiboom, L. C. Snyder, A. E. Meixner and R. E. Dietz, *J. Am. Chem. Soc.* **90**, 3567 (1968).
16 M. Schadt and W. Helfrich, *Appl. Phys. Letters*, **18**, 127 (1971).
17 C. Z. Van Doorn, *Phys. Letters*, **42***A*, 537 (1973).
18 The first practical field effect device was made by Fergason in 1971: see, for example, S. L. Arora and J. L. Fergason, in *Proceedings of the International Liquid Crystals Conference*, Bangalore, December 1973, *Pramana supplement I*, p. 520.
See also A. Sussman, *IEEE Transactions*, PHP-8, 24 (1972).
19 T. J. Scheffer, *J. Appl. Phys.* **44**, 4799 (1973).
S. Kobayashi and T. Shimomura, in *Proceedings of the International Liquid Crystals Conference*, Bangalore, December 1973, *Pramana supplement I*, p. 530.
20 G. Baur, A. Steib and G. Meier, in *Liquid Crystals and Ordered Fluids*, vol. 2 (eds. J. F. Johnson and R. S. Porter), p. 645, Plenum (1974).
21 D. W. Berreman, *Appl. Phys. Letters*, **25**, 12 (1974).
A more complete calculation taking into account backflow effects has been made by C. Z. Van Doorn, *J. de Physique C*1-261 (1975).
22 E. Sackmann and J. Voss, *Chem. Phys. Letters*, **14**, 528 (1972).
23 G. S. Ranganath, S. Chandrasekhar, U. D. Kini, K. A. Suresh and S. Ramaseshan, *Chem. Phys. Letters*, **19**, 556 (1973).
24 G. S. Ranganath, K. A. Suresh, S. R. Rajagopalan and U. D. Kini, in *Proceedings of the International Liquid Crystals Conference*, Bangalore, December 1973, *Pramana supplement I*, p. 353.
25 S. Chandrasekhar, G. S. Ranganath and K. A. Suresh, in *Proceedings of the International Liquid Crystals Conference*, Bangalore, December 1973, *Pramana supplement I*, p. 341.
26 R. Nityananda, U. D. Kini, S. Chandrasekhar and K. A. Suresh, in *Proceedings of the International Liquid Crystals Conference*, Bangalore, December 1973, *Pramana supplement I*, p. 325.
27 C. G. Darwin, *Phil. Mag.* **27**, 315, 675 (1914); *ibid.*, **43**, 800 (1922).
28 I. S. Gradshteyn and I. M. Ryzhik, *Tables of Integrals, Series and Products*, p. 27, Academic (1965).
29 R. Dreher, G. Meier and A. Saupe, *Mol. Cryst. Liquid Cryst.* **13**, 17 (1971).
30 S. Chandrasekhar and J. Shashidhara Prasad, *Mol. Cryst. Liquid Cryst.* **14**, 115 (1971).
Similar measurements have been reported recently by J. C. Martin and R. Cano, *C. R. Acad. Sci. B***278**, 219 (1974).
31 G. Borrmann, *Physik. Z.* **42**, 157 (1941).
P. P. Ewald, *Rev. Mod. Phys.* **37**, 46 (1965).
32 K. A. Suresh, *Mol. Cryst. Liquid Cryst.* **35**, 267 (1976).
33 E. I. Kats, *Sov. Phys. – JETP*, **32**, 1004 (1971).
34 R. Nityananda, *Mol. Cryst. Liquid Cryst.* **21**, 315 (1973).
35 G. Joly, 'Contribution à l'étude de la propagation de certaines ondes éléctromagnetiques dans les piles de reusch', thesis, University of Science and Technology, Lille, France (1972).
36 M. Aihara and H. Inaba, *Optics Commun.* **3**, 77 (1971).
A. S. Marathay, *J. Opt. Soc. Am.* **61**, 1363 (1971).
37 R. Nityananda and U. D. Kini, in *Proceedings of the International Liquid*

Crystals Conference, Bangalore, December 1973, *Pramana supplement I*, p. 311.
38 R. Nityananda, 'Study of some novel optical effects in periodic structures', thesis, Bangalore University, India (1975).
39 See, for example, M. Born and E. Wolf, *Principles of Optics*, Pergamon (1959).
40 An analytical solution in the two-wave approximation has been given by V. A. Belyakov and V. E. Dmitrienko, *Sov. Phys. – Solid State*, **15**, 1811 (1974).
V. E. Dmitrienko and V. A. Belyakov, *Sov. Phys. – Solid State*, **15**, 2365 (1974).
41 D. Taupin, *J. de Physique*, **30**, *C*4-32 (1969).
42 D. W. Berreman and T. J. Scheffer, *Phys. Rev. Letters*, **25**, 577 (1970).
43 See, for example, R. W. Wood, *Physical Optics*, 3rd edition, Macmillan (1934).
44 S. Chandrasekhar and J. Shashidhara Prasad, in *Physics of the Solid State*, (eds. S. Balakrishna, M. Krishnamurthi and B. R. Rao), p. 77, Academic (1969).
45 F. Grandjean, *C. R. Acad. Sci.* **172**, 71 (1921).
46 R. Cano, *Bull. Soc. Franc. Miner. Crist.* **91**, 20 (1968).
47 P. G. de Gennes, *C. R. Acad. Sci.* **266***B*, 571 (1968).
48 T. J. Scheffer, *Phys. Rev. A***5**, 1327 (1972).
49 Orsay Liquid Crystals Group, *J. de Physique*, **30**, *C*4-38 (1969).
50 M. Kléman and J. Friedel, *J. de Physique*, **30**, *C*4-43 (1969).
51 J. Rault, *Sol. State Commun.* **9**, 1965 (1971).
52 Y. Bouligand and M. Kléman, *J. de Physique*, **31**, 1041 (1970).
53 A. Saupe, *Mol. Cryst. Liquid Cryst.* **21**, 211 (1973).
54 J. Rault, *Mol. Cryst. Liquid Cryst.* **16**, 143 (1972); in *Liquid Crystals and Ordered Fluids*, vol. 2 (eds. J. F. Johnson and R. S. Porter), p. 677, Plenum (1974).
55 F. M. Leslie, *Proc. Roy. Soc. A***307**, 359 (1968).
56 O. Lehmann, *Ann. Physik*, **4** 649 (1900).
57 R. S. Porter, E. M. Barrall and J. F. Johnson, *J. Chem. Phys.* **45**, 1452 (1966).
58 S. Candau, P. Martinoty and F. Debeauvais, *C. R. Acad. Sci. B***277**, 769 (1973).
59 W. Helfrich, *Phys. Rev. Letters*, **23**, 372 (1969).
60 U. D. Kini, G. S. Ranganath and S. Chandrasekhar, *Pramana*, **5**, 101 (1975).
61 F. M. Leslie, *Mol. Cryst. Liquid Cryst.* **7**, 407 (1969).
62 U. D. Kini (unpublished).
63 J. J. Wysocki, J. Adams and W. Haas, *Phys. Rev. Letters*, **20**, 1024 (1968).
H. Baessler and M. M. Labes, *Phys. Rev. Letters*, **21**, 1791 (1968).
64 P. G. de Gennes, *Sol. State Commun.* **6**, 163 (1968).
65 R. B. Meyer, *Appl. Phys. Letters*, **12**, 281 (1968).
66 R. B. Meyer, *Appl. Phys. Letters*, **14**, 208 (1969).
67 G. Durand, L. Léger, F. Rondelez and M. Veyssie, *Phys. Rev. Letters*, **22**, 227 (1969).

The experiment has been done using electric fields by F. J. Kahn, *Phys. Rev. Letters*, **24**, 209 (1970).
68 F. M. Leslie, *Mol. Cryst. Liquid Cryst.* **12**, 57 (1970).
69 W. Helfrich, *J. Chem. Phys.* **55**, 839 (1971).
70 F. Rondelez and J. P. Hulin, *Sol. State Commun.* **10**, 1009 (1972).
71 F. Rondelez and H. Arnould, *C. R. Acad. Sci. B***273**, 549 (1971).
72 C. Gerritsma and P. Van Zanten, *Phys. Letters*, *A***37**, 47 (1971).
See also T. J. Scheffer, *Phys. Rev. Letters*, **28**, 593 (1972).
73 F. Rondelez, 'Contribution a l'étude des effets de champ dans les cristaux liquides nématiques et cholestériques', thesis, University of Paris (1973).
74 J. P. Hurault, *J. Chem. Phys.* **59**, 2068 (1973).
75 H. Arnould-Netillard and F. Rondelez, *Mol. Cryst. Liquid Cryst.* **26**, 11 (1974).
76 G. H. Heilmeier and J. E. Goldmacher, *Appl. Phys. Letters*, **13**, 132 (1968).
77 W. E. Haas, J. E. Adams, G. A. Dir and C. W. Mitchell, *Proceedings of the SID* 14/4, Fourth quarter, 121 (1973).
78 J. Cheng and R. B. Meyer, *Phys. Rev. A***9**, 2744 (1974).
79 P. G. de Gennes, *Mol. Cryst. Liquid Cryst.* **12**, 193 (1971).
80 P. N. Keating, *Mol. Cryst. Liquid Cryst.* **8**, 315 (1969).
81 See, for example, C. Kittel, *Introduction to Solid State Physics* 3rd edition, p. 184, John Wiley (1968).
82 J. L. Fergason, N. N. Goldberg and R. J. Nadalin, *Mol. Cryst.* **1**, 309 (1966).
83 J. L. Fergason, *Scientific American*, **211**, 77 (1964).
84 O. S. Selawry, H. S. Selawry and J. F. Holland, *Mol. Cryst.* **1**, 495 (1966).
M. Gautherie, *J. de Physique*, **30**, *C*4-122 (1969).
W. E. Woodmansee, *Appl. Optics*, **7**, 1721 (1968).
85 J. Hansen, J. L. Fergason and A. Okaya, *Appl. Opt.* **3**, 987 (1964).
F. Keilmann, *Appl. Opt.* **9**, 1319 (1970).
R. D. Ennulat and J. L. Fergason, *Mol. Cryst. Liquid Cryst.* **13**, 149 (1971).
86 K. Iizuka, *Electronics Letters*, **5**, 26 (1969).
87 P. Pollmann and H. Stegemeyer, *Chem. Phys. Letters*, **20**, 87 (1973).
88 P. H. Keyes, H. T. Weston and W. B. Daniels, *Phys. Rev. Letters*, **31**, 628 (1973).
89 R. Shashidhar and S. Chandrasekhar, *J. de Physique*, **36**, *C*1-49 (1975).
90 G. Friedel, *Ann. Physique*, **18**, 273 (1922).
91 J. E. Adams and W. E. L. Haas, *Mol. Cryst. Liquid Cryst.* **15**, 27 (1971).
92 J. E. Adams, W. E. L. Haas and J. J. Wysocki, *Phys. Rev. Letters*, **22**, 92 (1969).
93 T. Nakagiri, H. Kodama and K. K. Kobayashi, *Phys. Rev. Letters*, **27**, 564 (1971).
94 F. D. Saeva and J. J. Wysocki, *J. Am. Chem. Soc.* **93**, 5928 (1971).
95 A. D. Buckingham, G. P. Ceaser and M. B. Dunn, *Chem. Phys. Letters*, **3**, 540 (1969).
H. Stegemeyer, K. J. Mainusch and E. Steigner, *Chem. Phys. Letters*, **8**, 425 (1971).
96 F. D. Saeva, *Mol. Cryst. Liquid Cryst.* **23**, 171 (1973).
97 W. J. A. Goossens, *Mol. Cryst. Liquid Cryst.* **12**, 237 (1971).
98 A. Wulf, *J. Chem. Phys.* **59**, 1487 (1973).

99 D. Coates and G. W. Gray, *Mol. Cryst. Liquid Cryst.* **24**, 163 (1973).
100 *Interatomic Distances Supplement*, Special publication no. 18, The Chemical Society, London, p. 5 (1965).
101 S. Chandrasekhar and B. R. Ratna, *Mol. Cryst. Liquid Cryst.* **35**, 109 (1976).

Chapter 5

1 H. Sackmann and D. Demus, *Mol. Cryst. Liquid Cryst.* **21**, 239 (1973).
For a discussion of the principles underlying the calculation of phase diagrams for liquid crystalline mixtures see M. Domon and J. Billard, in *Proceedings of the International Liquid Crystals Conference*, Bangalore, December 1973, *Pramana supplement I*, p. 131.
2 B. K. Vainshtein and I. G. Chistyakov, in *Proceedings of the International Liquid Crystals Conference*, Bangalore, December 1973, *Pramana supplement I*, p. 79.
3 A. de Vries, in *Proceedings of the International Liquid Crystals Conference*, Bangalore, December 1973, *Pramana supplement I*, p. 93.
4 A. Levelut and M. Lambert, *C. R. Acad. Sci. B***272**, 1018 (1971).
J. Doucet, A. M. Levelut, M. Lambert, L. Liebert and L. Strzelecki, *J. de Physique*, **36**, *C*1-13 (1975).
5 A. de Vries, *Mol. Cryst. Liquid Cryst.* **20**, 119 (1973).
6 D. Demus and R. Rurainski, *Z. Phys. Chem.* (Leipzig), **253**, 53 (1973).
7 A. de Vries and D. L. Fishel, *Mol. Cryst. Liquid Cryst.* **16**, 311 (1972).
8 T. R. Taylor, S. L. Arora and J. L. Fergason, *Phys. Rev. Letters*, **25**, 722 (1970).
9 R. A. Wise, D. H. Smith and J. W. Doane, *Phys. Rev. A***7**, 1366 (1973).
10 D. Demus, H. König, D. Morzotko and R. Rurainski, *Mol. Cryst. Liquid Cryst.* **23**, 207 (1973).
11 D. Demus and H. Sackmann, *Z. Phys. Chem.* (Leipzig), **222**, 127 (1963).
12 G. R. Luckhurst and F. Sundholm, *Mol. Phys.* **21**, 349 (1971).
13 W. Helfrich and C. S. Oh, *Mol. Cryst. Liquid Cryst.* **14**, 289 (1971).
14 W. Z. Urbach and J. Billard, *C. R. Acad. Sci. B***274**, 1287 (1972).
15 D. Demus, G. Kunicke, J. Neelsen and H. Sackmann, *Z. Naturforsch.* **23***a*, 84 (1968).
16 S. Diele, P. Brand and H. Sackmann, *Mol. Cryst. Liquid Cryst.* **17**, 163 (1972).
17 D. Coates, K. J. Harrison and G. W. Gray, *Mol. Cryst. Liquid Cryst.* **22**, 99 (1973).
18 D. Demus, S. Diele, M. Klapperstuck, V. Link and H. Zaschke, *Mol. Cryst. Liquid Cryst.* **15**, 161 (1971).
19 P. C. Martin, O. Parodi and P. S. Pershan, *Phys. Rev. A***6**, 2401 (1972).
20 P. G. de Gennes and G. Sarma, *Phys. Letters*, **38***A*, 219 (1972).
21 Y. Liao, N. A. Clark and P. S. Pershan, *Phys. Rev. Letters*, **30**, 639 (1973).
22 R. E. Detjen, D. L. Uhrich and C. F. Sheley, *Phys. Letters*, **42***A*, 522 (1973).
23 W. L. McMillan, *Phys. Rev. A***4**, 1238 (1971).
24 J. G. Kirkwood and E. Monroe, *J. Chem. Phys.* **9**, 514 (1941).
25 K. Kobayashi, *Mol. Cryst. Liquid Cryst.* **13**, 137 (1971).
26 W. L. McMillan, *Phys. Rev. A***6**, 936 (1972).

27 J. W. Doane, R. S. Parker, B. Cvikl, J. L. Johnson and D. L. Fishel, *Phys. Rev. Letters*, **28**, 1694 (1972).
28 B. Cabane and W. G. Clark, *Sol. State Commun.* **13**, 129 (1973).
29 D. Djurek, J. Baturić-Rubčić and K. Franulović, in *Fifth International Liquid Crystals Conference*, Stockholm (1974).
F. Hardouin, H. Gasparoux and P. Delhaes, *J. de Physique*, **36**, *C*1-127 (1975).
S. Torza and P. E. Cladis, *Phys. Rev. Letters*, **32**, 1406 (1974).
30 W. L. McMillan, *Phys. Rev. A***7**, 1419 (1973).
31 B. I. Halperin, T. C. Lubensky and S. K. Ma, *Phys. Rev. Letters*, **32**, 292 (1974).
B. I. Halperin and T. C. Lubensky (to be published).
32 P. G. de Gennes, *J. de Physique*, **30**, *C*4-65 (1969).
33 G. Friedel, *Ann. Physique*, **18**, 273 (1922).
34 J. A. Geurst, *Phys. Letters*, **34***A*, 283 (1971).
35 W. H. Bragg, *Nature*, **133**, 445 (1934).
See also N. H. Hartshorne and A. Stuart, *Crystals and the Polarizing Microscope*, 2nd edition, Edward Arnold, London (1950).
36 S. Chandrasekhar and N. V. Madhusudana, *Acta Cryst. A***26**, 153 (1970).
37 F. Grandjean, *C. R. Acad. Sci.* **166**, 165 (1917).
38 S. Chandrasekhar, *Mol. Cryst.* **2**, 71 (1966).
39 L. D. Landau and E. M. Lifshitz, *Statistical Physics*, 2nd edition, Pergamon (1969).
40 M. Delaye, R. Ribotta and G. Durand, *Phys. Letters*, **44***A*, 139 (1973).
41 N. A. Clark and R. B. Meyer, *App. Phys. Letters*, **22**, 493 (1973).
42 G. Durand, in *Proceedings of the International Liquid Crystals Conference*, Bangalore, December 1973, *Pramana supplement I*, p. 23.
43 G. Durand, *C. R. Acad. Sci.* **275***B*, 629 (1972).
44 R. Ribotta, G. Durand and J. D. Litster, *Sol. State Commun.* **12**, 27 (1973).
45 N. A. Clark and P. S. Pershan, *Phys. Rev. Letters*, **30**, 3 (1973).
46 R. Ribotta, D. Salin and G. Durand, *Phys. Rev. Letters*, **32**, 6 (1974).
47 W. Helfrich, *Phys. Rev. Letters*, **23**, 372 (1969).
48 K. Sakamoto, R. S. Porter and J. F. Johnson, *Mol. Cryst. Liquid Cryst.* **8**, 443 (1969).
49 U. D. Kini, G. S. Ranganath and S. Chandrasekhar, *Pramana*, **5**, 101 (1975).
50 A. E. Lord, *Phys. Rev. Letters*, **29**, 1366 (1972).
51 K. Miyano and J. B. Ketterson, *Phys. Rev. Letters*, **31**, 1047 (1973).
52 See, for example, J. F. Nye, *Physical Properties of Crystals*, Oxford (1957).
53 See, for example, F. Brochard and P. G. de Gennes, in *Proceedings of the International Liquid Crystals Conference*, Bangalore, December 1973, *Pramana supplement I*, p. 1.
54 H. Gruler, *Z. Naturforsch.* **28***a*, 474 (1973).
55 P. G. de Gennes, *Sol. State Commun.* **10**, 753 (1972); *Mol. Cryst. Liquid Cryst.* **21**, 49 (1973).
56 See, for example, E. A. Lynton, *Superconductivity*, 2nd edition, Methuen (1964).
57 L. Cheung, R. B. Meyer and H. Gruler, *Phys. Rev. Letters*, **31**, 349 (1973).

58 M. Delaye, R. Ribotta and G. Durand, *Phys. Rev. Letters*, **31**, 443 (1973).
59 P. E. Cladis, *Phys. Rev. Letters*, **31**, 1200 (1973).
60 N. V. Madhusudana, P. P. Karat and S. Chandrasekhar, in *Proceedings of the International Liquid Crystals Conference*, Bangalore, December 1973, *Pramana supplement I*, p. 225.
61 L. Léger, *Phys. Letters*, **44***A*, 535 (1973).
62 D. d'Humières and L. Léger, *J. de Physique*, **36**, *C*1-113 (1975).
63 W. L. McMillan, *Phys. Rev. A***9**, 1720 (1974).
64 F. Brochard, *J. de Physique*, **34**, 411 (1973).
F. Jahnig and F. Brochard, *J. de Physique*, **35**, 301 (1974).
See also F. Jahnig, in *Proceedings of the International Liquid Crystals Conference*, Bangalore, December 1973, *Pramana supplement I*, p. 31.
65 S. Meiboom (unpublished).
66 F. Hardouin, M. F. Achard and H. Gasparoux, *Sol. State Commun.* **14**, 453 (1974).
67 C. C. Huang, R. S. Pindak, P. J. Flanders and J. T. Ho, *Phys. Rev. Letters*, **33**, 400 (1974).
68 R. A. Wise, A. Olah and J. W. Doane, *J. de Physique*, **36**, *C*1-117 (1975).
69 D. Salin, I. W. Smith and G. Durand, *J. de Physique Lettres*, **35**, *L*-165 (1974).
K. Chu and W. L. McMillan, *Phys. Rev. A***11**, 1059 (1975).
70 P. Kassubek and G. Meier, *Mol. Cryst. Liquid Cryst.* **8**, 305 (1969).
71 R. S. Pindak, C. C. Huang and J. T. Ho, *Sol. State Commun.* **14**, 821 (1974).
72 P. H. Keyes, H. T. Weston and W. B. Daniels, *Phys. Rev. Letters*, **31**, 628 (1973).
73 R. Shashidhar and S. Chandrasekhar, *J. de Physique*, **36**, *C*1-49 (1975).
74 R. B. Griffiths, *Phys. Rev. Letters*, **24**, 715 (1970).
75 A. Saupe, *Mol. Cryst. Liquid Cryst.* **7**, 59 (1969).
76 J. Nehring and A. Saupe, *J. Chem. Soc., Faraday Trans. 2*, **68**, 1 (1972).
77 A. Rapini, *J. de Physique*, **33**, 237 (1972).
78 D. W. Berreman, *Mol. Cryst. Liquid Cryst.* **22**, 175 (1973).
79 R. Nityananda, *Pramana*, **2**, 35 (1974).
80 Orsay Liquid Crystals Group, *Sol. State Commun.* **9**, 653 (1971).
81 Y. Galerne, J. L. Martinand, G. Durand and M. Veyssie, *Phys. Rev. Letters*, **29**, 562 (1972).
82 Z. Luz and S. Meiboom, *J. Chem. Phys.* **59**, 275 (1973).
83 R. Ribotta, R. B. Meyer and G. Durand, *J. de Physique* (to be published).

Index

Index